Resident Stem Cells and Regenerative Therapy

Resident Stem Cells and Regenerative Therapy

Edited by

Regina Coeli dos Santos Goldenberg

Carlos Chagas Filho Biophysics Institute, Federal University of Rio de Janeiro, Rio de Janeiro, Brazil

Antonio Carlos Campos de Carvalho

Federal University of Rio de Janeiro, National Cardiology Institute, Cell Therapy Network

ELSEVIER

AMSTERDAM • BOSTON • HEIDELBERG • LONDON
NEW YORK • OXFORD • PARIS • SAN DIEGO
SAN FRANCISCO • SINGAPORE • SYDNEY • TOKYO

Academic Press is an Imprint of Elsevier

Academic Press is an imprint of Elsevier
The Boulevard, Langford Lane, Kidlington, Oxford, OX5 1GB, UK
225 Wyman Street, Waltham, MA 02451, USA

First published 2013

British Library Cataloguing in Publication Data
A catalogue record for this book is available from the British Library

Library of Congress Cataloging-in-Publication Data
A catalog record for this book is available from the Library of Congress

ISBN: 978-0-12-416012-5

Printed and bound in the United States

12 13 14 15 10 9 8 7 6 5 4 3 2 1

Dedication

We dedicate this book to Professor Julio Cesar Voltarelli, who unfortunately left us prematurely this year. Julio was a Professor of Immunology at the Medical School of the University of São Paulo (USP) in Ribeirão Preto. He played a central role in developing cell-based therapies for autoimmune diseases in Brazil and worldwide. By using an ingenious strategy of hematopoietic stem collection purged of auto-reactive cells, bone marrow ablation, and marrow reconstitution he reasoned that he would reset the immune system of the patient and therefore attenuate or eradicate the auto-aggression mechanisms. He applied these therapies to many auto immune diseases, such as sclerodermia, systemic lupus erythematosus, multiple sclerosis, and type I diabetes. His clinical trials in diabetes were by far the most significant, leading to two articles published in the Journal of the American Medical Association (JAMA). Recognizing the limitations imposed by a therapy that required marrow ablation, Julio, in his never-ending search to benefit his patients, started a new protocol using allogeneic mesenchymal stem cells in type I diabetes patients. This trial is ongoing, now carried by the many disciples that Julio left during his prestigious career at USP. Julio's contribution to stem cell research and therapy in Brazil was paramount. He was a founding member of the Brazilian Association for Cell Therapy (ABTCEL) and an eloquent supporter of patients' rights. He had a brilliant academic career going back and forth between the basic laboratory and clinical practice. Besides all his academic and medical achievements, Julio was a simple man, always ready to establish a conversation (preferably over a cold beer) and to help colleagues and students alike. This simple and unpretentious style made him a very likeable person and his untimely passing left his many friends, among whom we include ourselves, orphan of a great scientist, teacher, colleague, and an enthusiastic supporter of the role that cell therapies will play in medicine in the near future.

Contents

Foreword

Resident stem cells, cells that reside in each organ and are ready to be activated at the time of injury or disease, hold the key to tissue and organ regeneration. These cells are continuously being activated as part of the normal cycle of cell proliferation and death. It is now well known that the tissues and organs in our bodies are constantly regenerating. For example, skin cells turn over completely about every 14 days throughout the body. The lining in the intestine turns over completely about every 7 days. Other cells turn over slowly, like bone cells (approximately every 10 years), or brain cells (approximately every 20 years). Nonetheless, every cell does turn over, and when it does, the resident stem cells in that specific tissue are involved. As the science of regenerative medicine advances, more knowledge is available regarding the supporting cells that are involved in the process of regeneration. Although the mechanisms are not yet entirely known, it is evident that bone marrow and vascular associated cells may also be involved in the process of natural cell turnover as well as regeneration at the time of injury. What is most remarkable about the field of resident stem cells is that most of the details we have today were not known even just two decades ago. Major work in the fields of cell and molecular biology has made these advances possible. It is indeed because of these major advances that the timeliness of this textbook is ideal. Dr. Regina Coeli dos Santos Goldenberg and Dr. Antonio Carlos Campos de Carvalho, currently Professors of Physiology at the Carlos Chagas Filho Biophysics Institute, Federal University of Rio de Janeiro in Brazil, took on the formidable task of assembling the current knowledge of the field in this comprehensive textbook. The contributors, all well-recognized researchers, cover the spectrum of topics that are most relevant to therapy. This body of work is certain to become the standard for the field of resident stem cells and regenerative therapy.

Anthony Atala MD,

Director and W.B. Boyce Professor, Wake Forest Institute for Regenerative Medicine, Wake Forest University, Winston-Salem, North Carolina, USA

List of Contributors

Mariana Alves Antunes
Instituto de Biofísica Carlos Chagas Filho; Programa de Terapia Celular e Bioengenharia; Laboratório de Investigação Pulmonar; Instituto Nacional de Ciência e Tecnologia de Biologia Estrutural e Bioimagem -INBEB; Universidade Federal do Rio de Janeiro, Rio de Janeiro, Brazil

Alex Balduino
Coordenação de Pesquisa, Universidade Veiga de Almeida (UVA), Rio de Janeiro, Brazil

Radovan Borojevic
Instituto de Ciências Biomédicas Universidade Federal do Rio de Janeiro and Excellion Serviços Biomédicos SA, Petrópolis, Rio de Janeiro

Giordano Wosgrau Calloni
Departamento de Biologia Celular, Embriologia e Genética, Centro de Ciências Biológicas, Universidade Federal de Santa Catarina, Campus Universitário-Trindade, 88040-900, Florianópolis, Santa Catarina, Brazil

Arnold I. Caplan
Skeletal Research Center, Department of Biology, Case Western Reserve University, Cleveland, Ohio, USA

Adriana B. Carvalho
Instituto de Biofísica Carlos Chagas Filho; Programa de Terapia Celular e Bioengenharia; Laboratório de Cardiologia Celular e Molecular; Instituto Nacional de Ciência e Tecnologia de Biologia Estrutural e Bioimagem -INBEB; Universidade Federal do Rio de Janeiro, Rio de Janeiro, Brazil

Antonio Carlos Campos de Carvalho
Instituto de Biofísica Carlos Chagas Filho; Programa de Terapia Celular e Bioengenharia; Laboratório de Cardiologia Celular e Molecular; Instituto Nacional de Ciência e Tecnologia de Biologia Estrutural e Bioimagem -INBEB; Universidade Federal do Rio de Janeiro, Rio de Janeiro, RJ, Instituto Nacional de Cardiologia, Rede de Terapia Celular, Brazil

Karen L. Christman
Department of Bioengineering, University of California, San Diego, La Jolla, California, USA

Grazielle Suhett Dias
Instituto de Biofísica Carlos Chagas Filho; Laboratório de Cardiologia Celular e Molecular; Instituto Nacional de Ciência e Tecnologia de Biologia Estrutural e Bioimagem -INBEB; Universidade Federal do Rio de Janeiro, Rio de Janeiro, Brazil

Milene R. Costa

Instituto de Biofísica Carlos Chagas Filho; Programa de Terapia Celular e Bioengenharia; Laboratório de Físico-Química Biológica Aída Hassón-Voloch; Instituto Nacional de Ciência e Tecnologia de Biologia Estrutural e Bioimagem -INBEB; Universidade Federal do Rio de Janeiro, Rio de Janeiro, Brazil

Maria Eugenia Leite Duarte

Centro de Pesquisa em Terapia Celular e Bioengenharia Ortopédica (CTCel), Instituto Nacional de Traumatologia e Ortopedia (INTO), Rio de Janeiro, Brazil; Instituto de Ciências Biomédicas (ICB), Universidade Federal do Rio de Janeiro (UFRJ), Rio de Janeiro, Brazil

Marcelo Einicker-Lamas

Instituto de Biofísica Carlos Chagas Filho; Programa de Terapia Celular e Bioengenharia; Laboratório de Físico-Química Biológica Aída Hassón-Voloch; Instituto Nacional de Ciência e Tecnologia de Biologia Estrutural e Bioimagem -INBEB; Universidade Federal do Rio de Janeiro, Rio de Janeiro, RJ, Brazil

Lanuza Alaby Pinheiro Faccioli

Instituto de Biofísica Carlos Chagas Filho; Laboratório de Cardiologia Celular e Molecular; Instituto Nacional de Ciência e Tecnologia de Biologia Estrutural e Bioimagem -INBEB; Universidade Federal do Rio de Janeiro, Rio de Janeiro, RJ, Brazil

Bernd K. Fleischmann

Institute of Physiology I, Life and Brain Center, University of Bonn, Bonn, Germany

Regina Coeli dos Santos Goldenberg

Instituto de Biofísica Carlos Chagas Filho; Programa de Terapia Celular e Bioengenharia; Laboratório de Cardiologia Celular e Molecular; Instituto Nacional de Ciência e Tecnologia de Biologia Estrutural e Bioimagem -INBEB; Universidade Federal doRio de Janeiro, Rio de Janeiro, RJ, Brazil

Fernanda Gubert

Instituto de Biofísica Carlos Chagas Filho; Programa de Terapia Celular e Bioengenharia; Laboratório de Neurobiologia Celular e Molecular; Instituto Nacional de Ciência e Tecnologia de Biologia Estrutural e Bioimagem -INBEB; Universidade Federal do Rio de Janeiro, Rio de Janeiro, RJ, Brazil

Patrícia M. Kossugue

Chemistry Institute, Biochemistry Department, Cell and Molecular Therapy Center (NUCEL), University of São Paulo, São Paulo, Brazil

Rafael S. Lindoso

Instituto de Biofísica Carlos Chagas Filho; Programa de Terapia Celular e Bioengenharia; Laboratório de Físico-Química Biológica Aída Hassón-Voloch; Instituto Nacional de Ciência e Tecnologia de Biologia Estrutural e Bioimagem -INBEB; Universidade Federal do Rio de Janeiro, Rio de Janeiro, RJ, Brazil

Fernando H. Lojudice

Chemistry Institute, Biochemistry Department, Cell and Molecular Therapy Center (NUCEL), University of São Paulo, São Paulo, Brazil

Rosalia Mendez-Otero
Instituto de Biofísica Carlos Chagas Filho; Programa de Terapia Celular e Bioengenharia; Laboratório de Neurobiologia Celular e Molecular; Instituto Nacional de Ciência e Tecnologia de Biologia Estrutural e Bioimagem -INBEB; Universidade Federal do Rio de Janeiro, Rio de Janeiro, RJ, Brazil

Lindolfo da Silva Meirelles
University of São Paulo, Ribeirão Preto, São Paulo, Brazil

Louise Mesentier-Louro
Instituto de Biofísica Carlos Chagas Filho; Laboratório de Neurobiologia Celular e Molecular; Programa de Terapia Celular e Bioengenharia; Instituto Nacional de Ciência e Tecnologia de Biologia Estrutural e Bioimagem -INBEB; Universidade Federal do Rio de Janeiro, Rio de Janeiro, RJ, Brazil

Luiz Fernando Quintanilha de Mesquita
Instituto de Biofísica Carlos Chagas Filho; Laboratório de Cardiologia Celular e Molecular; Instituto Nacional de Ciência e Tecnologia de Biologia Estrutural e Bioimagem -INBEB; Universidade Federal do Rio de Janeiro, Rio de Janeiro, RJ, Brazil; Division of Laboratory, Yamaguchi University Hospital, Yamaguchi University School of Medicine, Japan; Department of Gastroenterology and Hepatology, Yamaguchi University, Graduate School of Medicine Yamaguchi University, Japan

Marcelo Marcos Morales
Instituto de Biofísica Carlos Chagas Filho; Laboratório de Fisiologia Celular e Molecular; Programa de Terapia Celular e Bioengenharia; Instituto Nacional de Ciência e Tecnologia de Biologia Estrutural e Bioimagem -INBEB; Universidade Federal do Rio de Janeiro, Rio de Janeiro, Brazil

Nance Beyer Nardi
Laboratory of Stem Cells and Cell Therapy, Universidade Luterana do Brazil, Canoas, Rio Grande do Sul, Brazil

M.I. Nicolescu
Molecular Medicine Department, Victor Babeş, National Institute of Pathology, Bucharest, Romania and Division of Cellular and Molecular Medicine, Carol Davila, University of Medicine and Pharmacy, Bucharest, Romania

Bruno Diaz Paredes
Instituto de Biofísica Carlos Chagas Filho; Laboratório de Cardiologia Celular e Molecular; Programa de Terapia Celular e Bioengenharia; Instituto Nacional de Ciência e Tecnologia de Biologia Estrutural e Bioimagem -INBEB; Universidade Federal do Rio de Janeiro, Rio de Janeiro, RJ, Brazil

Paolo Pelosi
Department of Surgical Sciences and Integrated Diagnostics, University of Genoa, Genoa, Italy

Fernando Pitossi
Leloir Institute Foundation, Institute for Biochemical Investigations, CONICET, 1405 Buenos Aires, Argentina

L.M. Popescu
Division of Advanced Studies, Victor Babeş, National Institute of Pathology, Bucharest, Romania

Patricia Rieken Macêdo Rocco
Instituto de Biofísica Carlos Chagas Filho; Programa de Terapia Celular e Bioengenharia; Laboratório de Investigação Pulmonar; Instituto Nacional de Ciência e Tecnologia de Biologia Estrutural e Bioimagem -INBEB; Universidade Federal do Rio de Janeiro, Rio de Janeiro, RJ, Brazil

Isao Sakaida
Division of Laboratory, Yamaguchi University Hospital, Yamaguchi University School of Medicine, Japan; Department of Gastroenterology and Hepatology, Yamaguchi University, Graduate School of Medicine Yamaguchi University, Japan

Marcelo F. Santiago
Instituto de Biofísica Carlos Chagas Filho; Laboratório de Neurobiologia Celular e Molecular; Programa de Terapia Celular e Bioengenharia; Instituto Nacional de Ciência e Tecnologia de Biologia Estrutural e Bioimagem -INBEB; Universidade Federal do Rio de Janeiro, Rio de Janeiro, RJ, Brazil.

Eliana Scemes
The Dominick P. Purpura Department of Neuroscience, Albert Einstein College of Medicine, New York, New York, USA

Sonya B. Seif-Naraghi
Department of Bioengineering, University of California, San Diego, La Jolla, California, USA

Mari C. Sogayar
Instituto de Química, São Paulo, Brazil

Russell S. Taichman
Department of Periodontics and Oral Medicine, University of Michigan, Ann Arbor, Michigan, USA

Taro Takami
Division of Laboratory, Yamaguchi University Hospital, Yamaguchi University School of Medicine, Japan; Department of Gastroenterology and Hepatology, Yamaguchi University, Graduate School of Medicine Yamaguchi University, Japan

Christina M. Takiya
Instituto de Biofísica Carlos Chagas Filho; Laboratório de Patologia Celular; Programa de Terapia Celular e Bioengenharia; Instituto Nacional de Ciência e Tecnologia de Biologia Estrutural e Bioimagem -INBEB; Universidade Federal do Rio de Janeiro, Rio de Janeiro, RJ, Brazil

Shuji Terai
Division of Laboratory, Yamaguchi University Hospital, Yamaguchi University School of Medicine, Japan; Department of Gastroenterology and Hepatology, Yamaguchi University, Graduate School of Medicine Yamaguchi University, Japan

Andréa Gonçalves Trentin
Departamento de Biologia Celular, Embriologia e Genética, Centro de Ciências Biológicas, Universidade Federal de Santa Catarina, Campus Universitário-Trindade, 88040-900, Florianópolis, Santa Catarina, Brazil

Karine S. Verdoorn

Instituto de Biofísica Carlos Chagas Filho; Laboratório de Físico-Química Biológica Aída Hassón-Voloch; Programa de Terapia Celular e Bioengenharia; Instituto Nacional de Ciência e Tecnologia de Biologia Estrutural e Bioimagem -INBEB; Universidade Federal do Rio de Janeiro, Rio de Janeiro, RJ, Brazil

Hellen J. Vieira-Beiral

Instituto de Biofísica Carlos Chagas Filho; Laboratório de Físico-Química Biológica Aída Hassón-Voloch; Programa de Terapia Celular e Bioengenharia; Instituto Nacional de Ciência e Tecnologia de Biologia Estrutural e Bioimagem -INBEB; Universidade Federal do Rio de Janeiro, Rio de Janeiro, RJ, Brazil

Adalberto Vieyra

Instituto de Biofísica Carlos Chagas Filho; Laboratório de Físico-Química Biológica Aída Hassón-Voloch; Programa de Terapia Celular e Bioengenharia; Instituto Nacional de Ciência e Tecnologia de Biologia Estrutural e Bioimagem -INBEB; Universidade Federal do Rio de Janeiro, Rio de Janeiro, RJ, Brazil

Oliver Wessely

Cleveland Clinic, Lerner Research Institute, Department of Cell Biology, Cleveland, Ohio, USA

Camila Zaverucha-do-Valle

Instituto de Biofísica Carlos Chagas Filho; Laboratório de Neurobiologia Celular e Molecular; Programa de Terapia Celular e Bioengenharia; Instituto Nacional de Ciência e Tecnologia de Biologia Estrutural e Bioimagem -INBEB; Universidade Federal do Rio de Janeiro, Rio de Janeiro, RJ, Brazil

Chapter 1

Resident Stem Cells in Kidney Tissue

Milene R. Costa[*,†,‡,§,**], Karine S. Verdoorn[*,†,‡,§,**], Rafael S. Lindoso[*,†,‡,§,**], Marcelo Einicker-Lamas[*,†,‡,§,**], Hellen J. Vieira-Beiral[*,†,‡,§,**], Oliver Wessely[††], Adalberto Vieyra[*,†,‡,§,**]

[*] *Instituto de Biofísica Carlos Chagas Filho, Rio de Janeiro, Brazil*

[†] *Programa de Terapia Celular e Bioengenharia, Rio de Janeiro, Brazil*

[‡] *Laboratório de Físico-Química Biológica Aída Hassón-Voloch, Rio de Janeiro, Brazil*

[§] *Instituto Nacional de Ciência e Tecnologia de Biologia Estrutural e Bioimagem -INBEB, Rio de Janeiro, Brazil*

[**] *Universidade Federal do Rio de Janeiro, Rio de Janeiro, Brazil*

[††] *Cleveland Clinic, Lerner Research Institute, Department of Cell Biology, Cleveland, Ohio, USA*

Introductory Remarks

The mammalian kidney is recognized for its highly specialized structures and terminally differentiated cells arranged in an elaborate form to accomplish its function. Although these specialized cells have a low rate of renewal, the mammalian kidney possesses an inherent ability to regenerate or repair after ischemic or toxic injury, but the regenerative process is not fully understood. In the past few decades, investigations have concentrated on three main issues: (1) contribution of extrarenal stem cells (e.g., hematopoietic and mesenchymal bone marrow cells), (2) intrarenal regeneration by dedifferentiation of surviving tubular cells, and (3) intrarenal regeneration by recruitment of resident renal stem cells. The very low rate of extrarenal stem cell transdifferentiation into specialized kidney cells reinforces the intrinsic renal regenerative capacity. In parallel, the discovery of adult stem cells in many different tissues encouraged the search for renal resident stem/progenitor cells. This issue challenged well-established concepts and raised a new paradigm: renal

Resident Stem Cells and Regenerative Therapy. http://dx.doi.org/10.1016/B978-0-12-416012-5.00001-3

resident stem/progenitor cells exist and may contribute to tissue regeneration. Identification and characterization of these stem cells has presented an enormous challenge, given the complexity of the kidney structure. In this chapter, we will discuss the main findings to date and debate some of the open questions. The structural organization of the kidney and its multiple specialized cell types are important details when considering the susceptibility of kidney cells to injury and their ability to regenerate. We therefore present a brief overview in the following section of the structural characteristics of the kidney to give a better understanding of the issues that will be considered later in the chapter.

Brief Description of the Structural Organization of the Mammalian Adult Kidney

A longitudinal section of a mammalian kidney shows three clearly delimitated regions: cortex, external medulla, and internal medulla (Fig. 1.1). A close-up of the cortex moreover allows the recognition of two subregions, the more superficial

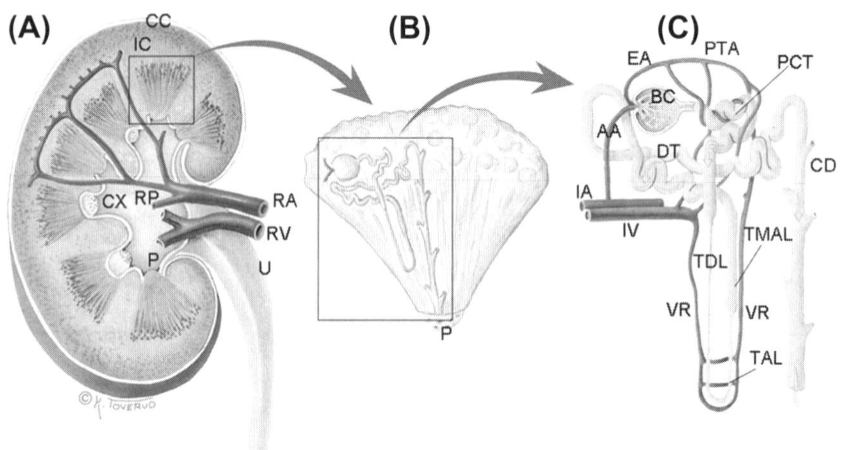

Figure 1.1 Structural organization of an adult kidney. **(A)** Longitudinal section of a mammalian kidney. CC, *cortex corticis*; IC, internal cortex; CX, calyx; P, papilla. To complete the description of the morphologic organization of an adult kidney, panel *A* also shows the renal artery (RA), the renal vein (RV), the renal pelvis (RP), and the ureter (U). **(B)** Enlargement of the area indicated in *A* shows the location of cortical and juxtamedullary nephrons. **(C)** Amplified picture of a nephron and its association with the vascular network. AA, afferent arteriole; BC, Bowman's capsule; EA, efferent arteriole; PTA, peritubular arteriole; PCT, proximal convoluted tubule; TDL, thin descending limb; TAL, thin ascending limb; TMAL, thick medullary ascending limb; DT, distal tubule; CD, collecting duct; IA, interlobular artery; IV, interlobular vein; VR, *vasa reta*. *(© Kari C. Toverud CMI.)*

cortex corticis (CC) and the internal cortex (IC) (Fig. 1.1A). The *cortex corticis* is particularly important in the context of this chapter because its cell population comprises >90% of the proximal tubule cells that are important targets in acute renal injury (AKI). The subregion called papilla (P) is found in the tip of the internal medulla (Fig. 1.1A and B), where a niche of stem/progenitor cells was described some years ago [1] (discussed later). The functional unit of the kidney, the nephron (more than 1 million in a human kidney; Fig. 1.1B and C), is a tubular structure, in which different segments can be distinguished by their specialized cell morphology and by their different permeability and transport properties. The nephron starts in the cortex with the glomerular or Bowman's capsule (BC) that receives the ultrafiltrate from a capillary network that originates from the afferent arteriole (AA), a vessel of the interlobular artery (IA) (Fig. 1.1B). As shown in Figure 1.2 the endothelial cells

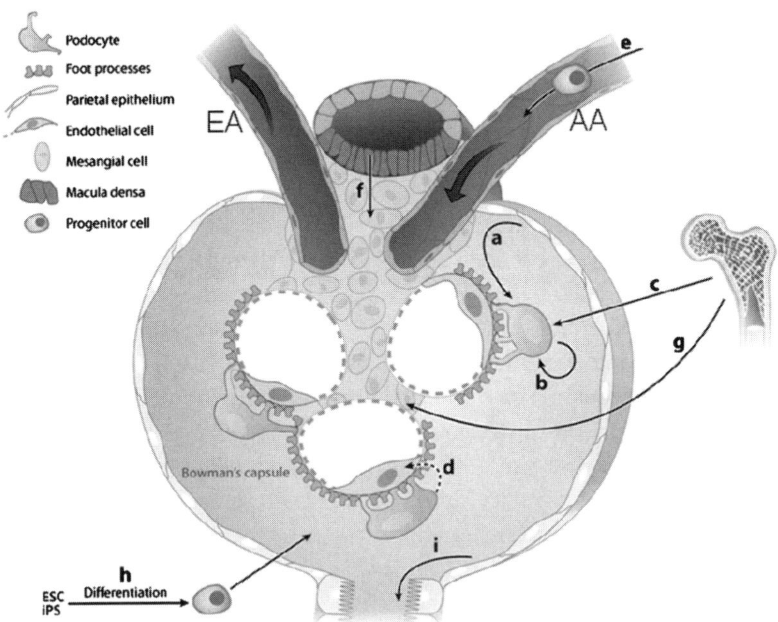

Figure 1.2 Detailed view of the glomerular compartment. AA, afferent arteriole; EA, efferent arteriole. The lowercase letters indicate possible pathways for cell replacement; *a-c*: podocyte replacement from the capsular epithelium, by self-renewal or by circulating stem cells; *d,e*: endothelial cell replacement from resident stem/progenitor cells or circulating cells; *f,g*: replacement of mesangial cells from *macula densa* or circulating cells; *h*: potential route of regeneration with the use of embryonic stem cells (ESC) or induced pluripotent stem (iPS) cells; *i*: possible regeneration of proximal tubule cells from parietal capsular epithelial cells. *(From [2], with permission; see this reference for details.)*

(EC) of the intertwined vessels are organized in a globular structure by two additional cell types, the podocytes (PD) and the mesangial cells (MC) [2]. Although the podocytes are critical structures for ultrafiltration—and are also especially sensitive to injuries—the mesangial cells are the more important targets for the action of stem cells in any potential regenerative process (Fig. 1.2).

After the capsular space, the next segment is the proximal convoluted tubule (PCT)[1] (Fig. 1.1B and C). Its terminal portion, *pars recta* (PR) dips into the outer medullary region. From here the tubule forms a hairpin structure (Henle's loop) (Fig. 1.1B and C), which is subdivided in three regions based on their differentiated structure and transport properties: the thin descending limb (TDL), the thin ascending limb (TAL), and the thick medullary ascending limb (TMAL). Reaching again the cortex, the distal tubule (DT) is the last nephron segment.[2] It fuses with the collecting duct (CD), which merges into the medulla and ends in the calyx (CX) (Fig. 1.1A). From a functional point of view, it is remarkable that specialized tubular cells in the region of the terminal part of the ascending limb and the early portion of the distal tubule (*macula densa* cells) (MD) (Fig. 1.2) form a complex (the juxtaglomerular apparatus) with cells of the afferent arteriole. This organized structure is key in the most important biochemical pathway of the kidney, the renin/angiotensin system (RAS), which plays a pivotal role in kidney disease and regeneration [5].

In respect to the vasculature, the glomerular capillary network described earlier converges to form the efferent arteriole (EA) (Fig. 1.1B and C; Fig. 1.2). This vessel exits the glomerulus and ramifies around the proximal and distal convoluted tubules, forming the peritubular arteriolar network (PTA) (Fig. 1.1C); it then fuses into small venules that drain into the interlobular veins (IV) (Fig. 1.1C). In the case of the juxtamedullary nephrons (i.e., the deeper nephrons below the *cortex corticis*) the efferent arteriole forms a straight-bore vessel that merges into the medulla (the descending arm of the *vasa reta*) (VR) in close proximity with the ascending limb of Henle's loop up to the papillary tip. Here it turns to ascend toward the external medulla as a venule (the ascending arm of the *vasa reta*) that finally ends into an interlobular vein [6]. Along their trajectories, the descending and ascending *vasa reta* anastomose with one another several times.

The special organization of the renal vasculature, with its intense blood flow in the cortex, and a lower oxygen and substrate supply in the medulla results in different metabolic characteristics to the three main kidney areas. Although in the cortex,

---

[1]The proximal convoluted tubule can be subdivided in segments with different morphologic and functional properties called S1, S2, and S3 [3]. Although they appear to have different regenerating capacity and different interactions with stem cells, this topic is beyond the scope of this chapter.

[2]Again, different subsegments have been described on the basis of their cellular composition and specialized transport properties [4], but these will not be considered in this chapter.

intense oxidative metabolism occurs (~10% of the oxygen consumption of the body at rest), the metabolism in the inner medulla is almost purely glycolytic [7]. The cells of the outer medulla show mixed metabolic characteristics depending on the blood flow and the local oxygen pressure. Interestingly, these differences in local oxygen pressure are important in AKI episodes and, possibly, in kidney regeneration.

Stem Cells in Kidney Organogenesis

The nephron epithelia and collecting system of the mammalian kidney derive from two precursor tissues, both of the intermediate mesoderm (Fig. 1.3) [8]. Early in the mouse embryogenesis at approximately day 9.5 (E9.5), the nephric duct (ND), or Wolffian duct, a simple epithelial tube, originates from a solid cord of cells from the dorsal intermediate mesoderm (IM). Interactions between the nephric duct and the undifferentiated ventral intermediate mesoderm, the nephrogenic cord, give rise to the pronephric and mesonephric kidneys. In mammals, these more primitive kidney forms are replaced soon thereafter by the adult metanephric kidney. At E10.5, mesenchymal-epithelial interactions between the nephric duct and the metanephric mesenchyme (MM)—a more specialized region of the intermediate mesoderm—will form the ureteric bud (UB) near its caudal end. The UB is an epithelial protrusion from the nephric duct into the MM. Reciprocal inductive signals from the UB and the MM induce growth and specialization of these two components. The UB will undergo elongation and repeated branching into the MM, giving rise to the collecting system, whereas the cells of the MM aggregate around the UB tips and undergo mesenchymal-epithelial transition to form a polarized renal vesicle (Fig. 1.4) [9]. This renal vesicle will contact the tip epithelium of the UB, generating the comma-shaped body, after which elongation of this structure forms a second cleft generating the S-shaped body. The distal end of the S-shaped body will subsequently fuse with the UB epithelium, while the proximal end will be invaded by blood vessels and stromal cells forming the glomerular tuft.

This sequence of events is coordinated by reciprocal signaling between the UB and MM cells and accompanied by multiple steps of differential gene expression and stem/progenitor cell maintenance or differentiation. The first precursor cell population that will give rise to all kidney structures is the intermediate mesoderm. Differentiation of these cells from the surrounding tissues begins with the expression of the genes that encode the two transcription factors *Lhx1/Lim1* (LIM homeobox 1) and *Osr1/Odd1* (odd-skipped related 1) [8,9]. *Lhx1* activity is required for the intermediate mesoderm differentiation, but its expression, like the expression of *Osr1*, also extends to the lateral plate mesoderm [10].

Once the nephric anlage has been specified, differentiation of the nephric duct from the intermediate mesoderm is induced by concentration gradients of bone

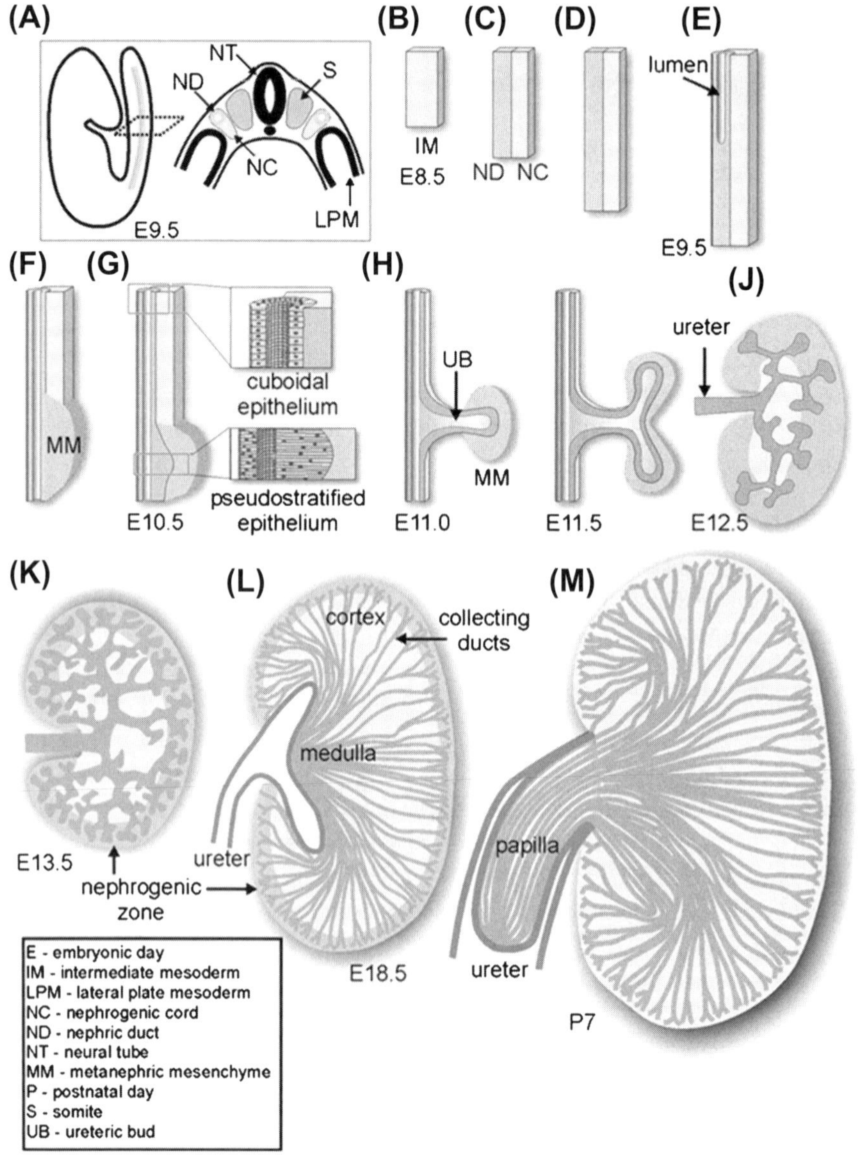

Figure 1.3 Renal organogenesis: origins of metanephric mesenchyme and ureteric bud. Mammalian metanephric kidneys develop from the intermediate mesoderm (IM) located bilaterally to the neural tube (NT), between the lateral plate mesoderm (LPM) and the somites (S) in the caudal region of the embryo **(A)**. The nephric duct (green, ND)—a simple epithelial tube originated from a solid cord of cells at the dorsal intermediate mesoderm **(C-E)**—interacts with the nephrogenic cord (blue, NC), the undifferentiated ventral IM. The nephric duct undergoes elongation and epithelialization **(D** and **E)**

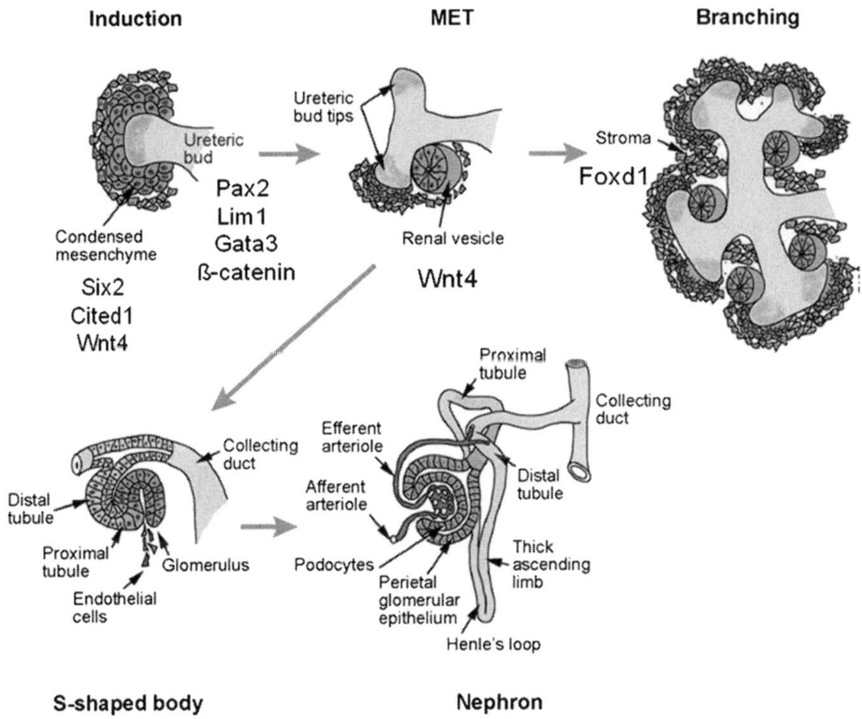

Figure 1.4 Renal organogenesis: specification of the metanephric mesenchyme and ureteric bud. The ureteric bud progenitor population is characterized by Pax2, Lim1, Gata3, and β-catenin expression. The metanephric mesenchyme is settled by different precursor populations: those that give rise to the stromal cells, which are Foxd1$^+$, and those that form the condensed mesenchyme, which are Six2$^+$. In the condensed mesenchyme, two subpopulations can be distinguished, those of the capping mesenchyme, which are Six2$^+$, Cited1$^+$, and Wnt4$^-$, and those of the induced mesenchyme, which are Six2$^+$, Cited1$^-$, and Wnt4$^+$. The cells of the induced mesenchyme form the renal vesicle via mesenchymal-epithelial transition (MET); they continue to express Wnt4 and give rise to all tubular epithelial cells. Differently, the cells of the capping mesenchyme stay undifferentiated and proliferate, providing nephron epithelial precursors until completion of renal development. *(From [9], with permission.)*

while the caudal end of the NC specializes into metanephric mesenchyme (pink, MM) **(F)**. The pseudo-stratified epithelium in the caudal ND gives rise to the ureteric bud (UB) **(G)**. Interactions between the UB and the MM promote the UB outgrowth **(H)**. and repeated branching within the MM **(I-K)**. The elongation of the collecting ducts forms the medulla **(L)** and papilla **(M)**. E, embryogenesis circa day, and P, postnatal day. *(From [8], with permission.)* **See Plate 1.**

morphogenetic proteins, activins, and retinoic acid, from the axial and paraxial mesoderm. With the specification of the nephric duct, *Lhx1* expression becomes restricted to this structure [11]. At this stage, co-expression of *Pax2* and *Pax8* becomes crucial for the continued expression of *Lhx1* and mesenchymal-epithelial transition. *Lhx1, Pax2,* and *Pax8* then induce the activation of the transcription factor *Gata3* and the signaling molecule *Wnt/β-catenin (Ctnnb1)* to guarantee appropriate UB growth and renal development. Mutations in any of the genes (*Pax2, Lhx1, Gata3,* and *β-catenin*) result in developmental defects. *Lhx1* is required for the correct nephric duct extension and patterning of all epithelial structures derived from the intermediate mesoderm [12]. Because *Pax2* and *Pax8* belong to the same class of transcription factors, they have redundant functions in the kidney. Whereas *Pax8* null mutant mice have no apparent phenotype, *Pax2* mutants develop a nephric duct that subsequently degenerates [13]. However, *Pax2/Pax8* double homozygous null mice have a pronounced kidney phenotype, being unable to generate any epithelial structure and completely failing to generate the nephric duct [14]. *Osr1* mutants can generate a nephric duct and express *Pax2,* but they do not develop mesonephric tubules or the metanephric kidney [15]. All these molecules are necessary for adequate proliferation and differentiation of cells that form the UB. These precursor cells have the potential to give rise only to the cells of the collecting system, whereas all the other cells of the nephron derive from the MM.

One interesting transcription factor in respect to multipotency is *Osr1*. It is detected in all renal precursors and also those from the adrenal gland and gonad. Moreover, it is clearly a marker of the stem/progenitor cells, and, as cell fates progress, it is no longer expressed. The nephric duct cells are the first to lose *Osr1* expression and because of that are considered the first committed renal cell population. *Osr1* activity is essential for the establishment of the stem/progenitor cells of the MM; its expression becomes restricted to this site at the very beginning of nephrogenesis (around E9.5) when the UB invades the MM [16]. Interstitial stromal cells (pericytes, mesangial, and capsule cells) are also derived from *Osr1*$^+$ precursors between E7.5 and E11. Although these epithelial and interstitial cells are believed to have a common precursor in the intermediate mesoderm, it is possible that their precursors commit early before E9.5 and constitute *Osr1*$^+$ heterogeneous population of cells.

The major role of *Osr1* in the MM is to induce and maintain *Pax2* and *Eya1* expression. These factors form a complex with *Hox11* and synergistically activate *Six2* and *Gdnf*. *Six2* activity is crucial for the cap mesenchyme to maintain the renal vesicle-forming progenitors in a self-renewing mesenchymal state [17,18]; *Gdnf* is required for the invasion of the UB within the MM and for its repeated branching [19]. The MM differentiates into a pool of highly active self-renewing cells ("capping mesenchyme"). These cells are characterized by *Six2* and *Cited1* expression and are *Wnt4* negative. They will differentiate into the so-called induced mesenchyme, when epithelial differentiation has already started, and are *Six2*$^+$, *Cited1*$^-$, and *Wnt4*$^+$.

*Cited*1 activity is responsible for the attenuation of canonical *Wnt* signaling that induces a nonrenewing and committed cell phenotype [20]. The cells of the "capping mesenchyme" might be the stem cells of the entire nephron epithelia. In fact, the ability of maintaining a stem cell-property among the epithelial precursors has been demonstrated only for cells expressing *Six2* [21,22]. In addition, all the renal tubular specialized epithelial cells originate from the *Six2*$^+$ stem/progenitor cell population of the MM, the interstitial stromal cells being derived from *Foxd*1 expressing cells.

The maintenance of the nephron stem/progenitor cell pool in the MM is accomplished by reciprocal inductive signals with the UB and the expression of several important molecules. *WT1* guarantees the differentiation competence of the progenitors and their survival [23]. *Sall1* expression provides the nephrogenic multilineage ability to the stem/progenitor cells [24]. Growth factors such as fibroblast growth factor (FGF) and bone morphogenetic protein (BMP) also play essential roles in the survival, renewal, and differentiation of stem/progenitor cells [9]. But among all these molecules, only *Six2* and *Wnt9b* have been directly associated to renal stem cell potential [25,26].

Based on the observation that the *Six2* expressing cells are "true" renal stem cells and the fact that *Six2* expression disappears after completion of kidney development, the existence of adult renal stem cells was contested [9,17,27]. However, other multipotent cell populations in the kidney exist and they express other marker genes that have been successfully used for identification purposes (discussed later).

Kidney Stem/Progenitor Cells: Niches and Isolation

As noted earlier, the lack of a specific marker for adult kidney stem/progenitor cells is one of the major challenges in the identification and isolation of these cells. So far, methodological approaches have relied on the detection of renal cells that exhibit characteristics of stem cells from different tissues (Table 1.1). These strategies have provided clear evidence that stem cells might exist in the adult kidney and that these cells are located in specific niches in the organ (for recent reviews see [28-30]).

One of the first approaches used to isolate kidney stem/progenitor cells was based on the finding that stem cells are slow-cycling cells characterized by low rates of DNA replication [28-30]. A common method to detect slow cycling cells consists of labeling cells with the synthetic nucleoside bromodeoxyuridine (BrdU) that incorporates into the newly synthesized DNA during replication. After consecutive cell divisions, the label gradually is diluted as new molecules of DNA are synthesized with unlabeled nucleosides. However, in the case of slow cycling cells, the BrdU labeling persists for long periods and can be used as a criterion for stem cell isolation. Using this approach Maeshima et al. [31] detected BrdU$^+$ cells in the renal tubules of normal rat kidney (Fig. 1.5). They also observed that these

TABLE 1.1 Isolation Methods and Niches of Kidney Adult Stem/Progenitor Cells

Reference	Niches	Isolation Method	Stem Cell Marker	Functional Tests for Stem Cell Attributes
Maeshima et al. (2003) [31]	Tubule	BrdU labeling	Vimentin	The cultured progeny of BrdU-labeled cells isolated after renal ischemia expressed vimentin and eventually E-cadherin
Iwatani et al. (2004) [34]	Not assigned	Hoechst 33342 extrusion	Sca-1, CD45, c-Kit	Wild-type adult rats that received intravenous injection of EGFP-labeled kidney derived Hoechst cells showed $EGFP^{+}$-labeled cells in skeletal muscle, liver, and bone marrow cells
Oliver et al. (2004) [1]	Papilla	BrdU labeling	α-smooth muscle actin	Formation of cellular spheres in cell culture, differentiation to myofibroblasts and neurons
Bussolati et al. (2005) [41]	Interstitium	Magnetic cell sorting	CD133 PAX2	Differentiation into epithelial or endothelial cells
Dekel et al. (2006) [39]	Interstitium	Magnetic cell sorting	Sca-1	Differentiation into myogenic, osteogenic, adipogenic, and neural lineage
Gupta et al. (2006) [47]	Tubule	Selective cell culture	Oct4, Pax2, and vimentin	Expression of endothelial, hepatocyte, and neural markers and differentiation into tubular cells
Maeshima et al. (2006) [32]	Tubule	BrdU labeling	None	Formed tubule-like or tubulocystic structures in response to growth factors when cultured in collagen gel
Sagrinati et al. (2006) [42]	Glomerulus	Magnetic cell sorting	CD133 and CD24	Self-renewal and high clonogenic potential, capacity to differentiate into tubular cells, osteogenic cells, and adipocytes
Bruno et al. (2009) [45]	Glomerulus	Magnetic cell sorting	CD116, CD25, and PAX2	Differentiation into osteogenic, chondrogenic, and adipogenic lineages and epithelial cells that express podocyte markers
Ronconi et al. (2009) [43]	Glomerulus	Magnetic cell sorting	CD133 and CD24	Differentiation into tubular cells and podocytes

cells became positive for proliferating cell nuclear antigen (PCNA) labeling after ischemia/reperfusion injury, which suggested that these cells underwent cell division and participated in renal repair during the recovery period. The same authors demonstrated that the BrdU-labeled cells isolated from tubules of rat adult

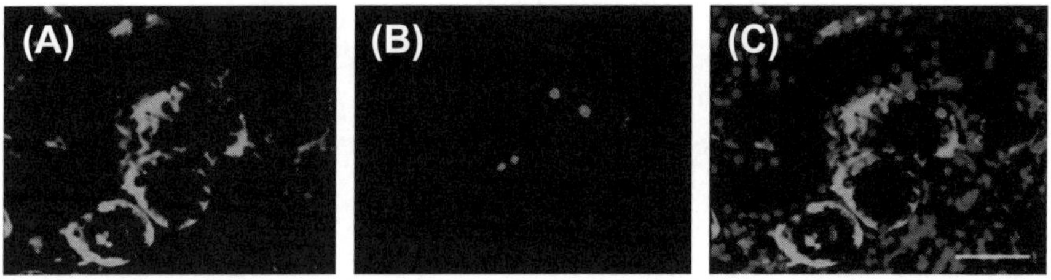

Figure 1.5 BrdU-positive cells in renal tubules of normal rat kidneys. **(A)** Kidney cells stained (green) for Lotus tetragonolobus agglutinin (a lectin that specifically labels proximal tubules). **(B)** BrdU labeling (red). **(C)** Merged images with DAPI blue nucleus. *(From [31], with permission.)* **See Plate 2.**

kidney are capable of forming tubulocystic structures in collagen gel culture and differentiate into fibroblast-like cells. Additionally, they found that these cells integrate into the proximal tubule and ureteric bud when they are transplanted into the metanephric kidney [32].

In a similar approach Oliver et al. [1] used BrdU labeling to demonstrate that the stem/progenitor cells reside in the renal papilla. This region is enriched with label-retaining cells in comparison to other regions of the kidney. They demonstrated that these cells quickly disappeared after transient renal ischemia, which might indicate that they began to proliferate during the repair phase after injury. Additionally, they showed that the label-retaining cells from the papilla display stem cell abilities, such as the formation of spheres in 3D culture and differentiation into other cell lineages, including myofibroblasts and neurons. They also showed that, as a homeostatic response, the label-retaining cells of the papilla migrate to the upper papilla to create a compartment of highly proliferative cells that could be a possible source of new renal cells after ischemia [33].

In a different approach Iwatani et al. [34] isolated stem/progenitor cells from kidney by the ability of stem cells to extrude Hoechst 33342 dye (Fig. 1.6) [35]. Those authors stained a cell suspension from adult rat kidney with Hoechst 33342 dye. Flow cytometric analysis showed that low Hoechst fluorescence cells had a frequency of 0.03% to 0.1% in the kidney preparation. They also investigated the ability of these cells to differentiate in vivo. They isolated low Hoechst fluorescence

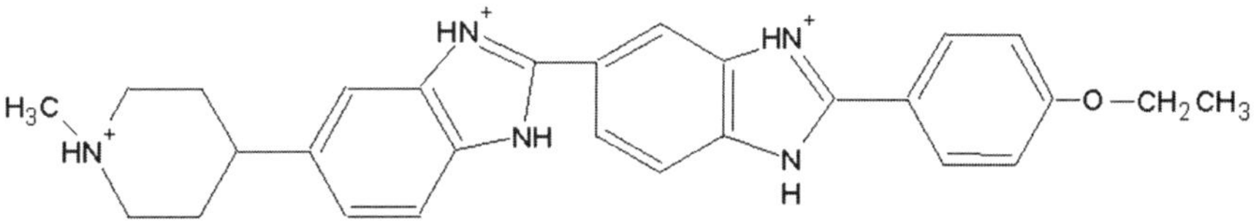

Figure 1.6 Chemical structure of Hoechst 33342 dye.

cells from the kidney of enhanced green fluorescent protein (EGFP) rats and transplanted them into wild-type rats. After 10 weeks, skeletal muscle, bone marrow, and liver cells expressing EGFP were detected in transplanted animals, suggesting that renal stem/progenitor cells were potentially capable of generating different cell lineages.

Recently, the identification of specific cell surface markers that specify stem cells—such as stem cell antigen-1 (Sca-1), CD133, CD24—has broadened the possibilities of isolating and identifying renal stem cells [36]. Sca-1 is an 18 kDa mouse glycosylphosphatidylinositol-anchored cell surface protein [37] expressed by different types of stem/progenitor cells in various tissues and organs [38]. Dekel et al. [39] used magnetic cell sorting to isolate nontubular cells that express Sca-1 in adult mouse kidney interstitium. The authors demonstrated that these cells were capable of differentiating into different cells types as myogenic, osteogenic, adipogenic, and neural lineages.

Another marker frequently used for isolating kidney stem cell is CD133 antigen, a glycosylated protein expressed on hematopoietic stem/progenitor cells, but also in embryonic kidneys [40]. Bussolatti et al. [41] used magnetic cell sorting to isolate $CD133^+$ cells from the tubular fraction of the cortex of human adult kidney. The authors found that these cells expressed the embryonic renal marker PAX2, which suggested their renal origin. $CD133^+$ cells could differentiate in vitro into epithelial or endothelial cells and, when injected in glycerol-induced tubulonecrotic mice, migrated and integrated in tubules, suggesting that $CD133^+$ cells are multipotent adult resident stem cells that participate in repair of renal injury [41].

CD133 has also been used in conjunction with other surface markers to identify specific kidney stem/progenitor cells. CD24 and CD133 co-expressing cells were isolated from a subset of parietal epithelial cells in the Bowman's capsule of adult human kidney [42]. CD24 is a surface molecule present in different types of human stem cells, but also in metanephric mesenchyme during kidney development [36]. These cells presented self-renewing ability and high cloning efficiency, both typical properties of stem cells. The cells could be induced to differentiate into tubular cells, osteogenic cells, and adipocytes. Also, injection of $CD133^+/CD24^+$ cells helped to regenerate tubular structures of the nephrons of severe combined immunodeficiency mice (SCID) after acute renal failure. Importantly, Ronconi et al. [43] have also identified a $CD133^+/CD24^+$ population of cells in the Bowman's capsule of human kidneys. These cells contributed to both podocyte and tubular epithelium regeneration when injected into SCID mice. Consistent with these findings, Appel et al. [44] showed that there are podocyte progenitor cells in the urinary pole of Bowman's capsule, which migrate into the glomerular tuft to replace damaged podocytes.

Interestingly, the cell surface marker CD146 defines another niche of resident stem cells within the glomerulus different from those of Bowman's capsule.

Bruno et al. [45] found that the human glomerulus deprived of Bowman's capsule contains a resident population of mesenchymal cells lacking CD133 but expressing CD146. CD146 is a 113 kDa membrane glycoprotein recognized as a marker of human bone mesenchymal cells [46]. Isolated $CD146^+/CD133^-$ cells expressed mesenchymal stem cell markers CD29, CD44, CD73, CD90, CD105, and CD166, and also renal stem cell markers, such as CD24 and Pax2. These cells showed self-renewal capability, osteogenic, adipogenic, and chondrogenic differentiation properties and differentiated into endothelial and epithelial cells. As an example, Bruno et al. [45] demonstrated that $CD146^+/CD133^-$ cells expressed α-smooth muscle antigen and angiotensin II type 1 receptors after multiple divisions in appropriate culture conditions.

A third approach for isolating stem/progenitor cells from rat kidney was used by Gupta et al. [47]. The technique is similar to the one used to isolate mesenchymal stem cells from bone marrow. It consists of harvesting, mincing, and digesting the kidney to obtain a single cell suspension. The suspension is then plated on fibronectin-coated flasks at low cellular density to avoid cell-to-cell contact and then cultured in the presence of epidermal growth factor (EGF), platelet-derived growth factor subunit B (PDGF-B), and leukemia inhibitory factor (LIF) at 37°C in the presence of 5% CO_2. After successive passages, the culture became monomorphic with spindle-shaped cells. The isolated cells were capable of long-term self-renewal and expressed the stem cell marker Oct4 and the transcription factor Pax2. Interestingly, putative $Oct4^+$ stem cells could be detected in the proximal tubules of rats. Finally, the authors demonstrated that these cells could differentiate into tubular epithelial cells after injection into normal kidneys or following ischemia-reperfusion injury.

Together the studies presented here support the idea that stem/progenitor cells exist in the adult kidney and can be isolated from different niches such as the renal papilla, the glomerulus, and the renal tubules. However, there are still many unanswered questions about the characterization of the niches as well as the specific markers that the cells express. According to Pleniceanu et al. [30], the criteria and methods used so far to isolate and identify renal stem/progenitor cells are often questionable regarding their specificity and efficiency. For example, there are cells other than stem cells that exhibit slow-cycling characteristics, including differentiated renal cells. Similarly, cell populations isolated by Hoescht 33342 extrusion are often heterogeneous, making this criterion inaccurate to isolate stem/progenitor cells. Finally, some of the surface markers used to isolate stem cells are also expressed in differentiated epithelia. However, despite such controversy, many research groups are concentrating their efforts on the study of renal resident stem/progenitor cells. This, to a large extent, is due to the therapeutic potential of stem cells in kidney repair as well as other renal pathologies that will be discussed in the following sections.

The Role of Stem/Progenitor Cells in Kidney Regeneration

As mentioned throughout the book, adult stem cells are involved in organ maintenance and repair after injury. The presence of these cells confers a rapid self-repair system as has been shown in several organs including the kidney [48,49]. Among the highly specialized different cell types present in the kidney (Fig. 1.1), the tubular epithelial cells are one of the most susceptible to ischemic and toxic insults [2]. This is due to the renal architecture and transport demand of tubular cells and depending on the severity, such insults can result in necrosis, apoptosis, and detachment of cells [50]. The second most affected region is the glomerulus where several insults are associated with the impairment of its function, such as intraglomerular hypertension, hyperfiltration, and loss of selective properties of the glomerular barrier [51]. The initial insults are often followed by excessive accumulation of ultrafiltered proteins in the lumen of tubules and Bowman's space, triggering inflammatory responses, and apoptotic pathways that will lead to progressive increase of the injured area. Therefore, the repair of the injured kidney relies on the rapid replacement of lost renal cells and the reestablishment of normal tubular transport properties.

Under normal physiologic conditions, adult kidney tissue has a low cell turnover, with an average of $<0.5\%$ of tubular cells proliferating. However, after kidney injury, the proliferation rates of different renal cells are significantly increased [52,53]. These new fully functional cells replace renal cells that were lost during injury. Although this process is clearly associated with an increase in proliferation, the origin of the new renal cells remains controversial. In this section, we explore how surviving cells and stem cells participate in kidney repair (Fig. 1.7). One of the proposed mechanisms of kidney regeneration is repopulation of the injured area by surviving differentiated cells [54].[3] The other mechanism is that resident stem/progenitor cell populations localized in specific renal niches [47,55] are mobilized to replace the lost cells. Thereby, the regeneration process can be mediated by surviving renal cells, stem/progenitor cells, or by a combination of all cell types (Fig. 1.7).

Renal tissue regeneration mediated by surviving epithelial cells relies on the capacity of the cells to dedifferentiate via epithelial-mesenchymal transition [56]. Once undifferentiated, these proliferate to replace the cells lost during injury and then redifferentiate into fully specialized adult renal cells culminating with the recovery of tissue function [54]. Indeed, histologic analyses of injured kidneys show

[3] At first, this mechanism appears not to involve progenitor cells. However, resident stem/progenitor cells can stimulate the proliferation of remnant surviving epithelial cells, as considered later.

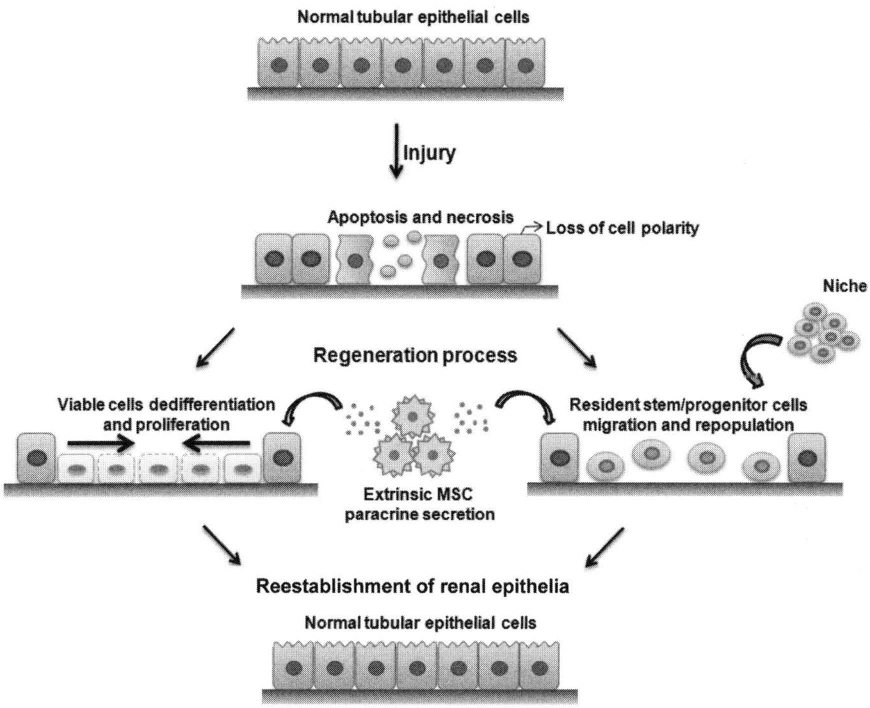

Figure 1.7 Possible regenerative mechanisms involving renal stem/progenitor cells and the role of extrinsic mesenchymal cells.

the presence of partially dedifferentiated cells that have both mesenchymal and tubular markers, reexpressing mesenchymal proteins—vimentin, neural cell adhesion molecule (NCAM), Pax2 [57,58]—and tubular epithelial proteins, such as megalin and Tamm-Horsfall glycoprotein (markers of proximal tubule and ascending limb, respectively) [59,60]. Similarly, glomerular parietal epithelial cells can dedifferentiate spontaneously, giving rise to immature glomeruli in a uninephrectomized mouse model [61]. The dedifferentiation could confer plasticity on terminally differentiated renal cells, resulting in functionally active progenitors (Figs. 1.2 and 1.7).

However, the capacity to dedifferentiate into an early renal cell phenotype may be restricted to some but not all tubular subpopulations. Langworthy et al. [62] have pointed to the existence of tubular cells that increase the expression of *Nfatc1* after injury; this transcription factor confers resistance to apoptosis and enhance proliferative capacity to these cells. Indeed, heterozygous mice for *Nfatc1* have a diminished regeneration capacity that results in the increase of fibrotic processes after tubule injury. Also noteworthy is the fact that some renal stem/progenitor cell identification studies do not clearly distinguish between adult renal stem/progenitors and surviving dedifferentiated epithelial cells. In this sense, Maeshima et al. [31] identified label-retaining cells among tubular epithelial cells and demonstrated that they have self-renewal capacity and are able to differentiate into tubular epithelial cells after injury. These label-retaining cells could be adult renal stem/progenitors but also early differentiated epithelial cells, besides extrarenal stem/progenitors. In addition, Gupta et al. [47] isolated $CD133^{-}/Oct4^{+}/Pax2^{+}$ renal cells that were able to differentiate into epithelial tubular cells. Once more it is difficult to identify the origin of these cells, and further studies are necessary to clearly address this issue.

Direct contribution of the identified kidney stem/progenitor cell populations (papilla, glomeruli, and proximal tubules) in the regeneration process remains to be demonstrated. Because these are undifferentiated cells per se, they could give rise to different renal cell types and replace the injured kidney cells [28,29,58]. Mechanistically these renal stem/progenitor cells need to be mobilized from their niche and undergo proliferation and differentiation into specialized renal cells, reestablishing thereby kidney function (Figs. 1.2 and 1.7). One important issue regarding the contribution of resident stem/progenitor cells relies on the mobilization of these cells to injured areas. During kidney injury, remnant-surviving renal cells secrete several chemokines, including the stromal cell-derived factor-1 (SDF-1), constitutively expressed in several organs including the kidney. This chemokine is known to recruit hematopoietic and mesenchymal stem cells from the bone marrow [63] and acts through specific receptors, CXCR4 and CXCR7, both of which are involved in the mobilization and targeting of stem cells [64]. Importantly, renal stem/progenitor cells express high levels of these SDF-1 receptors and can therefore be recruited from their niche by the higher levels of SDF-1 in the injured areas. Moreover, SDF-1 is essential for normal kidney development [65]. This supports the view that this chemokine plays an important role in the organization of the recovering process, mediating the migration, adhesion, and stimulating the proliferation of resident progenitor/stem cells [66]. In addition, SDF-1 has been implicated in the protection of resident stem/progenitors cells from ischemia-reperfusion injury, preventing apoptosis [67].

Paracrine Interactions in Renal Recovery

The recovery attempts are not only driven by differentiation of stem/progenitor cells. As for development and maintenance of renal homeostasis, interactions between the different cell types and structures in the kidney are essential for the repair process [68]. After injury, distal tubular epithelium contributes to the reepithelialization of proximal tubules through a differentiation-independent mechanism. The distal tubular cells are an important secretory source for several growth factors—such as EGF, HGF, and IGF-1—that act via autocrine and paracrine interactions to promote regeneration of tubular epithelial cells after injury [69]. Interestingly, although these factors are essential for renal regeneration, proximal tubular cells do not synthesize them, indicating that paracrine interactions are an essential aspect of kidney regeneration.

These paracrine interactions also play an important role in the regeneration process mediated by extrinsic stem cells [70]. Several studies showed improved renal regeneration after infusion of mesenchymal stem cells (MSC) [71-73]. MSCs, like the resident renal stem/progenitor cells, express the SDF-1 receptor CXCR4 [74], indicating that these cells are also susceptible to chemoattraction. In addition, the mobilization of MSC is mediated by the release of other cytokines by the injured kidney, such as granulocyte colony-stimulating factor (G-CSF), interleukin (IL)-1, and IL-6 [75-77]. Initial observations suggested that MSCs support regeneration by differentiation and replacement of renal cells [78-80]. However, subsequent studies showed that MSCs seldom persist in the injured kidney even though beneficial effects are observed, which reinforces the paracrine hypothesis. MSCs are known to secrete several growth factors that result in proliferative and antiapoptotic responses of renal cells [81]. MSCs, as other stem cells, can be recruited to the injured tissue and act as biologic minipumps, releasing several bioactive molecules such as chemokines, cytokines, growth factors, small RNAs, bioactive lipids, and peptides [4,49,73,82-86].

Tögel et al. [71] analyzed the conditioned medium of MSCs maintained in culture and found VEGF, IGF-1, and HGF to be present, indicating that these factors might be responsible for the paracrine actions of MSC. Later, another study from Tögel's group strengthened the importance of these factors, showing the decreased efficacy of MSCs in renal repair after VEGF knockdown by siRNA in the ischemic AKI rat model [82]. Therefore, the MSC can act through a paracrine mechanism, stimulating surviving renal cells to dedifferentiate and proliferate in order to repopulate the injured area [87]. Besides their potentially beneficial actions on surviving renal cells, MSCs have important immunomodulatory properties. These are mediated, at least in part, by secretion of prostaglandin E2 (PGE2) that inhibits secretion of pro-inflammatory cytokines (TNF-α and IFN-γ) by Th1, Th2, and natural killer (NK) cells [88]. Such modulation of the immune system promotes renoprotection and could also provide a microenvironment that favors resident renal stem/progenitor cells to migrate to and incorporate into injured renal tissue.

Based on such studies, it is evident that kidney regeneration is mediated by several mechanisms and also by different cell types (resident stem/progenitors and extrinsic cells). Such diversity in the regeneration process appears to be mandatory considering the different cell types and structures that compose this complex organ, as well as its richness in functions. Many questions, however, still remain open about the regulation and the role of resident stem cells in kidney regeneration. Understanding those in full will be essential for the development of new therapeutic strategies and treatment of kidney diseases.

Role of Kidney Resident Stem/Progenitor Cells in the Establishment and Progression of Renal Disorders

The overall processes that lead an increasing number of patients with chronic kidney disease to the end stage of renal failure is more complex than previously supposed. A pivotal role is played by the immune system because of the interaction of infiltrating inflammatory cells and renal resident stem/progenitor cells. This physiologic response results in the loss of renal cells, the recruitment of fibroblasts, and their replacement by the extracellular collagenous matrix (ECM) resulting in the scarring of the renal tissue [89]. Since the 1990s, new insights about the participation of adult renal stem/progenitor cells, as well as the involvement of extrinsic progenitor hematopoietic stem cells that migrate to the injured kidney, have emerged [2,70-73,78-80]. Because of the recruitment of extrinsic stem/progenitor cells from other niches (such as the bone marrow) it has proved very difficult to determine the actual contribution of the renal resident stem cells to kidney regeneration. Paradoxically, the same cell events that are related to kidney restoration upon injury are also implicated in some aspects that lead to the progression of kidney disease, as discussed later in a few representative examples. With this in mind, we will now highlight the role of the intrinsic renal stem/progenitor cells in the establishment and progression of different renal disorders.

Glomerularsclerosis

Over the past few decades, we have acquired a better understanding of glomerulosclerosis, which is scarring (sclerosis) of the glomeruli. It is caused by the interaction between resident glomerular cells and infiltrating inflammatory cells that are attracted to the glomeruli as a consequence of the release of different trophic factors, such as proinflammatory cytokines, chemokines, and growth factors, by the injured endothelial cells [90]. This response is triggered by changes in the endothelial cells, which reexpress molecules that favor the adhesion and anchoring of inflammatory infiltrating cells to the glomerular capillary wall [84]. As a consequence, a paracrine signaling cross talk between the glomerular and infiltrating cells develops. It triggers the synthesis and deposition of ECM and interferes with its breakdown, resulting in irreversible glomerularsclerosis [91]. However, glomerular injury results in a more complex response involving phenotypic adaptations that involve not only endothelial but also mesangial cells [92]. The mesangial cells "de-" and "trans-"differentiate in response to injury from a mature, adult pericyte-like phenotype, to an embryonic proliferative and contractile myofibroblast. This is accompanied by the expression of a range of cytoskeletal proteins, including α-smooth muscle actin [93,94]. Mesangial dedifferentiation into an embryonic phenotype favors healing and restoration of mesangial integrity.

However, the activated myofibroblast phenotype is also associated with changes in the ECM synthetic profile of mesangial cells. They start to produce and release collagen types I and III, instead of collagen type IV, to the interstitial space, a switch normally not detected within healthy glomeruli [95]. This switch in collagens enhances the progression of glomerularsclerosis. Because glomeruli are devoid of collagenases that are involved in the normal breakdown and remodeling of collagen types I and III, their glomerular deposition is augmented in an irreversible way.

It has been experimentally demonstrated that this phenotypic switch of mesangial cells can be obtained by their incubation with growth factors such as the transforming growth factor-β1 (TGF-β1) or by transfection of genes leading to the mesangial overexpression of TGF-β1 [95]. Once TGF-β1 is the main growth factor involved in the formation of glomerularsclerosis and it is expressed and released by stem/progenitor cells after injury [83], these cells could be responsible for one of the earliest events that culminate in renal interstitial fibrosis and consequently scarring.

Besides the contribution of endothelial and mesangial cells to glomerular injury, massive proliferation of glomerular epithelial cells resulting in crescent formation, and obliteration of Bowman's space is a third aspect. The molecular basis for the occurrence of these hyperplasic epithelial cells in different glomerular disorders is still a subject of controversy. Renal progenitors localized to the inner surface of Bowman's capsule are normally involved in the replacement of injured or dead podocytes. However, when this process occurs in a deregulated way, excessive proliferation of these progenitors contributes to formation of the so-called crescent [96]. This unobstructed proliferation of renal progenitors at different stages of their differentiation toward mature podocytes explains the pathogenesis of hyperplasic lesions in podocytopathies and "crescentic" glomerulonephritis [97].

Tubular Lesions

A key event in the recovery process of injured renal tubule epithelium is tubular cell dedifferentiation and subsequent proliferation that aims to repopulate the denuded basement membrane. However, recent experimental and clinical evidence strongly suggests that this phenomenon would not be sufficient for total renal recovery without the support of either extrinsic or intrinsic progenitors [2,98,99]. The main source for extrarenal progenitors is hematopoietic stem cells. Indeed, tubuloepithelial injury elicits the recruitment of bone marrow derived cells to the regenerating tubules [78,80]. Such recruited extrinsic progenitors acquire features of kidney resident cells and probably participate in the activation of the resident stem cells. One important aspect of kidney resident stem/progenitors is that they are thought to be primed to become mature renal cells and contribute to tissue recovery [2]. If any of these cells—dedifferentiated epithelial cells, extrinsic progenitors, and the resident stem/progenitor cells—fail to

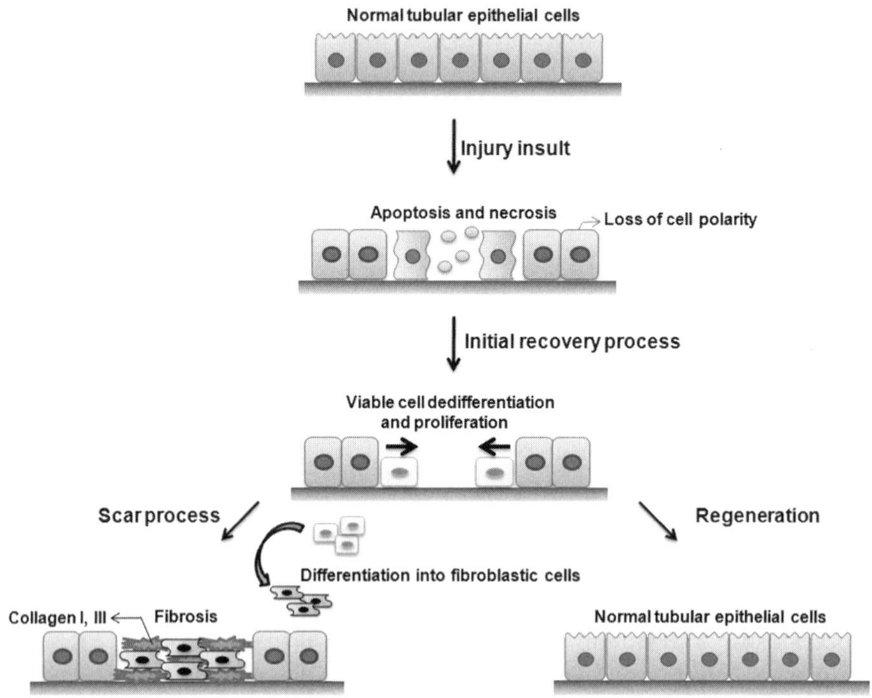

Figure 1.8 The dark side of renal remodeling. After injury, the surviving epithelial cells undergo dedifferentiation. These dedifferentiated cells may redifferentiate into functional epithelial cells leading to tubular regeneration or differentiate into fibroblasts leading to the scar process, characterized by deposition of collagen I and III.

mature into specialized epithelial cells, they can potentially be transformed into different cell types (Fig. 1.8). This aberrant process may be a consequence of the pathologic niche, where different bioactive molecules are acting together and can guide these cells toward fibroblastic differentiation and further fibrogenesis [93,94]. Unfortunately, the factors leading to a renal microenvironment that is beneficial for restoring epithelial differentiation versus one that causes harmful scarring fibroblastic phenotypes are still poorly understood. Improved understanding of these processes may lead to more innovative and original interventions that favor healing and prevent scarring.

Progenitor/Resident Stem Cells and Renal Cancer

One of the most intriguing and common diseases is cancer [100]. Although it affects virtually all types of tissues and organs, we will focus here on renal cancer and the contribution of the renal stem/progenitor cells to its progression. In recent years the cancer stem cell theory has emerged as a good hypothesis to explain many aspects of carcinogenesis [100]. It postulates that a subpopulation of cancer stem cells exists in different solid tumors as well as in cancers of hematopoietic origin. These cancer stem cells form a reservoir of self-sustaining cells with the exclusive ability to self-renew, allowing the maintenance of the tumor. Cancer stem cells have the capacity to both divide and expand the cancer stem cell pool, and they differentiate into the rather heterogeneous cancer cells that constitute the bulk of the tumor [101].

Cancer normally develops as a consequence to changes/errors in the cell-signaling program resulting in uncontrolled cell divisions. As a consequence, an abnormal cell mass develops that not only grows but also secretes and is susceptible to different bioactive molecules present in its microenvironment. As we have pointed out in this chapter, different nephropathies may arise as a consequence of erroneous differentiation of remnant kidney cells after the primary injury. This may result in abnormal proliferation or differentiation of nonkidney cell types. In a worst-case scenario, it is possible that, because of their pluripotency, different stem cells (recruited extrarenal or the intrinsic renal stem/progenitors) may lead to teratoma formation within the kidney. This is even more worrisome when stem cells are infused as in the diverse experimental cell therapeutic protocols (discussed later) and may start kidney neoplasia [100].

The problem of uncontrolled expansion of kidney progenitor cells is best exemplified with Wilms' tumor [102,103]. It is one of the most common childhood cancers and is classified as a primitive, multilineage malignancy of embryonic renal precursors that fail to terminally differentiate into epithelia and continue to proliferate [102]. Wilms' tumor cells express some molecular markers found in kidney embryonic cells, suggesting that the resident kidney stem/progenitor cells play an important role in its development.

Finally, it is noteworthy that kidney resident progenitors may also play another role in the development of renal tumors. As a consequence of their unique expression profile, they have the capacity to trigger angiogenesis and thereby contribute to tumor growth [104].

Stem/Progenitor Cells in Kidney: Features for Therapeutic Use after the Few Trials with Bone Marrow Cells?

As eluded to in the previous sections, many debates since the early 2000s have centered on the potential therapeutic role of resident stem/progenitor cells that reside within the injured kidney. These cells are essential for the spontaneous recovery of renal function that sometimes occurs in acute injury. Unfortunately, in many clinical cases this recovery does not take place [48]. It is exactly this imbalance between injurious mechanism and the regenerative capacity of the resident stem cell to repair the injured kidney that determines the progress of the disease toward chronic renal injury. As such the main intervention of cell therapy in the recovery of renal function could be the stimulation of endogenous repair by resident stem/progenitor cells in the kidney. However, only a small number of clinical trials have been conducted so far, and this review can only emphasize the limited progress that has been made and its potential in renoprotection and repair.

It is nevertheless stimulating that many studies have shown recovery or prevention of renal injury in experimental setups using treatment with exogenous stem cells in animal models and human patients (for excellent reviews see [49,83,105,106]). By way of a few significant examples, we note that in acute kidney injury, stem cells lead to the improvement of renal function in the ischemia/reperfusion model [82], experimental glomerulonephritis in rats [107], and acute renal failure induced by cisplatin in a mouse model [108]. In chronic kidney diseases, stem cells also show a powerful regenerative effect in the 5/6 nephrectomized rat [109,110], type-1 diabetes in the mouse [111], and in unilateral ureteral obstruction [53,85].

Among the different cells studied so far, the evidence highlights mesenchymal stem cells as an exogenous population with a great beneficial effect on the repair of renal injury [81,87,112,113]. Once transplanted, mesenchymal stem cells can migrate preferentially to sites where tissue is damaged or in response to an inflammatory process. As pointed out before, homing of administrated mesenchymal stem cells to the tubules, peritubular capillaries, glomeruli, and interstitium can achieve renal protection by multiple avenues: (1) stem cells incorporate in host tissue, (2) stem cells fuse with surviving differentiated mature cells, (3) stem cells release paracrine or endocrine factors to stimulate the mesenchymal stem cells themselves, and (4) stimulation of the resident stem/progenitor cells [114]. Based on recent studies, the two latter possibilities emerge as the more reliable mechanisms of renoprotection and repair, at least in kidney tissue; for this reason they deserve special attention in this chapter. The underlying mechanisms have been the focus of many clinical studies. Interestingly, cells administered at a local lesion are barely detectable, yet intense proliferation of tubular epithelial cells is observed. This was seen as evidence that exogenous stem cell administration can preserve the surviving renal cells and activate resident stem/progenitor cells [115,116].

Indeed, in other tissues, a mix of anti-inflammatory, antifibrotic, and angiogenic cytokines, as well as growth factors, act like a "multidrug system" that is required to effectively repair the damage in the different acute and chronic settings and in allograft rejection [49,117]. Therefore, cell-based therapies using mesenchymal stem cells could contribute to renal repair just by secreting a set of bioactive molecules that subsequently activate migration and proliferation of resident cells toward the damaged area. As already reviewed in this chapter, the kidney resident stem/progenitor cells express receptors for HGF and IGF-1 at their cell surface that can be activated by their specific ligands released from mesenchymal cells. Indeed, those ligands trigger signals that are of the utmost importance in stimulating endogenous renal stem/progenitor cells population, leading to cellular recovery and renal injury repair [118].

Although cell therapy in renal injury has shown beneficial effects in animal models, the clinical application of stem cells in renal disease has not yet been adequately explored. Perhaps because of the organ complexity—and despite the increasing number of patients waiting for renal transplantation or under dialysis—the number of clinical trials worldwide is relatively modest (discussed later). One of the greatest challenges for the cell-based therapies in the kidney—as in other organs—is not only to recover the architecture of the injured organ but also to make sure that the functionality of the organ itself is restored. An ideal cell therapy for kidney diseases should be able not only to reestablish the excretory functions—as does dialysis therapy—but also to improve other functions such as reabsorptive, secretory, metabolic, endocrine, immunologic, and synthetic functions. This would give patients with severe renal injuries or chronic kidney disease not only an increased life expectancy but also a higher quality of life. It would also alleviate some of the limitations of kidney transplantation as another strategy in the treatment of renal diseases, a problem that is beyond this chapter.

Obviously, several issues need to be addressed before this technology can be implemented into standard clinical care. Safety and efficacy of autologous and allogeneic bone marrow stem cells and mesenchymal stem cells for treating acute and chronic renal failure are being evaluated in eight ongoing clinical trials, with some of them still recruiting new participants. As already emphasized, this number is relatively low in comparison to clinical trials carried out to evaluate the impact of cell therapy in other organs [119-122; see also www.cellerix.com and www.osiristx.com]. Moreover, without partial and final results on studies of stem cell therapy in the kidney, it is difficult to predict whether treatment of renal injuries with stem cells will only be beneficial or will harbor some long-term side effects. Finally, we also need to know the extent to which resident cells can participate under these circumstances.

Another complication is that in most clinical trials, cells have been administered through a standard intravenous route (Table 1.2). It is important to emphasize that after intravenous administration of stem cells in a model of ischemia/reperfusion in mice, the majority of cells accumulated in the pulmonary capillaries [123], with reduced possibility of significant homing to the damaged kidney and, consequently, very low efficacy for recovery of renal function. In the same study, to circumvent this difficult, the vasodilator nitroprusside was intravenously administrated concomitantly with the stem cells. This maneuver increased the regenerative potential of these cells by allowing them access to the injured kidney. But these disadvantages of

TABLE 1.2 The Use of Stem Cells in Clinical Trails

Trial	Study Design	Delivery Route	Phase	Number of Patients	
Mesenchymal stem cell transplantation in the treatment of chronic allograft nephropathy	Historical control, safety/efficacy study, open label, treatment	Intravenous	I/II	20	Fuzhou General Hospital
Effect of mesenchymal stem cell transplantation for lupus nephritis	Historical control, safety/efficacy study, open label, treatment	Intravenous	I/II	20	Fuzhou General Hospital
Mesenchymal stem cells in cisplatin-induced acute renal failure in patients with solid organ cancers	Uncontrolled, safety/efficacy study, open label, treatment	Intravenous	I	9	Mario Negri Institute for Pharmacological Research
Mesenchymal stem cells transplantation for refractory systemic lupus erythematosus	Nonrandomized, safety/efficacy study, open label, treatment	Intravenous	I/II	20	Nanjing Medical University
Cotransplantation of islet and mesenchymal stem cell in type 1 diabetic patients	Historical control, safety/efficacy study, open label, treatment	Intravenous	I/II	30	Fuzhou General Hospital
Autologous bone marrow stem cells for treating chronic renal failure	Historical control, safety/efficacy study, open label, treatment	Intrarenal artery	I/II	5	International Stemcell Services Limited
Mesenchymal stem cells and subclinical rejection	Nonrandomized, safety/efficacy study, open label, treatment	Intravenous	I/II	15	Leiden University Medical Center
Allogeneic multipotent stromal cell treatment for acute kidney injury following cardiac surgery	Nonrandomized, open label, treatment	Intra-aortic	I	15	Intermountain Medical Center

systemic intravenous delivery of stem cells should be taken into account when considering the possibility of activation of resident stem/progenitor cells by administering (circulating) cells of either origin or type.

Nevertheless, at least two pioneering biotechnology companies, Osiris Therapeutics (www.osiristx.com) and Cellerix (www.cellerix.com) are now developing combined intravenous therapy, raising the expectations for successful cell therapy in renal diseases. Provacel, produced by Osiris Therapeutics, is a preparation of adult mesenchymal stem cells for intravenous administration. This approach has yielded encouraging results for different diseases [124-126], but clinical trials are still needed to test the safety and efficacy of either intravenous, intra-aortic, and intra-renal arterial routes with the use of this compound.

In summary, the discussions carried out in this chapter about the discovery of resident stem/progenitor cells in the kidney, the exciting molecular mechanisms underlying their actions, their successful participation in organ regeneration, and their possible stimulation by exogenous stem cells have provided new hope for successful therapies in patients with kidney diseases.

Financial support. This work was supported by grants from the Ministry of Health/SCTIE/DECIT, Ministry of Science and Technology/CNPq, Ministry of Education/CAPES, and FAPERJ (Brazil). OW is funded by NIH/NIDDK (7R01DK080745-03) (United States). Editing of the text by Biomedes (United Kingdom) is also acknowledged.

References

[1] Oliver JA, Maarouf O, Cheema FH, Martens TP, Al-Awqati Q. The renal papilla is a niche for adult kidney stem cells. J Clin Invest 2004;114:795–804.

[2] Guo JK, Cantley LG. Cellular maintenance and repair of the kidney. Annu Rev Physiol 2010;72:357–76.

[3] Venkatachalam MA, Bernard DB, Donohoe JF, Levinsky NG. Ischemic damage and repair in the rat proximal tubule: differences among the S1, S2, and S3 segments. Kidney Int 1978;14:31–49.

[4] Reilly RF, Ellison DH. Mammalian distal tubule: physiology, pathophysiology, and molecular anatomy. Physiol Rev 2000;80:277–313.

[5] Venkatachalam MA, Bernard DB, Donohoe JF, Levinsky NG, Lattanzio MR, Weir MR. Does blockade of the renin-angiotensin-aldosterone system slow progression of all forms of kidney disease? Curr Hypertens Rep 2010;12:369–77.

[6] Pitts RF. Physiology of the kidney and body fluids. 3rd ed. Chicago, USA: Year Book Medical Publishers Inc.; 1974.

[7] Klahr S, Hammerman M. Renal metabolism. In: Seldin DW, Giebisch G, editors. The kidney: physiology and pathophysiology. New York, USA: Raven Press; 1985.

[8] Costantini F, Kopan R. Patterning a complex organ: branching morphogenesis and nephron segmentation in kidney development. Dev Cell 2010;18:698–712.

[9] Dressler GR. The cellular basis of kidney development. Annu Rev Cell Dev Biol 2006;22:509–29.

[10] Tsang TE, Shawlot W, Kinder SJ, Kobayashi A, Kwan KM, Schughart K, et al. Lim1 activity is required for intermediate mesoderm differentiation in the mouse embryo. Dev Biol 2000; 223:77–90.

[11] Barak H, Rosenfelder L, Schultheiss TM, Reshef R. Cell fate specification along the anterior-posterior axis of the intermediate mesoderm. Dev Dyn 2005;232:901–14.

[12] Pedersen A, Skjong C, Shawlot W. Lim 1 is required for nephric duct extension and ureteric bud morphogenesis. Dev Biol 2005;288:571–81.

[13] Torres M, Gómez-Pardo E, Dressler GR, Gruss P. Pax-2 controls multiple steps of urogenital development. Development 1995;121:4057–65.

[14] Bouchard M, Souabni A, Mandler M, Neubüser A, Busslinger M. Nephric lineage specification by Pax2 and Pax8. Genes Dev 2002;16:2958–70.

[15] Wang Q, Lan Y, Cho ES, Maltby KM, Jiang R. Odd-skipped related 1 (Odd 1) is an essential regulator of heart and urogenital development. Dev Biol 2005;288:582–94.

[16] Mugford JW, Sipilä P, McMahon JA, McMahon AP. Osr1 expression demarcates a multi-potent population of intermediate mesoderm that undergoes progressive restriction to an Osr1-dependent nephron progenitor compartment within the mammalian kidney. Dev Biol 2008;324:88–98.

[17] Self M, Lagutin OV, Bowling B, Hendrix J, Cai Y, Dressler GR, et al. Six2 is required for suppression of nephrogenesis and progenitor renewal in the developing kidney. EMBO J 2006;25:5214–28.

[18] Kobayashi A, Valerius MT, Mugford JW, Carroll TJ, Self M, Oliver G, et al. Six2 defines and regulates a multipotent self-renewing nephron progenitor population throughout mammalian kidney development. Cell Stem Cell 2008;3:169–81.

[19] Costantini F. Renal branching morphogenesis: concepts, questions, and recent advances. Differentiation 2006;74:402–21.

[20] Mugford JW, Yu J, Kobayashi A, McMahon AP. High-resolution gene expression analysis of the developing mouse kidney defines novel cellular compartments within the nephron progenitor population. Dev Biol 2009;333:312–23.

[21] Boyle S, Shioda T, Perantoni AO, de Caestecker M. Cited1 and Cited2 are differentially expressed in the developing kidney but are not required for nephrogenesis. Dev Dyn 2007;236:2321–30.

[22] Boyle S, Misfeldt A, Chandler KJ, Deal KK, Southard-Smith EM, Mortlock DP, et al. Fate mapping using Cited1-CreERT2 mice demonstrates that the cap mesenchyme contains self--renewing progenitor cells and gives rise exclusively to nephronic epithelia. Dev Biol 2008;313: 234–45.

[23] Hartwig S, Ho J, Pandey P, Macisaac K, Taglienti M, Xiang M, et al. Genomic characterization of Wilms' tumor suppressor 1 targets in nephron progenitor cells during kidney development. Development 2010;137:1189–203.

[24] Nishinakamura R, Osafune K. Essential roles of Sall family genes in kidney development. J Physiol Sci 2006;56:131–6.

[25] Carroll TJ, Park JS, Hayashi S, Majumdar A, McMahon AP. Wnt9b plays a central role in the regulation of mesenchymal to epithelial transitions underlying organogenesis of the mammalian urogenital system. Dev Cell 2005;9:283–92.

[26] Karner CM, Das A, Ma Z, Self M, Chen C, Lum L, et al. Canonical Wnt9b signaling balances progenitor cell expansion and differentiation during kidney development. Development 2011;138: 1247–57.

[27] Costantini F. GDNF/Ret signaling and renal branching morphogenesis: From mesenchymal signals to epithelial cell behaviors. Organogenesis 2010;6:252–62.

[28] Gupta S, Rosenberg ME. Do stem cells exist in the adult kidney? Am J Nephrol 2008; 28:607–13.

[29] Little MH, Bertram JF. Is there such a thing as a renal stem cell? J Am Soc Nephrol 2009;10:2112–7.
[30] Pleniceanu O, Harari-Steinberg O, Dekel B. Concise review: kidney stem/progenitor cells: differentiate, sort out, or reprogram? Stem Cells 2010;28:1649–60.
[31] Maeshima A, Yamashita S, Nojima Y. Identification of renal progenitor-like tubular cells that participate in the regeneration processes of the kidney. J Am Soc Nephrol 2003;14:3138–46.
[32] Maeshima A, Sakurai H, Nigam SK. Adult kidney tubular cell population showing phenotypic plasticity, tubulogenic capacity, and integration capability into developing kidney. J Am Soc Nephrol 2006;17:188–98.
[33] Oliver JA, Klinakis A, Cheema FH, Friedlander J, Sampogna RV, Martens TP, et al. Proliferation and migration of label-retaining cells of the kidney papilla. J Am Soc Nephrol 2009;20:2315–27.
[34] Iwatani H, Ito T, Imai E, Matsuzaki Y, Suzuki A, Yamato M, et al. Hematopoietic and non-hematopoietic potentials of Hoechst(low)/side population cells isolated from adult rat kidney. Kidney Int 2004;65:1604–14.
[35] Goodell MA, Brose K, Paradis G, Conner AS, Mulligan RC. Isolation and functional properties of murine hematopoietic stem cells that are replicating in vivo. J Exp Med 1996;183:1797–806.
[36] Metsuyanim S, Harari-Steinberg O, Buzhor E, Omer D, Pode-Shakked N, Ben-Hur H, et al. Expression of stem cell markers in the human fetal kidney. PLoS One 2009;4:e6709.
[37] Spangrude GJ, Heimfeld S, Weissman IL. Purification and characterization of mouse hematopoietic stem cells. Science 1998;241:58–62.
[38] Holmes C, Stanford WL. Concise review: stem cell antigen-1: expression, function, and enigma. Stem Cells 2007;25:1339–47.
[39] Dekel B, Zangi L, Shezen E, Reich-Zeliger S, Eventov-Friedman S, Katchman H, et al. Isolation and characterization of nontubular sca-1+lin– multipotent stem/progenitor cells from adult mouse kidney. J Am Soc Nephrol 2006;17:3300–14.
[40] Corbeil D, Röper K, Hellwig A, Tavian M, Miraglia S, Watt SM, et al. The human AC133 hematopoietic stem cell antigen is also expressed in epithelial cells and targeted to plasma membrane protrusions. J Biol Chem 2000;275:5512–20.
[41] Bussolati B, Bruno S, Grange C, Buttiglieri S, Deregibus MC, Cantino D, et al. Isolation of renal progenitor cells from adult human kidney. Am J Pathol 2005;166:545–55.
[42] Sagrinati C, Netti GS, Mazzinghi B, Lazzeri E, Liotta F, Frosali F, et al. Isolation and characterization of multipotent progenitor cells from the Bowman's capsule of adult human kidneys. J Am Soc Nephrol 2006;17:2443–56.
[43] Ronconi E, Sagrinati C, Angelotti ML, Lazzeri E, Mazzinghi B, Ballerini L, et al. Regeneration of glomerular podocytes by human renal progenitors. J Am Soc Nephrol 2009;20:322–32.
[44] Appel D, Kershaw DB, Smeets B, Yuan G, Fuss A, Frye B, et al. Recruitment of podocytes from glomerular parietal epithelial cells. J Am Soc Nephrol 2009;20:333–43.
[45] Bruno S, Bussolati B, Grange C, Collino F, di Cantogno LV, Herrera MB, et al. Isolation and characterization of resident mesenchymal stem cells in human glomeruli. Stem Cells Dev 2009;18:867–80.
[46] Delorme B, Ringe J, Gallay N, Le Vern Y, Kerboeuf D, Jorgensen C, et al. Specific plasma membrane protein phenotype of culture-amplified and native human bone marrow mesenchymal stem cells. Blood 2008;111:2631–5.
[47] Gupta S, Verfaillie C, Chmielewski D, Kren S, Eidman K, Connaire J, et al. Isolation and characterization of kidney-derived stem cells. J Am Soc Nephrol 2006;17:3028–40.
[48] Bellomo R, May C, Wan L. Acute renal failure and sepsis. N Engl J Med 2004;351:2347–9.
[49] Bani-Hani AH, Campbell MT, Meldrum DR, Meldrum KK. Cytokines in epithelial-mesenchymal transition: a new insight into obstructive nephropathy. J Urol 2008;180:461–8.

[50] Bonventre JV, Weinberg JM. Recent advances in the pathophysiology of ischemic acute renal failure. J Am Soc Nephrol 2003;14:2199–210.

[51] Hostetter TH. Progression of renal disease and renal hypertrophy. Annu Rev Physiol 1995;57:263–78.

[52] Nadasdy T, Laszik Z, Blick KE, Johnson DL, Burst-Singer K, Nast C, et al. Human acute tubular necrosis: a lectin and immunohistochemical study. Hum Pathol 1995;26:230–9.

[53] Barreira AL, Takiya CM, Castiglione RC, Maron-Gutierrez T, Barbosa CM, Ornellas DS, et al. Bone marrow mononuclear cells attenuate interstitial fibrosis and stimulate the repair of tubular epithelial cells after unilateral ureteral obstruction. Cell Physiol Biochem 2009;24:585–94.

[54] Benigni A, Morigi M, Remuzzi G. Kidney regeneration. Lancet 2010;375:1310–7.

[55] Humphreys BD. Slow-cycling cells in renal papilla: stem cells awaken? J Am Soc Nephrol 2009;20:2277–9.

[56] Bonventre JV. Dedifferentiation and proliferation of surviving epithelial cells in acute renal failure. J Am Soc Nephrol 2003;14:S55–61.

[57] Witzgall R, Brown D, Schwarz C, Bonventre JV. Localization of proliferating cell nuclear antigen, vimentin, c-Fos, and clusterin in the postischemic kidney. Evidence for a heterogenous genetic response among nephron segments, and a large pool of mitotically active and dedifferentiated cells. J Clin Invest 1994;93:2175–88.

[58] Villanueva S, Céspedes C, Vio CP. Ischemic acute renal failure induces the expression of a wide range of nephrogenic proteins. Am J Physiol Regul Integr Comp Physiol 2006;290:R861–70.

[59] Xie Y, Nishi S, Iguchi S, Imai N, Sakatsume M, Saito A, et al. Expression of osteopontin in gentamicin-induced acute tubular necrosis and its recovery process. Kidney Int 2001;59:959–74.

[60] Lin F, Moran A, Igarashi P. Intrarenal cells, not bone marrow-derived cells, are the major source for regeneration in postischemic kidney. J Clin Invest 2005;115:1756–64.

[61] Swetha G, Chandra V, Phadnis S, Bhonde R. Glomerular parietal epithelial cells of adult murine kidney undergo EMT to generate cells with traits of renal progenitors. J Cell Mol Med 2009;15:396–413.

[62] Langworthy M, Zhou B, de Caestecker M, Moeckel G, Baldwin HS. NFATc1 identifies a population of proximal tubule cell progenitors. J Am Soc Nephrol 2009;20:311–21.

[63] Lapidot T, Kollet O. The essential roles of the chemokine SDF-1 and its receptor CXCR4 in human stem cell homing and repopulation of transplanted immune-deficient NOD/SCID and NOD/SCID/B2m(null) mice. Leukemia 2002;16:1992–2003.

[64] Mazzinghi B, Ronconi E, Lazzeri E, Sagrinati C, Ballerini L, Angelotti ML, et al. Essential but differential role for CXCR4 and CXCR7 in the therapeutic homing of human renal progenitor cells. J Exp Med 2008;205:479–90.

[65] Takabatake Y, Sugiyama T, Kohara H, Matsusaka T, Kurihara H, Koni PA, et al. The CXCL12 (SDF-1)/CXCR4 axis is essential for the development of renal vasculature. J Am Soc Nephrol 2009;20:1714–23.

[66] Kucia M, Jankowski K, Reca R, Wysoczynski M, Bandura L, Allendorf DJ, et al. CXCR4-SDF-1 signalling, locomotion, chemotaxis and adhesion. J Mol Histol 2004;35:233–45.

[67] Kunduzova OR, Escourrou G, De La Farge F, Salvayre R, Séguélas MH, Leducq N, et al. Involvement of peripheral benzodiazepine receptor in the oxidative stress, death-signaling pathways, and renal injury induced by ischemia-reperfusion. J Am Soc Nephrol 2004;15:2152–60.

[68] Schedl A, Hastie ND. Cross-talk in kidney development. Curr Opin Genet Dev 2000;10:543–9.

[69] Gobe GC, Johnson DW. Distal tubular epithelial cells of the kidney: Potential support for proximal tubular cell survival after renal injury. J Biochem Cell Biol 2007;39:1551–61.

[70] Bao J, Tu Z, Sun H, Luo G, Yang L, Song J, et al. R2: identification of renal potential progenitor/stem cells that participate in the renal regeneration processes of kidney allograft fibrosis. Nephrology 2008;2008(13):500–7.

[71] Tögel F, Weiss K, Yang Y, Hu Z, Zhang P, Westenfelder C. Vasculotropic, paracrine actions of infused mesenchymal stem cells are important to the recovery from acute kidney injury. Am J Physiol Renal Physiol 2007;292:F1626–35.
[72] Zhuo W, Liao L, Xu T, Wu W, Yang S, Tan J. Mesenchymal stem cells ameliorate ischemia-reperfusion-induced renal dysfunction by improving the antioxidant/oxidant balance in the ischemic kidney. Urol Int 2010. Epub ahead of print, doi:10.1159/000319366.
[73] Gatti S, Bruno S, Deregibus MC, Sordi A, Cantaluppi V, Tetta C, et al. Microvesicles derived from human adult mesenchymal stem cells protect against ischaemia-reperfusion-induced acute and chronic kidney injury. Nephrol Dial Transplant 2011. Epub ahead of print, doi:10.1093/ndt/gfr015.
[74] Tögel F, Isaac J, Hu Z, Weiss K, Westenfelder C. Renal SDF-1 signals mobilization and homing of $CXCR4^+$ cells to the kidney after ischemic injury. Kidney Int 2005;67:1772–84.
[75] Haq M, Norman J, Saba SR, Ramirez G, Rabb H. Role of IL-1 in renal ischemic reperfusion injury. J Am Soc Nephrol 1998;9:614–9.
[76] Zhang Y, Woodward VK, Shelton JM, Richardson JA, Zhou XJ, Link D, et al. Ischemia-reperfusion induces G-CSF gene expression by renal medullary thick ascending limb cells in vivo and in vitro. Am J Physiol Renal Physiol 2004;286:F1193–201.
[77] Patel NS, Chatterjee PK, Di Paola R, Mazzon E, Britti D, De Sarro A, et al. Endogenous interleukin-6 enhances the renal injury, dysfunction, and inflammation caused by ischemia/reperfusion. J harmacol Exp Ther 2005;312:1170–8.
[78] Gupta S, Verfaillie C, Chmielewski D, Kim Y, Rosenberg ME. A role for extrarenal cells in the regeneration following acute renal failure. Kidney Int 2002;62:1285–90.
[79] Fang TC, Alison MR, Cook HT, Jeffery R, Wright NA, Poulsom R. Proliferation of bone marrow-derived cells contributes to regeneration after folic acid-induced acute tubular injury. J Am Soc Nephrol 2005;16:1723–32.
[80] Poulsom R, Forbes SJ, Hodivala-Dilke K, Ryan E, Wyles S, Navaratnarasah S, et al. Bone marrow contributes to renal parenchymal turnover and regeneration. J Pathol 2001;195:229–35.
[81] Humphreys BD, Bonventre JV. Mesenchymal stem cells in acute kidney injury. Annu Rev Med 2008;59:311–25.
[82] Tögel F, Zhang P, Hu Z, Westenfelder C. VEGF is a mediator of the renoprotective effects of multipotent marrow stromal cells in acute kidney injury. J Cell Mol Med 2009;13:2109–14.
[83] Basile DP. The transforming growth factor beta system in kidney disease and repair: Recent progress and future directions. Curr Opin Nephrol Hypertens 1999;8:21–30.
[84] Anders HJ, Vielhauer V, Schlondorff D. Chemokines and chemokine receptors are involved in the resolution or progression of renal disease. Kidney Int 2003;63:401–15.
[85] Verdoorn KS, Lindoso RS, Lowe J, Lara LS, Vieyra A, Einicker-Lamas M. Bone marrow mononuclear cells shift bioactive lipid pattern in injured kidney towards tissue repair in rats with unilateral ureteral obstruction. Nephrol Dial Transplant 2010;25:3867–74.
[86] Sze SK, de Kleijn DP, Lai RC, Khia Way Tan E, Zhao H, Yeo KS, et al. Elucidating the secretion proteome of human embryonic stem cell-derived mesenchymal stem cells. Mol Cell Proteomics 2007;6:1680–9.
[87] Lindoso RS, Araujo DS, Adão-Novaes J, Verdoorn KS, Fragel-Madeira L, Caruso-Neves C, et al. Paracrine interaction between bone marrow-derived stem cells and renal epithelial cells. Cell Physiol and Biochem 2011 [in press].
[88] Aggarwal S, Pittenger MF. Human mesenchymal stem cells modulate allogeneic immune cell responses. Blood 2005;105:1815–22.
[89] El Nahas AM. Plasticity of kidney cells: Role in kidney remodeling and scarring. Kidney Int 2003;64:1553–63.

[90] Johnson RJ. The glomerular response to injury: progression or resolution? Kidney Int 1994;45: 1769–82.

[91] Eddy AA. Molecular basis of renal fibrosis. Pediatr Nephrol 2000;15:290–301.

[92] Savage COS. The biology of the glomerulus: Endothelial cells. Kidney Int 1994;45:314–9.

[93] Johnson RJ, Floege J, Yoshimura A, Iida H, Couser WG, Alpers CE. The activated mesangial cell: Aglomerular "myofibroblast." J AmSoc Nephrol 1992;2:S190–7.

[94] El Nahas AM, Muchaneta-Kubara EC, Essawy M, Soylemezoglu O. Renal fibrosis: Insights into pathogenesis and treatment. Int J Biochem Cell Biol 1997;29:55–62.

[95] Imai E, Isaka Y, Fujiwara Y, Kaneda Y, Kamada T, Ueda N. Introduction of a foreign gene into the kidney in vivo: Development of glomerulosclerosis by the transfection of gene for PDGF and TGF-beta. Contrib Nephrol 1994;107:205–15.

[96] Wiggins RC. The spectrum of podocytopathies: A unifying view of glomerular diseases. Kidney Int 2007;71:1205–14.

[97] Smeets B, Angelotti ML, Rizzo P, Dijkman H, Lazzeri E, Mooren F, et al. Renal progenitor cells contribute to hyperplastic lesions of podocytopathies and crescentic glomerulonephritis. J Am Soc Nephrol 2009;20:2593–603.

[98] Cantley LG. Adult stem cells in the repair of the injured renal tubule. Nat Clin Pract Nephrol 2005;1:22–32.

[99] Duffield JS, Park KM, Hsiao LL, Kelley VR, Scadden DT, Ichimura T, et al. Restoration of tubular epithelial cells during repair of the post ischemic kidney occurs independently of bone marrow–derived stem cells. J Clin Invest 2005;115:1743–55.

[100] Weissman IL. Normal and neoplastic stem cells. Novartis Found Symp 2005;265:35–50.

[101] Clarke MF, Dick JE, Dirks PB, Eaves CJ, Jamieson CH, Jones DL, et al. Cancer stem cells–perspectives on current status and future directions: AACR Workshop on cancer stem cells. Cancer Res 2006;66:9339–44.

[102] Rivera MN, Haber DA. Wilms' tumour: Connecting tumorigenesis and organ development in the kidney. Nat Rev Cancer 2005;5:699–712.

[103] Aiden AP, Rivera MN, Rheinbay E, Ku M, Coffman EJ, Truong TT, et al. Wilms tumor chromatin profiles highlight stem cell properties and a renal developmental network. Cell Stem Cell 2010;4:591–602.

[104] Bruno S, Bussolati B, Grange C, Collino F, Graziano ME, Ferrando U, et al. $CD133^+$ renal progenitor cells contribute to tumor angiogenesis. Am J Pathol 2006;169:2223–35.

[105] Watorek E, Klinger M. Stem cells in nephrology: present status and future. Arch Immunol Ther Exp (Warsz) 2006;54:45–50.

[106] Burt RK, Loh Y, Pearce W, Beohar N, Barr WG, Craig R, et al. Clinical applications of blood-derived and marrow-derived stem cells for nonmalignant diseases. JAMA 2008;299: 925–36.

[107] Kunter U, Rong S, Boor P, Eitner F, Müller-Newen G, Djuric Z, et al. Mesenchymal stem cells prevent progressive experimental renal failure but maldifferentiate into glomerular adipocytes. J Am Soc Nephrol 2007;18:1754–64.

[108] Imberti B, Morigi M, Tomasoni S, Rota C, Corna D, Longaretti L, et al. Insulin-like growth factor-1 sustains stem cell mediated renal repair. J Am Soc Nephrol 2007;18:2921–8.

[109] Cavaglieri RC, Martini D, Sogayar MC, Noronha IL. Mesenchymal stem cells delivered at the subcapsule of the kidney ameliorate renal disease in the rat remnant kidney model. Transplant Proc 2009;41:947–51.

[110] Semedo P, Donizetti-Oliveira C, Burgos-Silva M, Cenedeze MA. Avancini Costa Malheiros DM, Pacheco-Silva A, Câmara NO. Bone marrow mononuclear cells attenuate fibrosis development after severe acute kidney injury. Lab Invest 2010;90:685–95.

[111] Guillot PV, Cook HT, Pusey CD, Fisk NM, Harten S, Moss J, et al. Transplantation of human fetal mesenchymal stem cells improves glomerulopathy in a collagen type I alpha 2-deficient mouse. J Pathol 2008;214:627–36.

[112] Herrera MB, Bussolati B, Bruno S, Fonsato V, Romanazzi GM, Camussi G. Mesenchymal stem cells contribute to the renal repaisr of acute tubular epitelial injury. Int J Mol Med 2004;14: 1035–41.

[113] Tögel F, Hu Z, Weiss K, Isaac J, Lange C, Westenfelder C. Administered mesenchymal stem cells protect against ischemic acute failure through differentiation-independent mechanisms. Am J Physiol Renal Physiol 2005;289:F31–42.

[114] Asanuma H, Meldrum DR, Meldrum KK. Therapeutic applications of mesenchymal stem cells to repair kidney injury. J Urol 2010;184:26–36.

[115] Bi B, Schmitt R, Israilova M, Nishio H, Cantley LG. Stromal cells protect against acute tubular injury via an endocrine effect. J Am Soc Nephrol 2007;18:2486–96.

[116] Tögel F, Cohen A, Zhang P, Yang Y, Hu Z, Westenfelder C. Autologous and allogeneic marrow stromal cells are safe and effective for the treatment of acute kidney injury. Stem Cells Dev 2009;18:475–85.

[117] Reinders ME, Fibbe WE, Rabelink TJ. Multipotent mesenchymal stromal cell therapy in renal disease and kidney transplantation. Nephrol Dial Transplant 2010;25:17–24.

[118] Bussolati B, Tetta C, Camussi G. Contribution of stem cells to kidney repair. Am J Nephrol 2008;28:813–22.

[119] Battistella V, de Freitas GR, da Fonseca LM, Mercante D, Gutfilen B, Goldenberg RC, et al. Safety of autologous bone marrow mononuclear cell transplantation in patients with nonacute ischemic stroke. Regen Med 2011;6:45–52.

[120] Dohmann HF, Silva SA, Sousa AL, Braga AM, Branco RV, Haddad AF, et al. Multicenter double blind trial of autologous bone marrow mononuclear cell transplantation through intracoronary injection post acute myocardium infarction – MiHeart/AMI Study. Trials 2008;9:41.

[121] Couto BG, Goldenberg RC, da Fonseca LM, Thomas J, Gutfilen B, Resende CM, et al. Bone marrow mononuclear cell therapy for patients with cirrhosis: a Phase 1 study. Liver Int 2011;31: 391–400.

[122] Martino HF, Oliveira PS, Souza FC, Costa PC, Assunção E, Silva E, et al. A safety and feasibility study of cell therapy in dilated cardiomyopathy. Braz J Med Biol Res 2010;43:989–95.

[123] Schrepfer S, Deuse T, Reichenspurner H, Fischbein MP, Robbins RC, Pelletier MP. Stem cell transplantation: the lung barrier. Transplant Proc 2007;39:573–6.

[124] Weiss DJ, Kolls JK, Ortiz LA, Panoskaltsis-Mortari A, Prockop DJ. Stem cells and cell therapies in lung biology and lung diseases. Proc Am Thorac Soc 2008;5:637–67.

[125] Paul D, Samuel SM, Maulik N. Mesenchymal stem cell: Present challenges and prospective cellular cardiomyoplasty approaches for myocardial regeneration. Antioxid Redox Signal 2009;11: 1841–55.

[126] Undale AH, Westendorf JJ, Yaszemski MJ, Khosla S. Mesenchymal stem cells for bone repair and metabolic bone diseases. Mayo Clin Proc 2009;84:893–902.

Chapter 2

Pancreatic Stem Cells

Fernando H. Lojudice*, Patrícia M. Kossugue*, Mari C. Sogayar†

* *Chemistry Institute, Biochemistry Department, Cell and Molecular Therapy Center (NUCEL), University of São Paulo, São Paulo, Brazil*
† *Instituto de Química, São Paulo, Brazil*

Introduction

The main metabolic source of energy used by most human tissues is glucose, an essential metabolite for some cell types, such as neurons and erythrocytes. Maintenance of normoglycemia is crucial for correct metabolic function, and is regulated by several physiologic processes, including hormonal mechanisms, mainly orchestrated by the pancreas.

Two different tissues constitute the pancreas: the exocrine pancreas, which secretes digestive enzymes into the intestine, and the endocrine pancreas, which secretes hormones into the bloodstream. The endocrine cells are clustered, forming the islets of Langerhans, which are basically formed by four cell types: beta cells, which secrete insulin, C-peptide, and amylin; alpha cells, which secrete glucagon; delta cells, which secrete somatostatin; and PP cells, which secrete the pancreatic polypeptide.

Embryogenesis and Organogenesis

After gastrulation, all three germ layers are developed: endoderm, mesoderm, and ectoderm. The primitive gut develops from the endoderm, and, depending on the

Resident Stem Cells and Regenerative Therapy. http://dx.doi.org/10.1016/B978-0-12-416012-5.00002-5

signal that each part of this primitive gut receives, it gives rise to different organs, such as the liver, pancreas, or intestine.

In the mouse, the pancreatic development starts at embryonic day 8.5 (E8.5), with the expression of Pdx1 (*pancreas-duodenum homeobox*) and repression of Shh (*sonic hedgehog)* expression in a region located at the posterior foregut. Later on, two buds arise from this region, the ventral and dorsal buds. Some factors, such as Ptf1a (*pancreas transcription factor*), are necessary for ventral pancreas development, whereas others, such as Hlxb9 (*homeobox gene Hb9*), are necessary for dorsal pancreas formation.

The ventral bud fuses to the dorsal bud, and some evaginations arise from the epithelium. Cells residing at the "tips" of the branching epithelium adopt an acinar fate, whereas cells in the core or "trunk" become restricted to a ductal or endocrine fate [1,2], followed by formation of other cell types. The progenitor cells express Hes1 (*hairy and enhancer of split homolog*), which promotes the undifferentiated state, whereas FGFs (*fibroblast growth factors*) stimulate cell proliferation. The differentiation of endocrine versus exocrine cells is accomplished by lateral inhibition within the ductal epithelium, mediated by the Notch pathway [3,4]. Ngn3 induces the expression of Notch ligands, activating the Notch pathway in adjacent cells [5]. Downstream targets of the Notch pathway (e.g., hairy/enhancer-of-split 1, Hes1) repress Ngn3 expression, thereby inhibiting endocrine differentiation. Additionally, Hes1 represses the p57 cell cycle inhibitor, thus maintaining a proliferative pool of pancreatic progenitor cells within the embryonic ductal epithelium [6]. Therefore, cells in which the Notch signaling pathway is activated maintain their proliferative capacity, whereas cells in which this signaling is not activated express Ngn3, exit the cell cycle, and differentiate into endocrine cells—alpha, beta, delta, and PP cells—which produce glucagon, insulin, somatostatin, and pancreatic polypeptide, respectively. Subsequently, development of the endocrine pancreas is controlled by transcription factors such as PDX-1, Neurogenin 3, PAX-4 e PAX-6, ISL-1, BETA2-NEUROD1, MAF, NKX2.2, and NKX6.1. These transcription factors regulate migration and proliferation during pancreas development. The endocrine tissue is located in small aggregates known as *islets of Langerhans*, which can be isolated from cadaveric donors and used in islet cell transplantation (Fig. 2.1).

Expansion versus Neogenesis

After organogenesis, new beta cells may be originated, which is evidenced by islet cell mass increase after birth and the discovery of islet cells and precursors near the pancreatic ductal structures. It is also known that a significant regeneration of pancreatic tissue, including beta-cell mass increase, occurs in rats subsequent to a 90% pancreatectomy. Two main theories have been postulated: the expansion and the neogenesis theories. In the former, new beta cells are originated via replication of

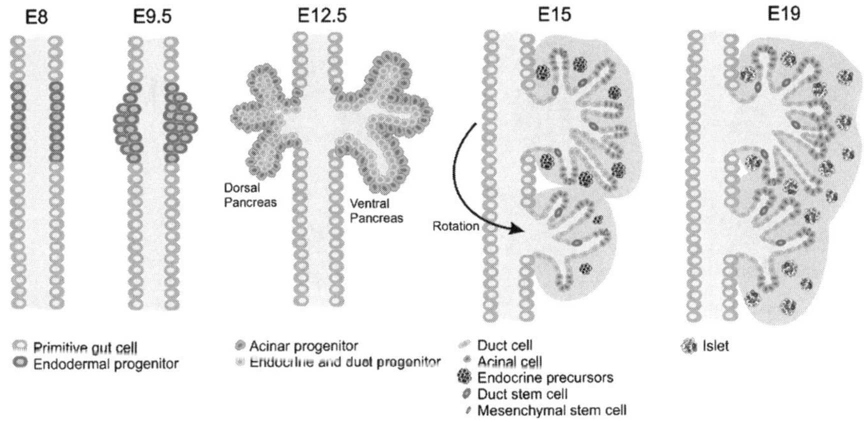

Figure 2.1 Embryonic development of the mouse pancreas. The pancreas organogenesis begins around embryonic day 9.5 (E9.5), when a small group of cells in the primitive gut initiate their differentiation. From E12.5 two buds arise: the ventral and the dorsal pancreas. On E15, the dorsal bud rotates and fuses with the ventral bud. From E19, the endocrine progenitors initiate their final differentiation in the islets of Langerhans.

preexisting beta cells and in the latter, new beta cells are differentiated from pancreatic precursor stem (progenitor) cells. The absence of Ngn3 expression in the adult pancreas supports the expansion theory, whereas the presence of small beta cell clusters or solitary beta cells adjacent to ducts are often interpreted as products of neogenesis.

The mechanism postulated to replace lost beta cells in adult individuals depends mainly on the expansion of preexisting beta cells rather than differentiation of stem cell precursors. Many growth factors and hormones have been tested in the search for inducers of pancreatic beta cells growth [7]. Also, it is well known that when the pancreatic duct is ligated, neogenesis occurs [8], because Ngn3+ precursor cells arise from the ducts and differentiate into new beta cells.

Pancreatic Stem Cells

Pancreatic stem cells have been studied for many years because of their advantageous characteristics, namely their high proliferation rate and natural commitment to the endocrine lineage. Utilization of these cells would be a better strategy to obtain beta cells when compared to other stem cell sources, because they may be stimulated to induce compensatory growth in vivo [9].

At least two different stem cell populations have been described in the pancreas (Fig. 2.2). The first group consists of ductal and acinar cells, which have been suggested to reside in the ductal pancreatic epithelium and to express the CK-19 and Pdx1 ductal markers [10]. These cells can proliferate and differentiate into endocrine

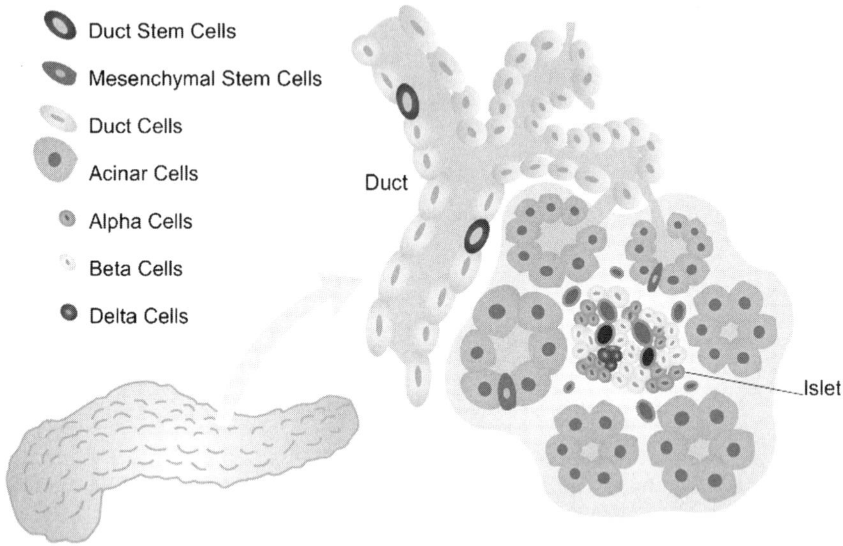

Figure 2.2 Different cell types and putative stem cell populations residing inside the pancreatic tissue. **See Plate 3.**

cells [11]. The second group consists of a population derived from the islets. Some reports show insulin-positive cells arising inside islets after total destruction of beta cells [9]. In a previous study, our group reported the existence of islet-progenitor/islet-stem cells arising from the islet itself [12] (Fig. 2.3).

For many years, nestin-positive cells were considered to be stem cells derived from pancreatic islets [13]. Nestin-positive cells express the ATP-binding cassette transporter (a stem cell marker) [14], display a high proliferative rate in vivo, and may differentiate in vitro into cells with characteristic phenotypes of hepatic, exocrine, ductal, and endocrine pancreatic cells [13]. Nestin+ cells are present not only in the pancreas but also in the ducts and display beta cell precursor characteristics, such as high Pdx1 levels [10, 15]. Some reports showed that nestin is only expressed in cells that are marked not to turn into beta cells [16-18]. However, other reports suggest that there are different populations of nestin+ cells in the pancreas and that nestin seems to be a functional protein that may be expressed not only in beta cell precursors but also in other cell types [19]. Therefore, it is important to study the role of nestin+ cells in precursor beta cells because nestin expression may play an essential function during precursor maturation.

Another pancreatic cell subpopulation expresses the HGF receptor (HGFr/c-Met). It has been suggested that HGF is important for regulating the function, proliferation, and differentiation of pancreatic beta cells [14]. In neonate mice pancreas, c-Met+ cells are present in the acinar tissue, but not in the islets. Some of

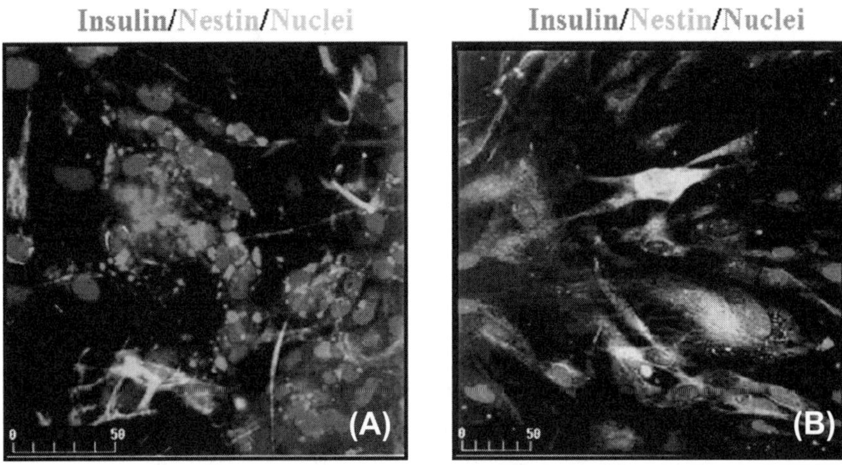

Figure 2.3 Confocal microscopy of triple-labeled immunofluorescence for insulin (red), nestin (green), and DAPI for nuclei (blue). Human pancreatic islet cell cultures were maintained for 24 **(A)** and 60 **(B)** days on uncoated plastic. Whereas the insulin signal tended to decrease with time, nestin-positive cells constitute the great majority of cells in long-term culture and co-localization of insulin and nestin is observed (B). Scale bars in μm. *(From [12].)* **See Plate 4**.

these cells are also present in ducts, but they do not assume acinar or endocrine phenotypes. In the adult pancreas, c-Met is mainly expressed in and adjacent to the pancreatic ducts and vascular endothelium, and adjacent to acinar tissue [20]. In addition, cells expressing endocrine markers are observed in clusters derived from c-Met+ cells, which can differentiate into endocrine cells in vivo.

Yet another pancreatic cell subpopulation is represented by the c-Kit-positive cells. The c-Kit is an universal stem cell marker. Binding of stem cell factor (SCF) to c-Kit results in cell proliferation, survival, and chemotaxis. The c-Kit receptor is expressed in the pancreas before and after birth. Co-localization studies have showen that c-Kit is expressed adjacent to islets, co-localizing with glucagon-positive cells [21]. Although the role of c-Kit in beta cell development is still unknown, some reports indicate that c-Kit+ stem cells display similar characteristics to endocrine precursor cells, being present in islets.

DCAMKL-1 is also a putative marker for pancreatic stem cells because cells expressing this marker were observed in the newborn and in adult mouse pancreatic epithelial cells and islets, but not within acinar cells. Co-expression was observed with somatostatin, NGN3, and nestin, but not with glucagon or insulin. Isolated DCAMKL-1-positive cells form spheroids in suspension culture and induce nodule formation in isotransplantation assays. Analysis of nodules demonstrated PDX-1, CK-14, and Ep-CAM markers, and islet-like structures (somatostatin and secretin positive) [22].

Studies using purified native cardiomyocyte and hematopoietic progenitor/stem cell lineages have contributed new data to generate these cells from other stem cells. For this reason, several groups are trying to understand the different phenotypes, gene expression profiles, and cell surface markers of native pancreatic cells, aiming at obtaining better results in the pancreatic cells differentiation field. Unfortunately, not much is known about the developing and postnatal mouse pancreas, and most of the available data focus on whole pancreas, islets, or immortalized cell lines. Moreover, some reports may have been based on cells obtained from "contaminated" cultures, a mix of different cell types [23].

There are some evidence that, purified progenitor cell cultures obtained by FACS sorting, may contribute to obtain large amounts of insulin-producing cells, almost without other cell types. Chemokine (C-X-C motif) receptor 4 (CXCR4), which is already known as a definitive endoderm-specific marker, has been used to select for definitive endoderm progenitors (CXCR4-positive cells). When these definitive endoderm progenitors are subjected to differentiation, 95% of the cells turn into PDX-1-positive cells, compared to less than 10% in the CXCR4-negative population [24].

Although several reports describe the isolation of pancreatic stem cells, and these cells seem to have a higher capacity to differentiate into insulin-producing cells when compared to other stem cells, none of the reports show their efficient differentiation into functional cells.

Transdifferentiation and Differentiation

The differentiation of a cell that is terminally differentiated into another unrelated cell lineage is possible, in theory, occurring through the process of transdifferentiation or by dedifferentiation followed by differentiation. Dedifferentiation is characterized as a multistep process that involves the loss of specific cell characteristics and reentrance into the cell cycle. For this reason, other pancreatic cells, which do not belong to the endocrine lineage, are possible sources of new beta cells [25-27], as both endocrine and exocrine cells have the same embryonic origin [4,28,29]. In vitro, endocrine, and exocrine cells dedifferentiate and express the same ductal marker, namely CK-19, indicating the same pattern of dedifferentiation.

It is important to mention that insular stem cells have yet to be fully characterized. In vitro co-localization experiments indicate that the ductal epithelium, exocrine tissue, and beta cell replication contribute to produce new insulin-positive cells [8,30,31].

It has also been shown that extrapancreatic tissue may be modified to express insulin, including the hepatic tissue [32], bone marrow [10,33], and monocytes [34]. The liver has an endodermic origin, as does the pancreas, with some reports indicating that oval hepatic cells can be manipulated to simultaneously co-express insulin and glucagon [35]. Another alternative is the use of bone marrow stem cells to repopulate the pancreas. However, the mechanisms involved seem to be more related to neoformation of pancreatic vasculature than transdifferentiation of bone marrow stem cells [36].

Stem cells from several sources, including embryonic stem cells, have been subjected to differentiation protocols hoping to achieve insulin-producing cells. Some reports, including from our own group [37-41], show that murine embryonic stem cells (mESCs) are able to form islet-like structures, but their insulin production levels are insufficient to render diabetic mice normoglycaemic.

Another strategy to obtain beta cells is based on cellular reprogramming by overexpression of defined factors in an adult organ. By reexpressing key developmental regulators (Ngn3, Pdx1, and Mafa) in vivo, it has been shown that it is possible to reprogram differentiated pancreatic exocrine cells into cells that closely resemble the beta cells [42].

As a result of important ethical and technical issues, the use of embryonic stem cells (ESCs) and molecular manipulation of these and other stem cells are still far from being applicable to patients, but ESCs have greatly contributed to elucidate some molecular aspects of pancreatic organogenesis and pancreatic islet biology.

Islet Cell Transplantation as an Attractive Treatment for Type 1 Diabetes Mellitus

Although studies related to stem cells and other alternative cell sources for pancreatic beta cells are only beginning, isolated pancreatic islets have been exhaustively studied for many years. Islet cell transplantation has been performed worldwide since the early 2000s, but it was introduced in Brazil by our group in 2002 [43]. Human islets may be isolated from pancreas obtained from cadaveric donors by enzymatic (collagenase) digestion, purified by density gradient, and implanted into the liver of hyperlabile type 1 diabetic patients (Fig. 2.4). Even though this procedure is much less invasive than total pancreas transplantation, it is still necessary to administer immunosuppressors to the patient in order to prevent immune rejection of the islet tissue. For that reason, some groups, including ours, have developed a technique in which islets are encapsulated with alginate or other biopolymers, such as the one developed by our group (patent pending [44]), showing that these encapsulated islets are able to revert type 1 diabetes induced in Balb/c mice by streptozotocin injection and protect these cells from the immune system, without interfering with the islet function [45]. This encapsulation technique may also be useful to protect other cell types from the immune rejection in other cell therapy protocols. We are aiming at encapsulation of insulin-producing cells, obtained by differentiation of ESCs or of adult progenitor cells, or by prolactin-induced beta cell mass increase [7] as a means to overcome the problems of organ shortage for islet isolation and the dramatic increase in the incidence of diabetes mellitus worldwide.

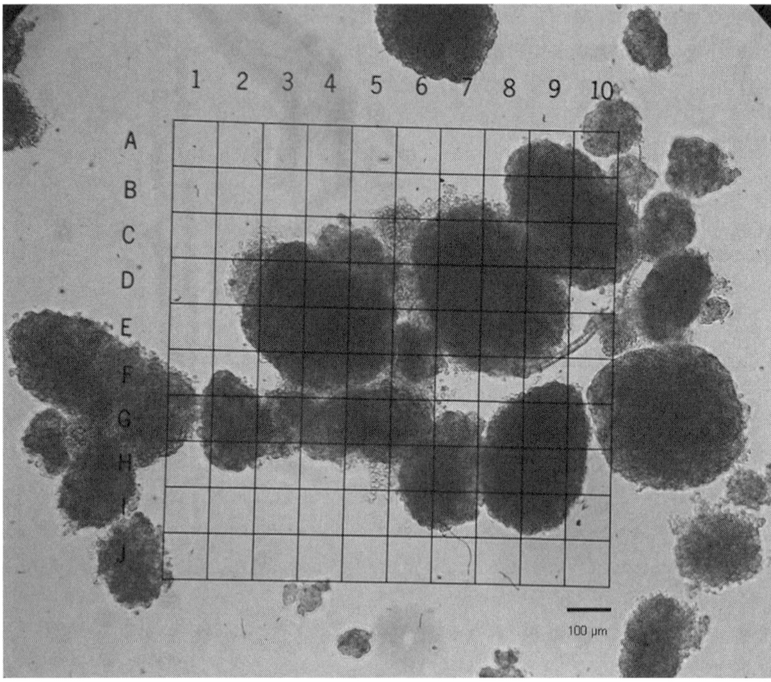

Figure 2.4 Purified human pancreatic islet. Human pancreas are harvested from cadaveric donors and digested through mechanic and enzymatic action, and the islets are purified in a continuous Ficoll density gradient, yielding extremely pure and viable islets, which are identified by staining with the zinc-chelating dithizone. **See Plate 5**.

References

[1] Solar M, et al. Pancreatic exocrine duct cells give rise to insulin-producing beta cells during embryogenesis but not after birth. Dev Cell 2009;17(6):849–60.

[2] Zhou Q, et al. A multipotent progenitor domain guides pancreatic organogenesis. Dev Cell 2007; 13(1):103–14.

[3] Jensen J, et al. Independent development of pancreatic alpha- and beta-cells from neurogenin3-expressing precursors: a role for the notch pathway in repression of premature differentiation. Diabetes 2000;49(2):163–76.

[4] Apelqvist A, et al. Notch signalling controls pancreatic cell differentiation. Nature 1999; 400(6747):877–81.

[5] Heremans Y, et al. Recapitulation of embryonic neuroendocrine differentiation in adult human pancreatic duct cells expressing neurogenin 3. J Cell Biol 2002;159(2):303–12.

[6] Georgia S, et al. p57 and Hes1 coordinate cell cycle exit with self-renewal of pancreatic progenitors. Dev Biol 2006;298(1):22–31.

[7] Labriola L, et al. Beneficial effects of prolactin and laminin on human pancreatic islet-cell cultures. Mol Cell Endocrinol 2007;263(1-2):120–33.

[8] Bonner-Weir S, et al. In vitro cultivation of human islets from expanded ductal tissue. Proc Natl Acad Sci USA 2000;97(14):7999–8004.
[9] Guz Y, Nasir I, Teitelman G. Regeneration of pancreatic beta cells from intra-islet precursor cells in an experimental model of diabetes. Endocrinology 2001;142(11):4956–68.
[10] Kim SY, et al. Activation of nestin-positive duct stem (NPDS) cells in pancreas upon neogenic motivation and possible cytodifferentiation into insulin-secreting cells from NPDS cells. Dev Dyn 2004;230(1):1–11.
[11] Ramiya VK, et al. Reversal of insulin-dependent diabetes using islets generated in vitro from pancreatic stem cells. Nat Med 2000;6(3):278–82.
[12] Maria-Engler SS, et al. Co-localization of nestin and insulin and expression of islet cell markers in long-term human pancreatic nestin-positive cell cultures. J Endocrinol 2004;183(3):455–67.
[13] Zulewski H, et al. Multipotential nestin-positive stem cells isolated from adult pancreatic islets differentiate ex vivo into pancreatic endocrine, exocrine, and hepatic phenotypes. Diabetes 2001;50(3):521–33.
[14] Lechner A, et al. No evidence for significant transdifferentiation of bone marrow into pancreatic beta-cells in vivo. Diabetes 2004;53(3):616–23.
[15] Nagasao J, et al. Expression of nestin and IGF-1 in rat pancreas after streptozotocin administration. Anat Histol Embryol 2004;33(1):1–4.
[16] Delacour A, et al. Nestin expression in pancreatic exocrine cell lineages. Mech Dev 2004;121(1):3–14.
[17] Street CN, et al. Heterogenous expression of nestin in human pancreatic tissue precludes its use as an islet precursor marker. J Endocrinol 2004;180(2):213–25.
[18] Treutelaar MK, et al. Nestin-lineage cells contribute to the microvasculature but not endocrine cells of the islet. Diabetes 2003;52(10):2503–12.
[19] Wang R, et al. Nestin expression and clonal analysis of islet-derived epithelial monolayers: insight into nestin-expressing cell heterogeneity and differentiation potential. J Endocrinol 2005;184(2):329–39.
[20] Suzuki A, et al. Liver repopulation by c-Met-positive stem/progenitor cells isolated from the developing rat liver. Hepatogastroenterology 2004;51(56):423–6.
[21] Yashpal NK, Li J, Wang R. Characterization of c-Kit and nestin expression during islet cell development in the prenatal and postnatal rat pancreas. Developmental dynamics: an official publication of the American Association of Anatomists 2004;229(4):813–25.
[22] May R, et al. Identification of a novel putative pancreatic stem/progenitor cell marker DCAMKL-1 in normal mouse pancreas. Am J Physiol Gastrointest Liver Physiol 2010;299(2):G303-10.
[23] McKnight KD, Wang P, Kim SK. Deconstructing pancreas development to reconstruct human islets from pluripotent stem cells. Cell Stem Cell 2010;6(4):300–8.
[24] Cai J, et al. Generation of homogeneous PDX1(+) pancreatic progenitors from human ES cell-derived endoderm cells. J Mol Cell Biol 2010;2(1):50–60.
[25] Arias AE, Bendayan M. Differentiation of pancreatic acinar cells into duct-like cells in vitro. Lab Invest 1993;69(5):518–30.
[26] De Lisle RC, Logsdon CD. Pancreatic acinar cells in culture: expression of acinar and ductal antigens in a growth-related manner. Eur J Cell Biol 1990;51(1):64–75.
[27] Hall PA, Lemoine NR. Rapid acinar to ductal transdifferentiation in cultured human exocrine pancreas. J Pathol 1992;166(2):97–103.
[28] Edlund H. Transcribing pancreas. Diabetes 1998;47(12):1817–23.
[29] St-Onge L, Wehr R, Gruss P. Pancreas development and diabetes. Curr Opin Genet Dev 1999;9(3):295–300.
[30] Dor Y, et al. Adult pancreatic beta-cells are formed by self-duplication rather than stem-cell differentiation. Nature 2004;429(6987):41–6.

[31] Baeyens L, et al. In vitro generation of insulin-producing beta cells from adult exocrine pancreatic cells. Diabetologia 2005;48(1):49–57.

[32] Zalzman M, Anker-Kitai L, Efrat S. Differentiation of human liver-derived, insulin-producing cells toward the beta-cell phenotype. Diabetes 2005;54(9):2568–75.

[33] Tang DQ, et al. In vivo and in vitro characterization of insulin-producing cells obtained from murine bone marrow. Diabetes 2004;53(7):1721–32.

[34] Ruhnke M, et al. Human monocyte-derived neohepatocytes: a promising alternative to primary human hepatocytes for autologous cell therapy. Transplantation 2005;79(9):1097–103.

[35] Yang L, et al. In vitro trans-differentiation of adult hepatic stem cells into pancreatic endocrine hormone-producing cells. Proc Natl Acad Sci USA 2002;99(12):8078–83.

[36] Hess D, et al. Bone marrow-derived stem cells initiate pancreatic regeneration. Nat Biotechnol 2003;21(7):763–70.

[37] Lumelsky N, et al. Differentiation of embryonic stem cells to insulin-secreting structures similar to pancreatic islets. Science 2001;292(5520):1389–94.

[38] Kahan BW, et al. Pancreatic precursors and differentiated islet cell types from murine embryonic stem cells: an in vitro model to study islet differentiation. Diabetes 2003;52(8):2016–24.

[39] Assady S, et al. Insulin production by human embryonic stem cells. Diabetes 2001;50(8):1691–7.

[40] Schroeder IS, et al. Differentiation of mouse embryonic stem cells to insulin-producing cells. Nat Protoc 2006;1(2):495–507.

[41] Lojudice FH, Kossugue PM, Fortuna VA, Sogayar MC. Unique genes identified during insulin-producing cells differentiation of murine embryonic stem cell line. in prep.

[42] Gu G, Dubauskaite J, Melton DA. Direct evidence for the pancreatic lineage: NGN3+ cells are islet progenitors and are distinct from duct progenitors. Development 2002;129(10):2447–57.

[43] Eliaschewitz FG, et al. First Brazilian pancreatic islet transplantation in a patient with type 1 diabetes mellitus. Transplant Proc 2004;36(4):1117–8.

[44] Mares-Guia TR, Campos-Lisboa ACV, Campanha-Rodrigues AL, Grazioli G, Labriola L. and Sogayar Mari C. Biopolymeric composite for encapsulation of cells, method for production of a biopolymeric composite for cells encapsulation, method for promoting cells cytoprotection and use of a biopolymeric composite for cells encapsulation. Patent requested to: Brazilian National Industrial Property Institute (INPI) PI 1008258-1 (deposited Dec 16th/2010), International PCT/BR2011/000479 (deposited Dec 16th, 2011).

[45] Maria-Engler SS, et al. Microencapsulation and tissue engineering as an alternative treatment of diabetes. Braz J Med Biol Res 2001;34(6):691–7.

Chapter 3

Tissue Engineering and the Role of Biomaterial Scaffolds

The Evolution of Cardiac Tissue Engineering

Sonya B. Seif-Naraghi, Karen L. Christman
Department of Bioengineering, University of California, San Diego, La Jolla, California, USA

Introduction to Tissue Engineering

History

Tissue engineering is a growing field with multidisciplinary players in engineering, biology, and medicine. The general goal of tissue engineering is to restore, maintain, or enhance tissue or organ function. The first use of the term *tissue engineering* to describe the research efforts toward this goal can be traced to Professor Y.C. Fung, of the University of California, San Diego. In 1985, Professor Fung submitted a proposal to the National Science Foundation requesting funding for a research center titled "Center for the Engineering of Living Tissues." The proposal was rejected, but 2 years later at a proposal review panel meeting there was a discussion focused on solidifying the concept of this quickly growing field, and Professor Fung volunteered the phrase again [1]. The field was further established within the

Resident Stem Cells and Regenerative Therapy. http://dx.doi.org/10.1016/B978-0-12-416012-5.00003-7

scientific community when Drs. Langer and Vacanti published the first review of tissue engineering in *Science* in 1993 [2]. This collaboration is also well known for their seminal publication in which an ear-shaped polymeric template was seeded with chondrocytes and subcutaneously implanted in a small animal [3]. Although this was a very early example of tissue engineering and achieved no functional outcome, it launched the field into popular view.

Though the field was not formally defined until 1987, there were prior examples of what would now be considered tissue engineering. Primitive vascular and skin grafts emerged in the early 20th century, and as cell culture techniques improved these technologies advanced. Preliminary work with cells and materials aimed at mimicking the function of the kidney, the pancreas/islet cells, and the liver started emerging in the wake of World War II. Research on growing bone and cartilage developed shortly thereafter, facilitated by the discovery of factors involved in bone induction. Studies representing the pre-1987 body of work were reviewed extensively in a report on the Emergence of Tissue Engineering as a Research Field, produced by the National Science Foundation (NSF) in 2003 [1].

In the years following the Langer/Vacanti review, articles citing tissue engineering increased exponentially. Combinations of cells, biomaterials, and exogenous factors aimed at repairing or reproducing every tissue and organ system have emerged. Investigated biomaterials span the natural and synthetic spectra; cell types vary from mature adult cells to embryonic stem cells (ESCs) and induced pluripotent stem cells; exogenous factors include angiogenic growth factors such as vascular endothelial growth factor (VEGF) and basic fibroblast growth factor (bFGF), bone inductive factors such as bone morphogenic protein 2 (BMP-2), and steroids and hormones that can be used to manipulate certain cell types. The classical tissue-engineering paradigm involves cell-seeded constructs, which are cultured in vitro, prior to implantation. More recently, in situ engineered and acellular scaffold approaches have been explored. A secondary application for the developing therapeutic systems was identified in the area of model systems for in vitro testing—toxicity, pathogenicity, and drug metabolism and uptake. While these applications arguably utilize tissue engineering principles, this chapter focuses on therapeutic systems, specifically tissue engineering approaches to cardiac repair postmyocardial infarction. Several other excellent resources exist that extensively detail the history and progress of tissue engineering [4,5].

Scope

Given the interdisciplinary nature of tissue engineering, success in the relevant areas of medicine, biology, and engineering all provide impetus for advancing the field. This is seen clearly with the advancement of techniques for the culture and differentiation of stem cells as well as the introduction of induced pluripotent stem cells. Additionally, as the technology used to develop bioreactors improves, so will the ability of tissue engineers to manipulate their constructs. Finally, each new insight into the mechanisms of disease provides an opportunity for a tissue engineer to incorporate that understanding into the design of relevant therapies. Thus, as these areas advance, their success has an immediate impact on tissue engineering. Similarly, achievements within the field of tissue engineering have widespread effects in related areas. If the goals of repair and regeneration of damaged or diseased organs are accomplished or if functional tissue can be engineered ex vivo for the replacement of damaged organs, the implications of that success will be innumerable.

Principles

When discussing principles of tissue engineering, it is prudent to define what exactly we mean by the term. Langer and Vacanti defined tissue engineering in the following way:

> *Tissue engineering is an interdisciplinary field that applies the principles of engineering and the life sciences toward the development of biological substitutes that restore, maintain, or improve tissue function. [2]*

A widely accepted umbrella used to conceptualize what constitutes "biological substitutes" is the tissue engineering triad (Fig. 3.1), which consists of cells, biomaterials, and exogenous bioactive factors. In general, tissue engineering approaches involve at least two of the three constituents of the triad. Traditionally, this is represented by a biomaterial scaffold seeded with a relevant cell type and manipulated with an exogenous biomolecule (e.g., VEGF or BMP-2) in order to

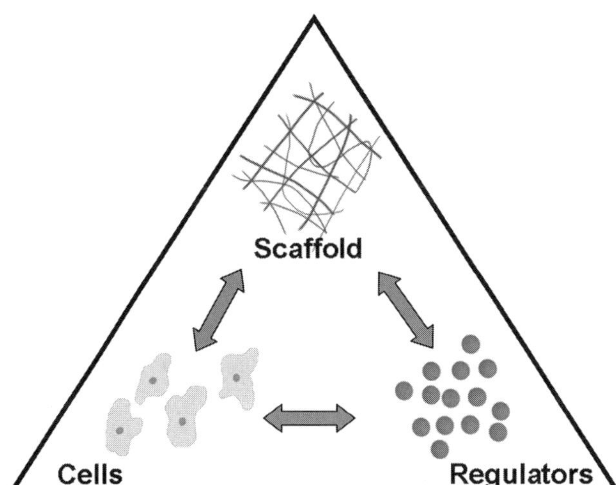

Figure 3.1 Tissue engineering triad. The umbrella of tissue engineering traditionally includes systems with at least two components of the tissue engineering triad, which is comprised of a biomaterial scaffold, cells, and regulators (e.g., exogenous growth factors).

reproduce or achieve some target function of the given tissue or organ. This definition can be relaxed to include biomaterial-only scaffolds that are designed to elicit a cellular response upon introduction to the host. In this case, although the "biological substitute" may include only one of the three components of the tissue-engineering triad, various cell types—including progenitor cells—can be recruited to the scaffold in vivo, a method of in situ tissue engineering.

Role of Biomaterials in Tissue Engineering

As an important component of the tissue-engineering triad, biomaterials have been used extensively to build both two-dimensional models and three-dimensional constructs. Here, the importance of the extracellular matrix as a biomaterial will be discussed, the range of naturally-derived and synthetic biomaterials used in tissue engineering will be outlined, and a set of examples of the applications for biomaterials as acellular, cellular, and growth factor delivery scaffolds will be presented. A discussion of the challenges specific to the individual areas of tissue engineering will be left to comprehensive references [4,5] and discipline-specific reviews, of which there are many.

Extracellular Matrix

The extracellular matrix (ECM), which is nature's scaffold, is a complex network of structural and fibrous proteins, proteoglycans, and glycosaminoglycans, which have been implicated in the regulation of cell communication, migration, proliferation, differentiation, adhesion, and signaling [6]. Specifically, the ECM is comprised primarily of collagen, with varying ratios of elastin, fibronectin, vitronectin, laminin, and other secreted proteins that are unique to each tissue [7]. In addition to these proteins, proteoglycans such as heparan sulfate proteoglycans (HSPGs), decorin, and dermatopontin allow for interactions between the cells and their ECM. Finally, glycosaminoglycans such as heparan sulfate, dermatan sulfate, and hyaluronic acid provide compressive strength as well as sequester and immobilize growth factors. Mimicking attributes of the architecture or composition of native ECM may allow for better manipulation of the cellular components in tissue-engineered systems. Additionally, presenting an environment that closely resembles native ECM and is therefore familiar to the local cell population could promote cell infiltration and endogenous repair. For these reasons, many of the naturally derived biomaterials commonly used in tissue engineering are ECM components.

Naturally Derived Biomaterials

Collagen, the primary component of most ECMs, is one of the most widely used biopolymers. Depending on the application, it can take various forms, including a soft hydrogel, a fibrous sponge, or an electrospun sheet [8]. ECM components such as elastin, laminin, heparan sulfate, and hyaluronic acid have also been used extensively as scaffold materials. Other popular natural materials include chitosan, alginate, and fibrin. These natural biopolymers have the general advantage of biocompatible degradation products, as well as being inherently bioactive. Unless chemically modified, these materials will all degrade relatively quickly—an advantage or a disadvantage depending on the application. Nonpolymeric natural materials such as hydroxyapatite or calcium carbonate are more resistant to degradation and have been utilized in applications that require stiff materials, such as bone tissue engineering [9].

Naturally derived biomaterials can be used in combination with one another, often in an attempt to more closely mimic the complex environment presented by the native ECM. One approach to harness the complexity of native ECM is to use the network secreted by cells in culture. In this way, a synthetic polymer scaffold can be used as a template on which cells can deposit ECM. If the original polymer template degrades, then eventually the construct will be remodeled to only consist of the cells and their ECM. Recently, decellularization techniques have provided the ability to start with native tissue, remove the cellular component, and then use the resulting ECM scaffold as a biomaterial scaffold. These materials have been shown to be biocompatible, as long as the decellularization protocol is effective in removal of the cellular antigens that may cause an immune response [10]. Complete decellularization of the heart with detergent solutions was first published by Ott et al. (Fig. 3.2) [11].

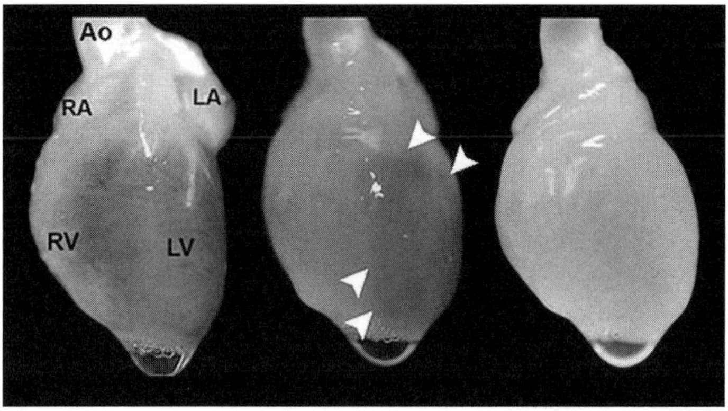

Figure 3.2 Whole heart decellularization. An adult rat heart is shown being decellularized with ionic detergents. The progression of decellularization is illustrated by the change in color from red/brown to a slightly translucent white, indicating the removal of blood and cellular content. The receding line of residual debris is indicated with arrows. *(Modified from [11] and reprinted with permission.)* **See Plate 6.**

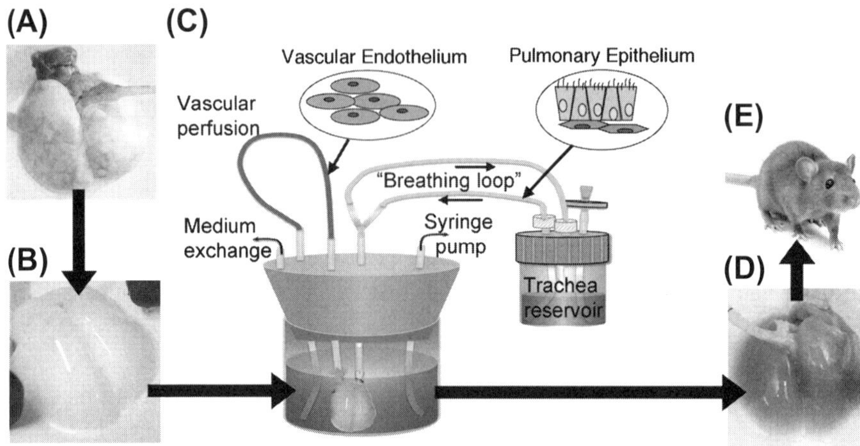

Figure 3.3 Whole lung engineering scheme. This scheme shows the fabrication of the scaffold **(A-B)**, seeding with relevant cells and culturing in a bioreactor **(C)**, and implantation of the in vitro tissue-engineered organ in a small animal model **(D-E)**. *(Modified from [13] and reprinted with permission.)*

This concept has subsequently been applied to nearly all organs and tissues [12]. Research utilizing decellularized scaffolds has involved seeding them with cells in an effort to repopulate the constructs with appropriate cell types and encourage either the organization and function of adult cells or the differentiation and maturation of stem/progenitor cells. In a recent example of this strategy, Petersen et al. decellularized adult rat lungs and showed that the hierarchic branching structures of airways and vasculature was retained. Pulmonary epithelial and vascular endothelial cells were able to successfully repopulate the scaffold, indicating the potential for using this tissue-engineered system for lung regeneration (Fig. 3.3) [13].

Not only can these decellularized ECM constructs be used as implantable, three-dimensional scaffolds, but they can also be processed into injectable materials that form gels in situ. Using methods modified from those first published by Freytes et al., a variety of tissues have been processed in this way, including the urinary bladder, myocardium, pericardium, adipose tissue, liver, small intestinal submucosa (SIS), and brain [14.20]. These injectable biomaterials are liquid at room temperature and form fibrous, porous hydrogels when brought to physiologic conditions of temperature, salt concentration, and pH. They are therefore potentially well-suited for minimally invasive delivery.

Synthetic Materials

Polymers represent the bulk of synthetic materials used in tissue engineering, ranging widely from poly(lactic acid) (PLA), poly(glycolic acid) (PGA), poly(ethylene glycol) (PEG), polyurethane, polytetrafluoroethylene (PTFE), and poly(N-isopropylacrylamide) (poly(NIPAAm)), among others. One of the advantages these materials carry over natural materials includes the ability to tailor properties (stiffness, porosity, degradation, ligand density, etc.) of the material for its intended purpose. For example, synthetic polymers can be designed to be either biodegradable, by including sites that allow for hydrolysis or enzymatic cleavage, or nonbiodegradable, in which case the scaffolds remain intact in the host. However, if the material does not degrade at all, there is potential for eliciting a foreign body reaction. On the other hand, another concern with synthetic polymers is the biocompatibility of degradation products.

Poly(N-isopropylacrylamide) (poly(NIPAAm)) is an interesting synthetic polymer because of its physiologically relevant lower critical solution temperature (LCST) of 32 to 33 °C. At room temperature, poly(NIPAAm) is hydrophilic and therefore soluble in aqueous solutions as well as resistant to protein adsorption and cell adhesion. At body temperature, however, poly(NIPAAm) is hydrophobic and therefore allows for protein adsorption and cell adhesion. Taking advantage of this property, Okano and colleagues pioneered cell sheet technology by culturing cells on plates coated in poly(NIPAAm) [21,22]. Cells are allowed to grow to confluency and produce their own extracellular matrix, at which point the temperature is dropped below the LCST. The change in hydrophobicity causes the adsorbed ECM proteins and associated cells to lift off of the material in a sheet. Stacked cell sheets have been produced and used as thin grafts; however, because of diffusion limitations, the maximum thickness that has been generated is approximately 600 μm [23,24]. Other applications utilizing poly(NIPAAm) are found in therapeutic embolization or as in situ tissue engineering scaffolds, as it can be kept in solution at room temperature, but when injected into the body, precipitates and forms a solid scaffold [25,26].

Growth Factor Delivery

In addition to using biomaterials as acellular scaffolds or cellular delivery vehicles, tissue engineering scaffolds can also be used to prolong the retention and release of bioactive moieties such as growth factors. Growth factors can be used to elicit a variety of biologic responses, for example, increased vascularization. Therapeutic angiogenesis via delivery of angiogenic factors such as VEGF and bFGF has specifically been investigated in a variety of disease models including myocardial and peripheral ischemia [27,28], and wound repair [29,30]; several reviews have been written on the topic [31-33].

Clinical Translation and Commercialization

In the wake of the initial tissue engineering boom in the early 1990s, many companies sprang up to commercialize the new products. Unfortunately, as a result of a series of underwhelming product launches, failed regulatory trials, and general lack of venture capital following the dot-com crash, tissue engineering suffered a severe downturn between 2000 and 2002. Remarkably, however, between 2002 and 2007, economic activity in the fields of tissue engineering, regenerative medicine, and stem cell therapeutics increased fivefold. By mid-2007 there were more than 100 development-stage companies with more than 55 products in preclinical and clinical trials [34]. Tissue-engineered products for repair of bone, cartilage, cornea, blood vessels, urinary structures, left mainstem bronchus, and skin have all proceeded to clinical trials in humans [35]. In a thorough review of the private sector involvement in the progression of tissue engineering, Lysaght et al. observes that contemporary commercial success results from identification of products that can be achieved given the current technology and regulatory guidelines [34]. It is concluded that the future of tissue engineering is bright, given the field's demonstrated resilience and the current diversity of research.

Future of Tissue Engineering

Advancement in both stem cell technologies and our understanding of disease states will inform the design of future tissue-engineered systems. As discussed, because the field has such strong ties in different areas of engineering, medicine, and biology, the improvement in knowledge in a variety of areas may provide impetus for advancing the field. As our understanding of how to best manipulate progenitor and stem cells increases, scaffolds will be able to be fabricated to better achieve goals with respect to cellular organization and differentiation. Advancements in the development of smart materials—which include an ever-growing arsenal of materials sensitive to temperature, salt concentration, and pH—will also accelerate the growth of tissue engineering. Additionally, although tissue engineering has historically been identified as applied research, a new focus on elucidating the mechanisms by which many of the tissue-engineered scaffolds are effective must be undertaken. Only then will scaffolds be able to be designed and tailored under an informed set of design criteria out of which novel, relevant, and clinically translatable approaches can emerge.

Cardiac Tissue Engineering

Introduction

The implications of cardiac tissue engineering are widespread; more than one in three deaths in the United States alone are attributed to cardiovascular disease, and heart failure following myocardial infarction is responsible for a majority of those deaths [36]. The discovery and investigation of novel approaches to treating this difficult population will benefit patients with little other recourse. The term *cardiac tissue engineering* will be used here to describe tissue-engineered strategies for the support, repair, regeneration, or replacement of myocardial tissue (cardiac muscle) as a treatment for myocardial infarction and the prevention of heart failure.

Disease State

To understand the rationale behind the cardiac tissue engineering designs that will be discussed in this chapter, it is useful to first understand the underlying disease state. Briefly, the occurrence of events in patients with ischemic cardiomyopathy involves an initial coronary event that injures downstream tissue. The ensuing inflammatory response leads to remodeling of the infarcted tissue, a process that can be beneficial at first but eventually results in congestive heart failure.

Myocardial Infarction and Left Ventricular Remodeling

More specifically, myocardial infarction (MI) refers to an ischemic injury caused by the blockage of a coronary artery. These vessels provide the blood supply to the cardiac muscle, and cessation of oxygen delivery causes cell death in the downstream tissue. After an initial inflammatory period, the ECM is remodeled by fibroblasts, which lay down collagenous scar tissue. Unfortunately, this noncontractile replacement cannot contribute to the pumping action of the heart. Initially, cells in the noninfarcted myocardium hypertrophy to compensate for the loss in function. This stiffens the muscle, increasing myocardial wall stress. Over time, this causes the infarct area to thin, further increasing wall stress. This accelerates the negative left ventricular (LV) remodeling process that leads to the development of heart failure. In this condition, the ability of the heart to pump blood, as often measured by ejection fraction, is severely reduced. Average mortality data indicate that 50% of end-stage heart failure patients die within 5 years of diagnosis; in patients with coronary disease or who are older than 55, mortality is significantly higher [37].

Current Treatment Options

Advancements in medical therapy and percutaneous interventions have improved the prognosis post-MI considerably, but the incidence of heart failure is only increasing, likely as a result of the increasing number of patients who now survive the initial attack. Currently, the only approved treatments for end-stage heart failure post-MI are left ventricular assist devices (LVADs) and total heart transplantation. The former is limited by the complications of a chronic external assist device and the inherently risky surgical procedure necessary to implant the device in a fragile patient population. Although total heart transplantation may provide a solution for a few patients, there is a severe dearth of donor hearts, and even for the lucky population who do receive a transplant, a lifetime of immunosuppressant drugs and the side effects thereof is not an ideal solution.

TISSUE ENGINEERING APPROACHES

Introduction

Initial experimental therapies involved the delivery of cells and or growth factors in simple liquid solutions. More recently, tissue engineering approaches have been explored. As discussed, classical tissue engineering strategies involve some combination of a biomaterial scaffold, a cellular component, and bioactive moieties designed to achieve some functional outcome. When applied to cardiac repair post-MI, the first application for biomaterials was as cell delivery scaffolds, utilized to enhance cell survival and retention upon implantation. This was first explored in the form of three-dimensional cardiac patches that were seeded with cells and delivered to the epicardial surface of the heart. In an effort to reduce the invasiveness of delivery, injectable biomaterials that gel in situ within the ventricular wall have also been investigated as scaffolds for the enhancement of cell delivery and retention. More recently, biomaterials in the form of both patches and injectable scaffolds have been investigated for use as acellular therapies, designed for in situ tissue engineering in which endogenous cellular infiltration and remodeling is encouraged. This discussion will outline scaffold design parameters, focus on the rationale for both cellular and acellular therapies, introduce the two primary tissue engineering approaches, and summarize seminal findings. For an exhaustive assessment of in vivo studies utilizing biomaterials in these applications, the reader is referred to two reviews on the use of biomaterials for MI [24,38].

Design Parameters

Cardiac tissue–engineered scaffolds intended for the treatment of ischemic cardiomyopathy should ideally do the following: mitigate negative left ventricular remodeling, salvage at-risk cells in borderzone myocardium, provide a template for cellular infiltration, promote angiogenesis to support the tissue remodeling process, and degrade in a biocompatible fashion. Thus, the biochemical, structural, and mechanical properties all need to be appropriate for the application. Ligand density and distribution can regulate cell attachment, and degradation products of scaffold materials can either promote or retard chemotaxis into the construct. Porosity and fiber diameter are two structural considerations that can change cellular attachment and migration; pore size of approximately 30 μm has been reported as ideal for cellular and vascular infiltration [39]. Material mechanics is another consideration—with respect to cardiac tissue engineering, these mechanics can be modeled after healthy myocardium (20 kPa), or individual ECM components (0.5 to 1 kPa). The mechanics of the scaffold will affect cellular behavior, so if stem or progenitor cells are incorporated, it is possible that the mechanics of the developing heart, which is significantly more compliant than adult tissue, would be appropriate.

When designing an in vitro engineered cardiac patch, perfusion is an important parameter and has historically been a limiting factor. If cells are seeded throughout, constraints of diffusion prevent patches of appreciable thickness (greater than 600 um) from being successful. For in situ engineered systems, it is essential to have a template appropriate for the survival and retention of cells incorporated into the system or for the recruitment and organization of cells infiltrating into the scaffold. For injectable systems, gelation kinetics are an important consideration as systems that gel too quickly may not be compatible with the minimally invasive, catheter delivery methods.

Cellular versus Acellular Scaffolds

Biomaterial scaffolds have been utilized extensively as cell delivery scaffolds, intended to provide structural support and improve survival and retention in the harsh post-MI environment [23]. In this context, a variety of biomaterials have been explored, including synthetic materials such as polyethylene glycol (PEG), poly(NIPAAm), and poly(glycerol-sebacate), as well as naturally derived materials such as collagen, Matrigel, gelatin, alginate, fibrin, and decellularized ECM. These biomaterials have served as delivery scaffolds for many cell types, including adult cells such as skeletal myoblasts, cardiomyocytes, and dermal fibroblasts as well as multi- and pluripotent cells such as bone marrow–derived mesenchymal progenitor cells and embryonic stem cells. Although a wide range of materials and cell types have been combined and explored, the general results of these studies confirm the utility of biomaterial scaffolds to promote cell survival and retention in vivo.

In addition to their use as cell delivery vehicles and scaffolds, biomaterials have been used extensively as acellular scaffolds. In this role, the materials provide a template for cellular infiltration and remodeling. As an in situ tissue engineering approach, this takes advantage of the resident progenitor and other cells already present in the given tissue. This approach may prove especially useful when the tissue has a high regenerative capacity and a larger population of progenitor cells—the urinary bladder, small intestinal submucosa (SIS), and the skin, for example—but the strategy can be applied in any organ, even one with a small population of resident progenitor cells, such as the heart. As noted, one of the challenges of this particular tissue engineering problem is that the heart has very little regenerative capacity. Until recently, it was thought that all the cells in the heart were terminally differentiated. Studies have now identified a population of progenitor cells in the heart, but they are in low numbers. These cells are identified as being negative for blood lineage markers (Lin-) and positive for stem cell surface markers such as c-kit, Sca-1, and MDR-1 [40]. Interestingly, in failing human hearts it was determined that more of these cells are actually present [41]. If these cells could be recruited to the infarct area and given an environment that encouraged differentiation and organization, then some endogenous repair could potentially be elicited. Thus, an acellular treatment that promotes cell infiltration and provides an appropriate environment within the harsh infarct milieu could also elicit regeneration by taking advantage of these resident progenitor cells. A handful of studies have used immunohistochemistry to identify the presence of progenitor cells in or near the treatment region or infarct. Examples include studies involving acellular injectable scaffolds of Matrigel [42], an emulsion of decellularized SIS [43], and matrix gels derived from pericardial [44] and myocardial tissue [45]. The presence of these cells indicates the potential for stem cell mobilization within the tissue.

Although the techniques and materials vary greatly, one remarkable aspect these approaches have in common is that, when applied in small animal MI models, most of these tissue-engineered systems show improvements in cardiac function, as measured by increases in ejection fraction (EF) or fractional shortening (FS). Other positive outcomes may include a decrease in infarct size or fibrosis, an increase in infarct wall thickness, and an improvement in contractility or elasticity.

Growth Factor Delivery

In addition to cellular delivery vehicles and acellular biomaterial-only scaffolds, cardiac tissue engineering scaffolds for growth factor delivery have been investigated. Inducing angiogenesis may preserve endogenous cardiomyocytes and functionally contractile myocardium post-MI [46-48]. To harness this potential, tissue engineering systems have been designed to deliver these proteins to infarcted tissue. Growth factors such as VEGF and bFGF have been covalently bound to synthetic polymers such as PLGA [49] and poly(NiPAAm) [50], as well as to naturally derived biomaterials such as collagen [51] and hyaluronic acid [31,32,52]. Bio-inspired self-assembling peptides [53] have also been used as growth factor delivery scaffolds. Other hybrid materials that incorporate either biomolecules that associate with growth factors natively, such as heparin or heparan sulfate [54,55], or derivatives that include highly sulfated sugars or heparin-like growth factor binding domains [57,58], have also proved useful. Immobilization on a biomaterial scaffold has been demonstrated to be more effective than physical entrapment or bolus injection, as it increases growth factor stability and localizes the effects to the site of treatment [59]. In vivo, growth factors interact with sulfated glycosaminoglycans (sGAGs) that associate with ECM proteins [60]. Thus, the ECM can sequester and present growth factors to the cells in the tissue microenvironment. Recently, an injectable, acellular pericardial matrix hydrogel was shown to allow for the natural incorporation of a heparin-binding growth factor, bFGF, for prolonged release and increased vascularization in a rat MI model (data unpublished). Delivery of other growth factors to reduce the degree of fibrosis or enhance cell survival or differentiation has also been investigated.

Patches

Cardiac patches, a primary tissue engineering approach for treating MI, are three-dimensional constructs fabricated in vitro and implanted over the infarcted tissue. In some cases, the patches are used to replace the damaged tissue after resection in a ventriculotomy. Li et al. was the first to deliver cells in a biomaterial patch; scaffolds improved cell survival and retention but showed no improvement in cardiac function [61]. Leor et al. then used alginate to transplant fetal cardiomyocytes in a rat MI model and successfully demonstrated improvement in both cell survival and cardiac function [62]. The first examples of cardiac patches used a variety of cell types, including stem cells and progenitor cells.

Stem cells appear in conjunction with a wide range of materials. For example, collagen scaffolds have been seeded with bone marrow–derived CD133+ cells [63], human mesenchymal stem cells [64], and human umbilical cord mononuclear cells [65] and applied in small animal infarct or cryoinjury models with varying degrees of success. In general, the application of these patches leads to increased vascularization and a maintenance or improvement in functional parameters such as EF and FS. An exciting development is the use of cell-seeded collagen patches in a phase 1 clinical trial, the Myocardial Assistance by Grafting a New Bioartificial Upgraded Myocardium (MAGNUM). In this study, a collagen patch seeded with autologous bone marrow cells was administered to 10 patients undergoing coronary artery bypass graft (CABG) surgery in conjunction with intramyocardial injection of bone marrow cells. Results from this study indicate that both cell injection alone and with the patch improve EF and the patient's New York Heart Association functional class at a 1-year follow-up. Improvements in LV end diastolic volume and scar thickness, however, were only observed in the patients who received both patch and injection and not those who only received the bolus injection [66].

Other naturally-derived materials that have been used in conjunction with stem cells include decellularized pericardial, SIS, and myocardial tissue. Mesenchymal stem cells have been seeded on decellularized pericardial tissue and used to replace tissue in a small animal ventriculotomy model with observed improvements in cardiac function [67]. Implantation of a decellularized SIS patch, with and without mesenchymal stem cells, improved LV contractility and microvessel density in a rabbit MI model [68]. In addition to stem cells, many studies have used progenitor cells and ESC-derived cells. For example, Xiong et al. seeded human ESC-derived endothelial cells and human ESC-derived smooth muscle cells onto a fibrin scaffold. When implanted in a porcine model, EF improved after 7 days and maintained the improvement over 4 weeks [69]. Godier-Furnemont et al. incorporated human mesenchymal progenitor cells in fibrin and seeded them on a scaffold made from a decellularized human myocardial sheet. The scaffold was implanted over infarcts in a rodent MI model, and the researchers observed a preservation of cardiac function and an increase in vascularization [70]. These results support the hypothesis that not only can a biomaterial scaffold support the survival and retention of the cells seeded within it, but that the application of the cell-seeded patches may provide therapeutic benefit.

Acknowledging the challenges of clinical translation for strategies that rely on a cellular component, there has been recent work focused on developing acellular cardiac patches. The following examples illustrate the breadth of these approaches. Callegari et al. applied a patch made from a simple collagen sponge and showed an increase in angiogenesis compared to controls in a cryoinjury small animal model up to 60 days postapplication [71]. An acellular synthetic scaffold composed of an elastic biodegradable polyester urethane urea (PEUU) developed by Fujimoto et al. (Fig. 3.4) was shown to increase fractional area change, wall thickness, and capillary density 8 weeks post-implantation when compared to the control group [72]. Another synthetic scaffold used is

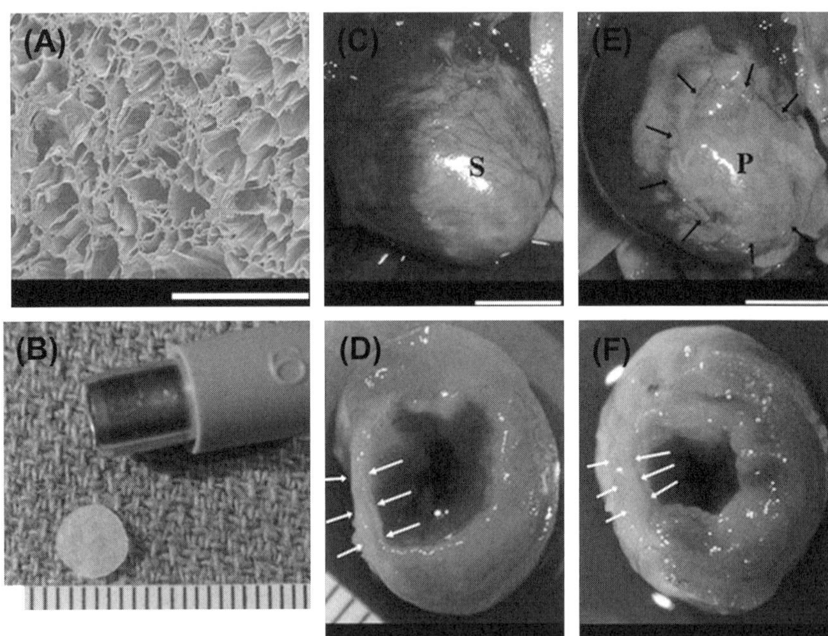

Figure 3.4 Cardiac patch application. Here, an elastic synthetic biodegradable cardiac patch is applied extramurally over the infarct. In **(C)**, the infarct of a control animal is shown and compared to the applied patch, which can be seen in **(D)**. After 8 weeks there is a marked difference in wall thickness between saline-injected control animals **(E)** and the patch-treated animals **(F)**. *(Modified from [72] and reprinted with permission.)* **See Plate 7**.

biodegradable poly(glycolide-co-caprolactone); when implanted both with and without bone marrow–derived mononuclear cells, this patch improved FS, ventricular diameters, and hemodynamics after 4 weeks when compared to the control [73].

Hybrid patches have been also been constructed; for example, Krupnick et al. designed a patch out of a combination of poly(lactic acid) mesh, collagen type I, and Matrigel, all of which was then reinforced with polytetrafluroethylene. They used this as both an acellular and cellular scaffold to replace ventricular tissue after ventriculotomy. Miyagi et al. designed another hybrid system for the delivery of bone marrow MSCs and cytokines by using a combination of synthetic and natural components. Using a gelatin sponge (Gelfoam) coated with poly ε-caprolactone, this group took advantage of the gelatin to promote cellular attachment as well as the synthetic component to provide structural support. When used in a small animal ventriculotomy model, the hybrid scaffold showed an improvement in cardiac function when compared to control animals and to unmodified gelatin sponge. This improvement was seen in all groups with the hybrid scaffold—alone, with or without cytokines, and with or without cells [74].

Thus, work with acellular patches as potential in situ tissue engineering approaches to cardiac repair has increased dramatically and given the promising results, this area is likely to continue to expand. There is, however, some dispute within the literature, as studies often have conflicting results. When investigating the effectiveness of direct cell injection compared to delivery in a poly(lactide-co-ε-caprolactone) cardiac patch, Jin et al. found that MSCs delivered via both routes were able to differentiate into cardiomyocytes. Additionally, both the direct cell injection and delivery in the patch showed similar increases in EF when compared to both saline and implantation of an acellular scaffold, which was insufficient for the preservation or improvement of cardiac function [71]. As the body of literature increases with respect to the number of studies that specifically investigate the differences in the effects of cellular and acellular patches, it may become possible to understand how and why they differ and if one can provide more benefit. To date, the literature disagrees as to the efficacy of acellular cardiac patches; these discrepancies may be due to differences unrelated to the scaffolds themselves—differences in the animal model, time of treatment, and length of study make it difficult to compare results across studies.

Injectable Scaffolds

Similar to cardiac patches, injectable scaffolds have been investigated for use as cell delivery vehicles, as well as for acellular treatments or the delivery of growth factors. A wide variety of injectable scaffolds have been investigated, including naturally-derived materials such as collagen, fibrin, ECM hydrogels, and alginate as well as synthetic materials such as various polyesters as well as PEG, and poly(NiPAAm).

Other materials used as injectable scaffolds for cardiac repair include self-assembling peptides and combination materials such as collagen-GAG hybrids. One of the most exciting advantages of developing an injectable scaffold for cardiac repair is that if the material is compatible with delivery via catheter, then the treatment could be administered via minimally invasive methods.

The potential for using an injectable biomaterial to enhance cell delivery was first demonstrated by Christman et al., by intramyocardial injection of fibrin glue, with and without skeletal myoblasts, into a small animal MI model [75,76]. Improvements in cardiac function as well as increases in cellular retention and survival were seen when compared to cells delivered in saline. One of the most interesting results of this study, however, was that injecting acellular fibrin demonstrated similar improvements in cardiac function. Since then, fibrin has been studied extensively as an acellular, cellular, and growth factor delivery scaffold. Marrow-derived stem cells with cardiomyogenic potential [77], adipose-derived stem cells [78,79], bone marrow mononuclear cells, bFGF, VEGF, hepatocyte growth factor (HGF), and transforming growth factor beta (TGFB) have all been incorporated into an injectable fibrin scaffold for cardiac repair. The results of these studies are varied. Danoviz et al. demonstrated that both an acellular injection of fibrin and an injection of adipose-derived stem cells in fibrin improved cardiac function compared to the control, but the cellular scaffold reduced scar area to a greater degree [78]. Similarly, Zhang et al. showed that although both fibrin alone and fibrin with cells improved function compared to the control, the cellular scaffold-treated animals were significantly improved compared to the acellular scaffold. As with patches, the disagreement in literature over whether or not cellular scaffolds provide additional benefit compared to acellular scaffolds has yet to be resolved.

Although Christman et al. laid the foundation with their initial work, fibrin is a two-component system comprised of thrombin and fibrinogen, which undergo a rapid gelation process when combined. To use this material in vivo, a direct injection with a Duploject syringe, or similar device, is required. As such, fibrin is not compatible with minimally invasive delivery methods—it would clog any catheter before reaching the myocardium. In the past few years, alginate has emerged as a biomaterial that can be compatible with catheter delivery. After the seminal work by Leor et al. that established alginate as a potential candidate for an injectable treatment for cardiac repair [6], it has been used extensively. Preservation or improvement in cardiac function has been observed in small animal models after injection of alginate in combination with fetal cardiomyocytes [80] and human mesenchymal stem cells [81], among others. Incorporating growth factors such as VEGF and platelet derived growth factor [82], insulin-like growth factor [83], and HGF [83] has been shown to increase neovascularization or attenuate infarct expansion and LV remodeling. Preclinical trials with alginate in a porcine MI model demonstrated that intracoronary injection of alginate alone could prevent or reverse negative LV remodeling. With this method, alginate is delivered into the coronary arteries and leaks through the vessels and into the myocardium [84]. Based on these successes, a "first-in-man" study was initiated to evaluate the safety and feasibility of intracoronary injection of alginate in acute MI patients [85].

Decellularized ECM hydrogels reappear under this category because they can also be used as injectable scaffolds. As discussed, via methods first outlined by Freytes et al., decellularized ECM materials can be processed into injectable scaffolds that are liquid at room temperature and form gels when brought to physiologic conditions. For cardiac applications, ECM hydrogels derived from porcine SIS, porcine myocardium, and porcine and human pericardium have been investigated. Okada et al. investigated the efficacy of two forms of SIS-ECM hydrogels and found that one form improved fractional area change compared to both the other ECM gel and the control at 2 weeks posttreatment and reduced infarct size and increased capillary density compared to the other ECM gel and the control at 6 weeks [86]. With the rationale that a tissue-specific ECM scaffold would provide a more appropriate combination of biochemical cues, Singelyn et al. developed a myocardial matrix material derived from porcine cardiac tissue, which was shown to self-assemble into a fibrous gel upon injection [20] and preserve function in a small animal MI model [45]. Interestingly, histologic staining of the self-assembled gel in vivo closely resembles the decellularized ECM prior to processing (Fig. 3.5). When

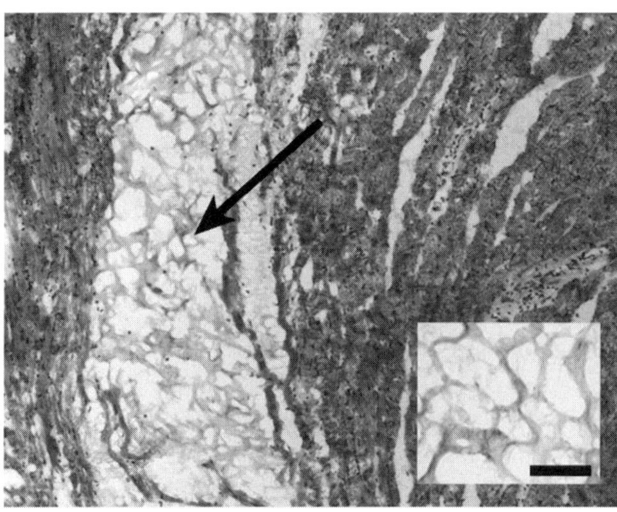

Figure 3.5 Injectable acellular biomaterial scaffold. Here, a ventricular ECM-derived injectable hydrogel has been injected in a healthy rat heart. Hematoxylin and eosin staining allows for visualization of the matrix injection *(arrow)*. The inset at right shows the decellularized ECM prior to processing, which has a strikingly similar structure to the self-assembled gel in vivo. *(Modified from [20] and reprinted with permission.)* **See Plate 8**.

scaled up to delivery in a porcine MI model, percutaneous transendocardial catheter delivery of the myocardial matrix was successful, a benchmark for translation potential [45]. Seif-Naraghi et al. developed a potentially autologous ECM hydrogel from decellularized pericardium, establishing proof of concept by injecting this acellular scaffold in healthy rats and observing cell infiltration, specifically vascular cells. The authors also identified the presence of c-kit+ cells in the injection region, indicating the potential for recruiting resident progenitor cells [44]. Similarly, when an SIS-ECM suspension was injected in a rat MI model, a population of c-kit positive cells was identified in the region. Additionally, the acellular injection alone promoted cell migration and improvement in cardiac function when compared to the saline control [43].

Other injectable scaffolds have been investigated; chitosan has been injected with and without cells [87,88] as well as with bFGF [89], Matrigel has been delivered alone [42], gelatin and bFGF have been co-injected [90,91], and bio-inspired self-assembling peptides have been used to deliver insulin-like growth factor 1 and cardiomyocytes [92,93] as well as autologous bone marrow mononuclear cells [94]; these peptides have also been used for dual delivery of platelet-derived growth factor and fibroblast growth factor [95].

All of the injectable scaffolds discussed thus far have been naturally-derived materials, but there have also been a number of interesting synthetic polymers used as injectable treatments. As mentioned, synthetic materials carry the advantage of being highly tunable with respect to degradation time, mechanical properties, and porosity. Additionally, the inherent batch-to-batch variability present in natural materials is significantly reduced with synthetic polymers. Some examples of successful applications of synthetic polymers for cardiac repair include combination scaffolds based on PEG and poly(NIPAAm). As discussed, the degradation of these materials is a consideration; many synthetic scaffolds are modified to be enzymatically degraded in vivo. For example, matrix metalloproteinase (MMP)-degradable PEG hydrogel with thymosin B-4 and human ESCs [96] and MMP-degradable PNIPAAM-co-acrylic acid hydrogels with bone marrow–derived mononuclear cells have both been shown to improve cardiac function in small animal MI models [97].

Although the exact mechanism by which these scaffolds are effective is unclear, a variety of hypotheses have been discussed, including that an appropriate degradation time may allow for resident cell in-growth and tissue remodeling, that having a fibrous, porous microstructure may support the cell infiltration process, that mechanical support may decrease stress on the ventricle, and, finally, that degradation products of the naturally derived materials discussed may offer biochemical cues for cellular migration and angiogenesis. Additionally, when these biomaterials serve as delivery vehicles for cellular cardiomyoplasty, there could be a synergistic effect that enhances benefit. Recently, attention has turned to elucidating these mechanisms. Ifkovits et al. investigated how material properties affect cardiac function and remodeling by injecting two different formulations of methacrylated hyaluronic acid (one at 8 kPa and one at 43 kPa) in an ovine MI model. Functional outcomes were not different between the two groups, but the stiffer hydrogel did reduce infarct area to a greater degree than the more compliant hydrogel [98].

Previously, computational modeling was used to try to predict how an injected material would affect the myocardial mechanics. Results of this work suggested that the increase in wall thickness provided by the presence of the bulk material would lead to a decrease in wall stress because of Laplace's law and a subsequent improvement in cardiac function [99]. However, recent work by Rane et al. established that the presence of a bio-inert material within the infarct wall did not prevent the negative remodeling process in a rat model and was not sufficient to improve cardiac function despite increasing wall thickness [100]. Dobner et al. also injected nondegradable PEG hydrogel in a rat model in order to investigate the effects on remodeling and found that by 13 weeks, pathologic progression was similar to that of saline [101]. Taken together, these results suggest that the structural support provided by a bulk material is insufficient to mitigate negative LV remodeling. Therefore, the biologic properties, degradation properties, or other aspects of the material that encourage cell infiltration may be the mechanisms by which these materials are efficacious.

These examples represent only a few of the many promising approaches, and given the advances made since the early 2000s, it is not surprising that materials have advanced both to preclinical large animal models and beyond to the initial stages of clinical trials [85]. Bearing in mind the goal of clinical translation, it is encouraging to see materials that are designed to be compatible with percutaneous transendocardial delivery via the catheters already in use. Alginate and the myocardial ECM hydrogel both have gelation properties conducive to these delivery methods [45,84,102], potentially circumventing the need for the surgical procedures required for the placement of patches or the direct injection of other materials.

Future of Cardiac Tissue Engineering

The early years of the 21st century have seen tremendous growth in the field of cardiac tissue engineering. Most recently, injectable biomaterials have emerged as a promising alternative therapy that could potentially be delivered via minimally invasive methods, eliminating the need for risky surgical procedures. Improvement in our mechanistic understanding of both the pathophysiology of heart disease and the effect of these biomaterial-based tissue-engineered approaches will drive the advancement of the field. As we begin to understand the mechanisms behind the success of these therapies, we will be able to establish true design criteria and start optimizing the strategies to meet relevant clinical needs.

References

[1] Viola J, Lal B, Grad O. The Emergence of Tissue Engineering as a Research Field. Nat Sci Found 2003.

[2] Langer R, Vacanti JP. Tissue engineering. Science 1993 May 14;260(5110):920–6.

[3] Cohen S, Bano MC, Cima LG, Allcock HR, Vacanti JP, Vacanti CA, et al. Design of synthetic polymeric structures for cell transplantation and tissue engineering. Clin Mater 1993;13(1-4):3–10.

[4] Atala A, Lanza RP. Methods of tissue engineering. San Diego, CA: Academic Press; 2001.

[5] Lanza RP, Langer RS, Vacanti J. Principles of tissue engineering. 3rd ed. Amsterdam; Boston: Elsevier/Academic Press; 2007.

[6] Leor J, Amsalem Y, Cohen S. Cells, scaffolds, and molecules for myocardial tissue engineering. Pharmacol Ther 2005 Feb;105(2):151–63.

[7] Lutolf MP, Hubbell JA. Synthetic biomaterials as instructive extracellular microenvironments for morphogenesis in tissue engineering. Nat Biotechnol 2005 Jan 1;23(1):47–55.

[8] Glowacki J, Mizuno S. Collagen scaffolds for tissue engineering. Biopolymers 2008 May;89(5):338–44.

[9] van Hout WM, Mink van der Molen AB, Breugem CC, Koole R, Van Cann EM. Reconstruction of the alveolar cleft: can growth factor-aided tissue engineering replace autologous bone grafting? A literature review and systematic review of results obtained with bone morphogenetic protein-2. Clin Oral Investig 2011 Jun;15(3):297–303.

[10] Robinson KA, Li J, Mathison M, Redkar A, Cui J, Chronos NA, et al. Extracellular matrix scaffold for cardiac repair. Circulation 2005 Aug 30;112(Suppl. 9):I135–43.

[11] Ott HC, Matthiesen TS, Goh SK, Black LD, Kren SM, Netoff TI, et al. Perfusion-decellularized matrix: using nature's platform to engineer a bioartificial heart. Nat Med 2008 Feb;14(2):213–21.

[12] Crapo PM, Gilbert TW, Badylak SF. An overview of tissue and whole organ decellularization processes. Biomaterials 2011 Apr;32(12):3233–43.

[13] Petersen TH, Calle EA, Zhao L, Lee EJ, Gui L, Raredon MB, et al. Tissue-engineered lungs for in vivo implantation. Science 2010 Jul 30;329(5991):538–41.

[14] Seif-Naraghi SB, Salvatore MA, Schup-Magoffin PJ, Hu DP, Christman KL. Design and characterization of an injectable pericardial matrix gel: a potentially autologous scaffold for cardiac tissue engineering. Tissue Eng Part A 2010 Jun;16(6):2017–27.

[15] Freytes DO, Martin J, Velankar SS, Lee AS, Badylak SF. Preparation and rheological characterization of a gel form of the porcine urinary bladder matrix. Biomaterials 2008 Apr;29(11): 1630–7.

[16] Young DA, Ibrahim DO, Hu D, Christman KL. Injectable hydrogel scaffold from decellularized human lipoaspirate. Acta Biomater 2011 Mar;7(3):1040–9.

[17] Dequach JA, Yuan SH, Goldstein LS, Christman KL. Decellularized porcine brain matrix for cell culture and tissue engineering scaffolds. Tissue Eng Part A 2011 Oct 17.

[18] Sellaro TL, Ranade A, Faulk DM, McCabe GP, Dorko K, Badylak SF, et al. Maintenance of human hepatocyte function in vitro by liver-derived extracellular matrix gels. Tissue Eng Part A 2010 Mar;16(3):1075–82.

[19] Okada M, Payne TR, Oshima H, Momoi N, Tobita K, Huard J. Differential efficacy of gels derived from small intestinal submucosa as an injectable biomaterial for myocardial infarct repair. Biomaterials 2010 Oct;31(30):7678–83.

[20] Singelyn JM, DeQuach JA, Seif-Naraghi SB, Littlefield RB, Schup-Magoffin PJ, Christman KL. Naturally derived myocardial matrix as an injectable scaffold for cardiac tissue engineering. Biomaterials 2009 Oct;30(29):5409–16.

[21] Okano T, Yamada N, Sakai H, Sakurai Y. A novel recovery system for cultured cells using plasma-treated polystyrene dishes grafted with poly(N-isopropylacrylamide). J Biomed Mater Res 1993 Oct;27(10):1243–51.
[22] Yamada N, Okano T, Sakai H, Karijusa F, Sawasaki Y, Sakurai Y. Thermo-responsive polymeric surfaces; control of attachment and detachment of culture cells. Macromol Chem Rapid Commun 1990;11:571–6.
[23] Sarig U, Machluf M. Engineering cell platforms for myocardial regeneration. Expert Opin Biol Ther 2011 Aug 1;11(8):1055–77.
[24] Christman KL, Lee RJ. Biomaterials for the treatment of myocardial infarction. J Am Coll Cardiol 2006 Sep 5;48(5):907–13.
[25] Stile RA, Burghardt WR, Healy KE. Synthesis and characterization of injectable poly(N-isopropylacrylamide)-based hydrogels that support tissue formation in vitro. Macromolecules 1999 Nov 2;32(22):7370–9.
[26] Lee BH, West B, McLemore R, Pauken C, Vernon BL. In-situ injectable physically and chemically gelling NIPAAm-based copolymer system for embolization. Biomacromolecules 2006 Jun;7(6):2059–64.
[27] Carmeliet P, Conway EM. Growing better blood vessels. Nat Biotechnol 2001 Nov;19(11): 1019–20.
[28] Asahara T, Bauters C, Zheng LP, Takeshita S, Bunting S, Ferrara N, et al. Synergistic effect of vascular endothelial growth factor and basic fibroblast growth factor on angiogenesis in vivo. Circulation 1995 Nov 1;92(Suppl. 9):II365–71.
[29] West DC, Hampson IN, Arnold F, Kumar S. Angiogenesis induced by degradation products of hyaluronic acid. Science 1985 Jun 14;228(4705):1324–6.
[30] Fournier N, Doillon CJ. Biological molecule-impregnated polyester: an in vivo angiogenesis study. Biomaterials 1996 Sep;17(17):1659–65.
[31] Soker S, Machado M, Atala A. Systems for therapeutic angiogenesis in tissue engineering. World J Urol 2000 Feb;18(1):10–8.
[32] Zisch AH, Lutolf MP, Hubbell JA. Biopolymeric delivery matrices for angiogenic growth factors. Cardiovasc Pathol 2003 Nov-Dec;12(6):295–310.
[33] Patel ZS, Mikos AG. Angiogenesis with biomaterial-based drug- and cell-delivery systems. J Biomater Sci Polym Ed 2004;15(6):701–26.
[34] Lysaght MJ, Jaklenec A, Deweerd E. Great expectations: private sector activity in tissue engineering, regenerative medicine, and stem cell therapeutics. Tissue Eng Part A 2008 Feb;14(2):305–15.
[35] Vacanti J. Tissue engineering and regenerative medicine: from first principles to state of the art. J Pediatr Surg 2010 Feb;45(2):291–4.
[36] Lloyd-Jones D, Adams RJ, Brown TM, Carnethon M, Dai S, De Simone G, et al. Heart disease and stroke statistics-2010 update: a report from the American Heart Association. Circulation 2010 Feb 23;121(7):E46–215.
[37] Strickman NE. The pathogenesis and prognosis of end-stage heart disease. Tex Heart Inst J 1987 Dec;14(4):346–50.
[38] Rane AaK LC. Biomaterials for the treatment of myocardial infarction: a five-year update. J Am Coll Cardiol 2011.
[39] Ratner BD. A paradigm shift: biomaterials that heal. Polym Int 2007;56(10):1183–5.
[40] Beltrami AP, Barlucchi L, Torella D, Baker M, Limana F, Chimenti S, et al. Adult cardiac stem cells are multipotent and support myocardial regeneration. Cell 2003 Sep 19;114(6):763–76.
[41] Kubo H, Jaleel N, Kumarapeli A, Berretta RM, Bratinov G, Shan XY, et al. Increased cardiac myocyte progenitors in failing human hearts. Circulation 2008 Aug 5;118(6):649–57.

[42] Ou L, Li W, Zhang Y, Wang W, Liu J, Sorg H, et al. Intracardiac injection of Matrigel induces stem cell recruitment and improves cardiac functions in a rat myocardial infarction model. J Cell Mol Med 2011 Jun 1;15(6):1310–8.

[43] Zhao ZQ, Puskas JD, Xu D, Wang NP, Mosunjac M, Guyton RA, et al. Improvement in cardiac function with small intestine extracellular matrix is associated with recruitment of C-kit cells, myofibroblasts, and macrophages after myocardial infarction. J Am Coll Cardiol 2010 Mar 23;55(12):1250–61.

[44] Seif-Naraghi SB, Salvatore MA, Schup-Magoffin PJ, Hu DP, Christman KL. Design and characterization of an injectable pericardial matrix gel: a potentially autologous scaffold for cardiac tissue engineering. Tissue Eng Pt A 2010:2017–27.

[45] Singelyn JM, Sundaramurthy P, Johnson T, Schup-Magoffin PJ, Hu DP, Faulk DM, et al. Catheter-deliverable hydrogel derived from decellularized ventricular extracellular matrix increases endogenous cardiomyocytes and preserves cardiac function post-myocardial infarction. J Am Coll Cardiol 2012 Feb 23;59(8):751–63.

[46] Lei Y, Haider H, Shujia J, Sim ES. Therapeutic angiogenesis. Devising new strategies based on past experiences. Basic Res Cardiol 2004 Mar;99(2):121–32.

[47] Henry TD, Annex BH, McKendall GR, Azrin MA, Lopez JJ, Giordano FJ, et al. The VIVA trial: Vascular endothelial growth factor in Ischemia for Vascular Angiogenesis. Circulation 2003 Mar 18;107(10):1359–65.

[48] Haider H, Ye L, Jiang S, Ge R, Law PK, Chua T, et al. Angiomyogenesis for cardiac repair using human myoblasts as carriers of human vascular endothelial growth factor. J Mol Med 2004 Aug;82(8):539–49.

[49] Richardson TP, Peters MC, Ennett AB, Mooney DJ. Polymeric system for dual growth factor delivery. Nat Biotechnol 2001 Nov;19(11):1029–34.

[50] Garbern JC, Minami E, Stayton PS, Murry CE. Delivery of basic fibroblast growth factor with a pH-responsive, injectable hydrogel to improve angiogenesis in infarcted myocardium. Biomaterials Mar;32(9):2407–2416.

[51] Shen YH, Shoichet MS, Radisic M. Vascular endothelial growth factor immobilized in collagen scaffold promotes penetration and proliferation of endothelial cells. Acta biomaterialia 2008 May;4(3):477–89.

[52] Goncalves LM. Angiogenic growth factors: potential new treatment for acute myocardial infarction? Cardiovasc Res 2000 Jan 14;45(2):294–302.

[53] Kim JH, Jung Y, Kim SH, Sun K, Choi J, Kim HC, et-al. The enhancement of mature vessel formation and cardiac function in infarcted hearts using dual growth factor delivery with self-assembling peptides. Biomaterials Sep;32(26):6080–6088.

[54] Sakiyama-Elbert SE, Hubbell JA. Controlled release of nerve growth factor from a heparin-containing fibrin-based cell ingrowth matrix. J Control Release 2000 Oct 3;69(1):149–58.

[55] Pieper JS, Hafmans T, van Wachem PB, van Luyn MJ, Brouwer LA, Veerkamp JH, et al. Loading of collagen-heparan sulfate matrices with bFGF promotes angiogenesis and tissue generation in rats. J Biomed Mater Res 2002 Nov;62(2):185–94.

[56] Freeman I, Kedem A, Cohen S. The effect of sulfation of alginate hydrogels on the specific binding and controlled release of heparin-binding proteins. Biomaterials 2008 Aug;29(22):3260–8.

[57] Sakiyama SE, Schense JC, Hubbell JA. Incorporation of heparin-binding peptides into fibrin gels enhances neurite extension: an example of designer matrices in tissue engineering. Faseb J 1999 Dec;13(15):2214–24.

[58] Sakiyama-Elbert SE, Hubbell JA. Development of fibrin derivatives for controlled release of heparin-binding growth factors. J Control Release 2000 Apr 3;65(3):389–402.

[59] Miyagi Y, Chiu LL, Cimini M, Weisel RD, Radisic M, Li RK. Biodegradable collagen patch with covalently immobilized VEGF for myocardial repair. Biomaterials Feb;32(5):1280-1290.

[60] Iozzo RV. Proteoglycans: structure, biology, and molecular interactions. New York: Marcel Dekker; 2000.

[61] Li RK, Jia ZQ, Weisel RD, Mickle DAG, Choi A, Yau TM. Survival and function of bioengineered cardiac grafts. Circulation 1999 Nov 9;100(19):63–9.

[62] Leor J, Aboulafia-Etzion S, Dar A, Shapiro L, Barbash IM, Battler A, et al. Bioengineered cardiac grafts - A new approach to repair the infarcted myocardium? Circulation 2000 Nov 7;102(19):56–61.

[63] Pozzobon M, Bollini S, Iop L, De Gaspari P, Chiavegato A, Rossi CA, et al. Human bone marrow-derived CD133(+) cells delivered to a collagen patch on cryoinjured rat heart promote angiogenesis and arteriogenesis. Cell Transplant 2010;19(10):1247–60.

[64] Simpson D, Liu H, Fan THM, Nerem R, Dudley SC. A tissue engineering approach to progenitor cell delivery results in significant cell engraftment and improved myocardial remodeling. Stem Cells 2007 Sep;25(9):2350–7.

[65] Cortes-Morichetti M, Frati G, Schussler O, Van Huyen JPD, Lauret E, Genovese JA, et al. Association between a cell-seeded collagen matrix and cellular cardiomyoplasty for myocardial support and regeneration. Tissue Eng 2007 Nov;13(11):2681–7.

[66] Chachques JC, Trainini JC, Lago N, Herreros J, Schussler O, Carpentier AF. Myocardial Assistance by Grafting a New Upgraded bioartificial Myocardium (MAGNUM Trial): clinical results at 2 years. Eur Heart J 2010 Sep;31. 941–941.

[67] Wei H, Chen C, Lee W, Chiu I, Hwang S, Lin W, et al. Bioengineered cardiac patch constructed from multilayered mesenchymal stem cells for myocardial repair. Biomaterials 2008 Sep 1;29(26):3547–56.

[68] Tan MY, Zhi W, Wei RQ, Huang YC, Zhou KP, Tan B, et al. Repair of infarcted myocardium using mesenchymal stem cell seeded small intestinal submucosa in rabbits. Biomaterials 2009:3234–40.

[69] Xiong Q, Hill KL, Li Q, Suntharalingam P, Mansoor A, Wang X, et al. A fibrin patch-based enhanced delivery of human embryonic stem cell-derived vascular cell transplantation in a porcine model of postinfarction left ventricular remodeling. Stem Cells 2011 Feb 1;29(2):367–75.

[70] Godier-Furnémont AF, Martens TP, Koeckert MS, Wan L, Parks J, Arai K, et al. Composite scaffold provides a cell delivery platform for cardiovascular repair. Proc Natl Acad Sci USA 2011 May 10;108(19):7974–9.

[71] Jin J, Jeong SI, Shin YM, Lim KS, Shin H, Lee YM, et al. Transplantation of mesenchymal stem cells within a poly(lactide-co-epsilon-caprolactone) scaffold improves cardiac function in a rat myocardial infarction model. Eur J Heart Fail 2009 Feb 1;11(2):147–53.

[72] Fujimoto KL, Tobita K, Merryman WD, Guan J, Momoi N, Stolz DB, et al. An elastic, biodegradable cardiac patch induces contractile smooth muscle and improves cardiac remodeling and function in subacute myocardial infarction. J Am Coll Cardiol 2007:2292–300.

[73] Piao H, Kwon J, Piao S, Sohn J, Lee Y, Bae J, et al. Effects of cardiac patches engineered with bone marrow-derived mononuclear cells and PGCL scaffolds in a rat myocardial infarction model. Biomaterials 2007:641–9.

[74] Miyagi Y, Zeng F, Huang X, Foltz WD, Wu J, Mihic A, et al. Surgical ventricular restoration with a cell- and cytokine-seeded biodegradable scaffold. Biomaterials 2010 Oct 1;31(30):7684–94.

[75] Christman KL, Vardanian AJ, Fang Q, Sievers RE, Fok HH, Lee RJ. Injectable fibrin scaffold improves cell transplant survival, reduces infarct expansion, and induces neovasculature formation in ischemic myocardium. J Am Coll Cardiol 2004 Aug 4;44(3):654–60.

[76] Christman KL, Fok HH, Sievers RE, Fang Q, Lee RJ. Fibrin glue alone and skeletal myoblasts in a fibrin scaffold preserve cardiac function after myocardial infarction. Tissue Eng 2004 Mar-Apr;10(3-4):403–9.

[77] Guo HD, Wang HJ, Tan YZ, Wu JH. Transplantation of marrow-derived cardiac stem cells carried in fibrin improves cardiac function after myocardial infarction. Tissue Eng Pt A 2011 Jan 1;17(1-2):45–58.

[78] Danoviz ME, Nakamuta JS, Marques FLN, dos Santos L, Alvarenga EC, dos Santos AA, et al. Rat adipose tissue-derived stem cells transplantation attenuates cardiac dysfunction post infarction and biopolymers enhance cell retention. PLoS ONE 2010:e12077.

[79] Zhang X, Wang H, Ma X, Adila A, Wang B, Liu F, et al. Preservation of the cardiac function in infarcted rat hearts by the transplantation of adipose-derived stem cells with injectable fibrin scaffolds. Exp Biol Med (Maywood) 2010 Dec 1;235(12):1505–15.

[80] Landa N, Miller L, Feinberg MS, Holbova R, Shachar M, Freeman I, et al. Effect of injectable alginate implant on cardiac remodeling and function after recent and old infarcts in rat. Circulation 2008 Mar 18;117(11):1388–96.

[81] Yu J, Du KT, Fang Q, Gu Y, Mihardja SS, Sievers RE, et al. The use of human mesenchymal stem cells encapsulated in RGD modified alginate microspheres in the repair of myocardial infarction in the rat. Biomaterials 2010:7012–20.

[82] Hao X, Silva EA, MÔnsson-Broberg A, Grinnemo KH, Siddiqui AJ, Dellgren G, et al. Angiogenic effects of sequential release of VEGF-A165 and PDGF-BB with alginate hydrogels after myocardial infarction. Cardiovasc Res 2007 Jul 1;75(1):178–85.

[83] Ruvinov E, Leor J, Cohen S. The promotion of myocardial repair by the sequential delivery of IGF-1 and HGF from an injectable alginate biomaterial in a model of acute myocardial infarction. Biomaterials 2011:565–78.

[84] Leor J, Tuvia S, Guetta V, Manczur F, Castel D, Willenz U, et al. Intracoronary injection of in situ forming alginate hydrogel reverses left ventricular remodeling after myocardial infarction in Swine. J Am Coll Cardiol 2009 Sep 8;54(11):1014–23.

[85] BioLineRX, Ltd. Safety and feasibility of the injectable BL-1040 implant. ClinicalTrials.gov NCT00557531 2007.

[86] Okada M, Payne TR, Oshima H, Momoi N, Tobita K, Huard J. Differential efficacy of gels derived from small intestinal submucosa as an injectable biomaterial for myocardial infarct repair. Biomaterials 2010:7678–83.

[87] Lu WN, Lü SH, Wang HB, Li DX, Duan CM, Liu ZQ, et al. Functional improvement of infarcted heart by co-injection of embryonic stem cells with temperature-responsive chitosan hydrogel. Tissue Eng Pt A 2009 Jun 1;15(6):1437–47.

[88] Lue S, Wang H, Lu W, Liu S, Lin Q, Li D, et al. Both the transplantation of somatic cell nuclear transfer- and fertilization-derived mouse embryonic stem cells with temperature-responsive chitosan hydrogel improve myocardial performance in infarcted rat hearts. Tissue Eng Pt A 2010:1303–15.

[89] Wang H, Zhang X, Li Y, Ma Y, Zhang Y, Liu Z, et al. Improved myocardial performance in infarcted rat heart by co-injection of basic fibroblast growth factor with temperature-responsive Chitosan hydrogel. J Heart Lung Transpl 2010:881–7.

[90] Shao ZQ, Takaji K, Katayama Y, Kunitomo R, Sakaguchi H, Lai ZF, et al. Effects of intramyocardial administration of slow-release basic fibroblast growth factor on angiogenesis and ventricular remodeling in a rat infarct model. Circ J 2006:471–7.

[91] Liu Y, Sun L, Huan Y, Zhao H, Deng J. Effects of basic fibroblast growth factor microspheres on angiogenesis in ischemic myocardium and cardiac function: analysis with dobutamine cardiovascular magnetic resonance tagging. Eur J Cardiothorac Surg 2006:103–7.

[92] Davis ME, Hsieh PCH, Takahashi T, Song Q, Zhang SG, Kamm RD, et al. Local myocardial insulin-like growth factor 1 (IGF-1) delivery with biotinylated peptide nanofibers improves cell therapy for myocardial infarction. Proc Natl Acad Sci USA 2006:8155–60.

[93] Dubois G, Segers VF, Bellamy V, Sabbah L, Peyrard S, Bruneval P, et al. Self-assembling peptide nanofibers and skeletal myoblast transplantation in infarcted myocardium. J Biomed Mater Res Part B Appl Biomater 2008 Oct 1;87(1):222–8.

[94] Lin Y, Yeh M, Yang Y, Tsai D, Chu T, Shih Y, et al. Intramyocardial peptide nanofiber injection improves postinfarction ventricular remodeling and efficacy of bone marrow cell therapy in pigs. Circulation 2010 Sep 14;122(Suppl. 11):S132–41.

[95] Kim JH, Jung Y, Kim SH, Sun K, Choi J, Kim HC, et al. The enhancement of mature vessel formation and cardiac function in infarcted hearts using dual growth factor delivery with self-assembling peptides. Biomaterials 2011 Sep 1;32(26):6080–8.

[96] Kraehenbuehl TP, Ferreira LS, Hayward AM, Nahrendorf M, van der Vlies AJ, Vasile E, et al. Human embryonic stem cell-derived microvascular grafts for cardiac tissue preservation after myocardial infarction. Biomaterials 2011:1102–9.

[97] Wall ST, Yeh CC, Tu RYK, Mann MJ, Healy KE. Biomimetic matrices for myocardial stabilization and stem cell transplantation. J Biomed Mater Res Part A 2010 Dec 15;95A(4):1055–66.

[98] Yoon SJ, Fang YH, Lim CH, Kim BS, Son HS, Park Y, et al. Regeneration of ischemic heart using hyaluronic acid-based injectable hydrogel. J Biomed Mater Res B 2009:163–71.

[99] Wall ST, Walker JC, Healy KE, Ratcliffe MB, Guccione JM. Theoretical impact of the injection of material into the myocardium: a finite element model simulation. Circulation 2006 Dec 12;114(24):2627–35.

[100] Rane AA, Chuang JS, Shah A, Hu DP, Dalton ND, Gu YS, et al. Increased infarct wall thickness by a bio-inert material is insufficient to prevent negative left ventricular remodeling after myocardial infarction. PloS one 2011 Jun 23;6(6).

[101] Dobner S, Bezuidenhout D, Govender P, Zilla P, Davies N. A synthetic non-degradable polyethylene glycol hydrogel retards adverse post-infarct left ventricular remodeling. J Cardiac Fail 2009 Sep;15(7):629–36.

[102] Singelyn JM, Christman KL. Injectable materials for the treatment of myocardial infarction and heart failure: the promise of decellularized matrices. J Cardiovasc Transl Res 2010 Oct; 3(5):478–86.

Chapter 4

Resident Neural Stem Cells

Camila Zaverucha-do-Valle[*,†,‡,§,**,††], Fernanda Gubert[*,†,‡,§,**,††], Louise Mesentier-Louro[†,‡,§,**,††], Eliana Scemes[‡‡], Fernando Pitossi[§§], Marcelo F. Santiago[†,‡,§,**,††], Rosalia Mendez-Otero[†,‡,§,**,††]

[†] *Instituto de Biofísica Carlos Chagas Filho, Rio de Janeiro, RJ, Brazil*

[‡] *Laboratório de Neurobiologia Celular e Molecular, Rio de Janeiro, RJ, Brazil*

[§] *Programa de Terapia Celular e Bioengenharia, Rio de Janeiro, RJ, Brazil*

[**] *Instituto Nacional de Ciência e Tecnologia de Biologia Estrutural e Bioimagem -INBEB, Rio de Janeiro, RJ, Brazil*

[††] *Universidade Federal do Rio de Janeiro, Rio de Janeiro, RJ, Brazil*

[‡‡] *The Dominick P. Purpura Department of Neuroscience, Albert Einstein College of Medicine, New York, New York, USA*

[§§] *Leloir Institute Foundation, Institute for Biochemical Investigations, CONICET, 1405 Buenos Aires, Argentina*

Neurogenesis in the Adult Brain

It was established that neurons were not able to proliferate and after their death they could not be replaced. This idea that new neurons could not be generated in the adult mammalian brain was accepted for most of the 20th century.

In 1962, however, Joseph Altman published a paper in *Science* that started to change this dogma. He performed a lesion in the lateral geniculate nucleus of adult rats and injected [^{3}H]-thymidine in the area of injury. The [^{3}H]-thymidine is incorporated during the S phase of the cell cycle, therefore labeling proliferating cells. He observed several labeled glial cells, and also described labeled neurons not necessarily related to the area of injury [1]. Nevertheless, although this work was published in a prestigious journal, it was not well accepted by the scientific community at that time. It was assumed that the available techniques at the time were

[*] These authors contributed equally to this work.

Resident Stem Cells and Regenerative Therapy. http://dx.doi.org/10.1016/B978-0-12-416012-5.00004-9

not sensitive enough to clearly distinguish dividing glia from dividing neurons (see [2] for revision).

In the 1970s, new data came when Kaplan and Hinds injected [^{3}H]-thymidine into adult rats and described labeled cells in the olfactory bulb and in the dentate gyrus of the hippocampus. Some of these cells were analyzed by electron microscopy and showed several morphologic characteristics of neurons [3]. After this report, Kaplan published a series of papers showing neurogenesis in the adult brain, but at the same time, Pasko Rakic analyzed the brain of rhesus monkeys injected with [^{3}H]-thymidine and found no labeled cells, concluding that there was no generation of new neurons in those brains [4] and therefore rejecting Kaplan's findings.

It was not until the 1980s that the scientific community started to accept the idea of new neurons being formed in the adult brain. Fernando Nottebohm and coworkers demonstrated in a series of papers the presence of new neurons around the higher vocal center in adult canaries. These cells originated in the ventricular zone and migrated to the higher vocal center [5]. Later, it was demonstrated that peaks of neurogenesis in male canaries were correlated with the seasonal song modification observed in this species [6].

At the beginning of the 1990s, neurogenesis was finally accepted in the adult rodent brain through the use of specific neuronal markers, such as the proteins MAP-2 (Microtubule-associated protein 2) and NSE (neuron specific enolase) [7,8]. The [^{3}H]-thymidine started to be replaced by BrdU (5-bromo-3'-deoxyuridine), which is also a thymidine analogue incorporated during the S phase of the cell cycle. BrdU labeling can be visualized by immunohistochemical techniques and can be combined with neuronal markers, which allows a more reliable identification of newly formed neurons.

At the end of the 1990s, Elisabeth Gould and coworkers injected adult monkeys with BrdU and found labeled cells in the subventricular zone [9]. Around this same time, Eriksson and coworkers demonstrated the generation of new neurons in the human brain [10], by analyzing the brains of five patients who received BrdU for cancer diagnosis. In all patients analyzed, labeled cells in the granular layer of the hippocampus were found. Also, the subventricular zone of all the examined patients contained BrdU$^+$ cells.

Nowadays, it is well accepted that new neurons are generated in at least two regions of the adult mammalian brain: the subventricular zone (SVZ) and the subgranular zone (SGZ) in the hippocampus. These neurogenic regions and the cells present in these areas are described in more detail in the following sections.

Neurogenesis during Development

During embryonic development, the neurogenic region is the ventricular zone, a pseudo-stratified epithelium next to the ventricles. In the beginning of the development, the ventricular zone is composed of a single layer of epithelial cells. These cells, called neuroepithelial cells, are the primary neural stem cells, which give rise to all cells of the brain [11].

For many years, it was believed that neurons and glia originated from separate precursors. However, since the early 2000s many groups have demonstrated that radial glia cells are able to generate both glia and neurons [12,13]. Nowadays the more accepted idea is that when cortical neurogenesis begins, there is a transition where the neuroepithelial cells start to acquire a glial morphology, differentiating into radial glia cells [14].

Radial glia cells have their soma in the ventricular zone and radial processes that extend toward the pial surface. These cells where first known as guides for neuroblast migration [15]. As neurogenesis proceeds, there is a progressive thickening of the cortex and the radial glia process elongates to accompany it [16].

The development of new techniques that allow the tracking and fate of each cell—especially techniques using virus vectors—led to the discovery that in addition to the function in neuronal guidance, radial glia cells also act as neural stem cells in the developing brain. These cells are able of self-renewal and they can generate neurons [12,17], oligodendrocytes [18], and ependymal cells [19]. Noctor and coworkers transduced the ventricular zone of embryonic animals (E17) using replication-incompetent enhanced GFP expressing retrovirus. These retroviruses infect only the proliferative cells and in the embryonic stage E17 96% of the proliferative cells are radial glia. The authors were able to follow the progeny of radial glia cells and observed that these cells give rise to the majority of cortical neurons [12,17].

At the end of development, radial glia cells retract their processes and differentiate into astrocytes [20]. Radial glia cells express some specific markers such as RC2, nestin, and vimentin [21] and share features that resemble astrocytes, such as the presence of glycogen granules and the expression of brain lipid-binding protein (BLBP), L-glutamate/L-aspartate transporter (GLAST), and, in some species, glial fibrillary acidic protein (GFAP).

A second germinative zone that is originated by asymmetrically dividing cells in the ventricular zone is the subventricular zone [22]. The subventricular zone contains rapidly proliferating cells—called intermediate progenitors or transit amplifying cells—and expands during the last third of rodent prenatal development. This process occurs in parallel with a progressive reduction of the ventricular zone (for review, see [23]). In adult animals, the ventricular zone disappears and the subventricular zone remains as a place where the neural stem cells are located.

In 2004, Merkle and co-workers labeled the radial glia retrogradely through their processes with adenovirus and, stained them with GFP or alkaline phosphatase. Using this technique, the authors were able to follow radial glia progenies. They concluded that in addition to exhibiting neural stem cells features during development, radial glia gave rise to the neural stem cells in the adult subventricular zone [18].

Neural Stem Cells

Neural stem cells are multipotent cells that are able to self-renew and to differentiate into the three major cellular types of the central nervous system: neurons, astrocytes, and oligodendrocytes. The neural progenitors, on the other hand, have limited proliferative capacity and are not able of self-renewal indefinitely. In addition, they can only differentiate into a more restricted variety of cell types.

Neural stem cells were first described by Reynolds and Weiss in 1992 [24]. After cultivating cells from adult mouse striatum using a nonadherent, serum-free medium containing epidermal growth factor (EGF), they noticed the presence of spheroid bodies in the culture dish. These spheroids were later termed neurospheres, and they are formed by the division of a single neural stem cell. They showed that the cells inside the neurospheres were positive for the protein nestin, and they also demonstrated that the cells in the spheres were able to generate secondary spheres when plated individually into a new dish. The ability to form secondary neurospheres demonstrates the neural stem cells' capacity of self-renewal [24,25].

When the neurospheres were transferred to poly-l-ornithine-coated coverslips, the researchers noticed that some cells started migrating away from the sphere core. In the absence of mitogens, the neurospheres adhere to the culture dish and cells differentiate into neurons and glia, demonstrating the multipotentiality capacity [24].

After this first report, Morshead and coworkers demonstrated that the EGF (epidermal growth factor)-responsive cells were in fact located in the subventricular zone around the lateral ventricles and not in the striatum. They showed that the neurospheres were only formed when subventricular zone cells were plated and demonstrated that the rostral region of the subventricular zone was the region that contained more sphere-generating cells [26].

It was then established that neural stem cells could be isolated from the subventricular zone of adult mammals. These cells can be kept *in vitro* in the presence of FGF-2 (basic fibroblast growth factor) and EGF, and they can proliferate and generate neurospheres. The neurospheres are, however, heterogeneous and each contains both stem cells and more committed progenitors. Only a small percentage of the neurosphere cells (less than 1%) can give rise to secondary neurospheres. In Figure 4.1 we can see the neurosphere assay and an example of a neurosphere derived from a GFP mouse.

It is possible to generate neurospheres from single cells derived from both the subventricular zone and the subgranular zone, but these neurospheres have different characteristics. Unlike the subventricular zone, the subgranular zone neurospheres are smaller and are not able to form secondary neurospheres. In addition, each subgranular zone neurosphere can only be differentiated into one cell type. This indicates that neurospheres derived from the subgranular zone are in fact formed by neural progenitors instead of neural stem cells [27].

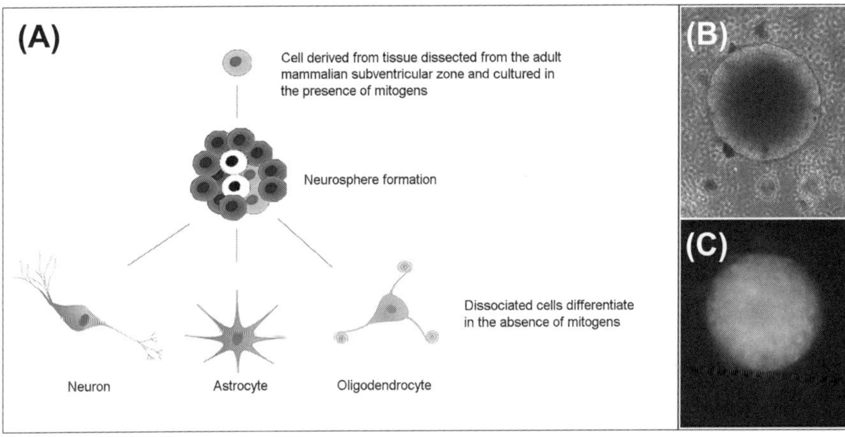

Figure 4.1 **(A)** In the neurosphere formation assay, cells are isolated from neurogenic areas, such as the subventricular zone. The tissue is dissociated into single-cell suspension and cultured in the presence of EGF and FGF-2, generating a cluster of cells called a neurosphere. A neurosphere can be dissociated and replated as single-cell suspension in the presence of mitogens to generate secondary neurospheres, demonstrating the self-renewal ability. Neurosphere cells can also be differentiated in the absence of mitogens, generating neurons, astrocytes, and oligodendrocytes, which demonstrate its multipotency. **(B-C)** Photomicrograph of a neurosphere generated from a beta-actin-GFP transgenic mouse. In (B) we can see the phase-contrast image and in (C) the epifluorescence photomicrograph. Scale bar: 100μm.

There are still many controversies in the literature concerning the presence of adult neural stem cells in other regions of the central nervous system. There was a report describing extremely low-level constitutive neurogenesis in the neocortex [9]; but this report was strongly contested primarily for methodological reasons [28]. It was argued that the BrdU used to identify proliferating cells could also label cells undergoing DNA repair, and that the use of conventional instead of confocal microscopy could make the distinction between proliferating glia and neurons difficult [29]. The group that reported neurogenesis in the neocortex finally modified their interpretation and stated that the proliferating neurons existed only transiently [30].

Many groups were also able to isolate cells with neural stem cells characteristics *in vitro* from non-neurogenic regions, such as the spinal cord, the amygdala, the cerebral cortex, and the substantia nigra (reviewed by [31]). Other groups, however, reported that they could only isolate neural stem cells from the regions that were proved to be neurogenic *in vivo* [32,33]. Overall, these isolated data remain controversial.

Several groups have also demonstrated the induction of low levels of neurogenesis in normally non-neurogenic regions in response to neuronal death, such as the neocortex, the non-neurogenic region of the hippocampus and the striatum (reviewed by [31]).

The Subventricular Zone

In adult mammals, the subventricular zone is one of the major neurogenic regions. New neurons formed in this region constantly migrate toward the olfactory bulb to replace interneurons [34,35], a process that also occurs in nonhuman primates [36] and infant humans [37]. In the olfactory bulb of adult rats, approximately 80,000 new granular neurons are formed each day from the subventricular zone progenitors [38].

The subventricular zone is composed mainly by three cell types: the A, B, and C cells. The type A cells are the migratory neuroblasts and correspond to approximately 33% of the subventricular zone cells. They express doublecortin (DCX) and PSA-NCAM, and they exhibit elongated cell body and one or two processes. They migrate through the rostral migratory stream (RMS) to the olfactory bulb where they differentiate into granular and glomerular neurons. B cells express GFAP, have characteristics of astrocytes, and correspond to around 23% of the subventricular zone cells [39]. B cells can be classified as B1 and B2 cells. B1 cells are in contact with the lateral ventricle lumen and have a small apical surface, a single basal body, and a single short primary cilium [39,40]. Finally, the C cells are the transit-amplifying cells and correspond to about 11% of the subventricular zone cells. They express the transcription factor Dlx2 [39].

Within the subventricular zone and the lateral ventricle lumen, there is a layer of ependymal cells, also called E cells. These cells have round nuclei and clear cytoplasm and express CD24, vimentin, and S100β. The ependymal cells can be classified as E1 or E2 cells. E1 cells have multiple long cilia, and E2 cells are often biciliated with long cilia [40]. In this region there are also other cellular types, such as mature neurons, microglia, and tanycytes [39].

It has been recently demonstrated that the presence of B1 cells with apical surface in contact with the ventricle is not as rare as previously thought and that B1 cells account for almost one third of the cells touching the ventricle [40]. Although E1 and E2 cells were found in all the ventricles analyzed (including the third ventricle and medial wall of the lateral ventricle), the apical surfaces of B1 cells were found only in neurogenic regions (the lateral wall and anterior medial wall of the lateral ventricles). The presence of apical surfaces of B1 cells alters the geometry of the E1 cells' apical surfaces in these regions, with one or more B1 cells surrounded by E1 cells in a pattern similar to a pinwheel. This pattern is observed in the neurogenic regions but not in non-neurogenic regions [40].

There are still some controversies in the field concerning the identity of the neural stem cells in the subventricular zone. However, nowadays the most accepted idea is that B cells are the neural stem cells, which are quiescent or have a slow cell cycle (around 14 days). These cells divide in an asymmetric way and give rise to C cells. These cells, in turn, are transit-amplifying cells and have a high proliferation rate, increasing the number of progenitor cells in the region and eventually differentiating into A cells that migrate tangentially through the RMS to the olfactory bulb [25]. In the RMS, A cells migrate along each other path, ensheathed by astrocytes and forming chains [41]. It is believed that only the activated B cells—the ones that are already programmed to generate type C cells—touch the ventricle with a cilium [42]. The presence of C cells to amplify the number of progenitors in the region is important because in this way the neural stem cells do not have to go through many cell cycles, thereby decreasing the chance of mutations [43].

The second line of thought proposes that the neural stem cells are in fact the ependymal cells [44]. Johansson and coworkers observed proliferating cells around the lateral ventricle. In another experiment, they injected the lipophilic tracer DiI in one of the lateral ventricles to label the cells around the ventricle wall and analyzed the contralateral hemisphere. When the region was dissociated and the cells were used in a neurosphere formation assay, 88% neurospheres were DiI^+. These neurospheres were able to form secondary neurospheres and differentiate into neurons, astrocytes, and oligodendrocytes [44].

Shortly after this report, other groups were unsuccessful to obtain similar results. Chiasson and coworkers, and Doetsch and coworkers did not find any $BrdU^+$ ependymal cells [25,45] and they speculated that the proliferating cells located in the lateral ventricles walls were in fact activated B cells [25,40,42]. This discussion recently re-emerged when Coskun and coworkers demonstrated that a subpopulation of ependymal $CD133^+$ cells was able to give rise to neurospheres and to differentiate into neurons, astrocytes, and oligodendrocytes [46]. Alvarez Buylla's group, however, contradicted these results and reported that B1 cells were also $CD133^+$, suggesting that the cells isolated from the ventricle wall were in fact B1 cells and not ependymal cells [40].

Recently, Frisén's group that published the first report identifying the ependymal cells as neural stem cells demonstrated that E cells were quiescent and only have proliferative capacity, giving rise to A cells and astrocytes in response to stroke [47]. They showed that ependymal cells did not contribute to adult neurogenesis under normal conditions and, therefore, they did not fulfill the criteria to be classified as neural stem cells [47].

Besides forming neuroblasts, it has been demonstrated that neural stem cells from the subventricular are also capable of forming oligodendrocytes *in vivo*. Labeling specifically the subventricular zone B cells, Menn and coworkers showed that these

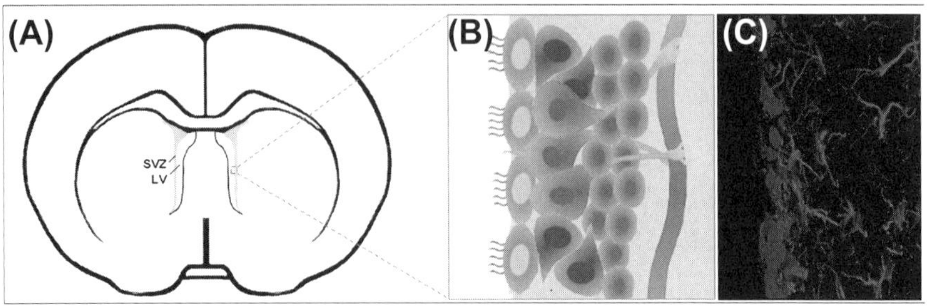

Figure 4.2 Subventricular zone of adult rodents. **(A)** Schematic drawing of a coronal section of the adult mouse brain showing the subventricular zone (SVZ) and the lateral ventricle lumen (LV). The boxed area is enlarged in (B). **(B)** Cytoarchitecture of the SVZ. In brown are the multiciliated ependymal cells placed in the ventricle wall. In blue are the B cells, some of which have a single cilium that contacts the ventricle lumen. In green are the C cells, also called transit-amplifying cells, and in red are the A cells, which are the neuroblasts that migrate to the olfactory bulb. **(C)** Photomontage of confocal-projection images taken from an adult rat brain coronal section showing GFAP$^+$ cells (in blue), PSA-NCAM$^+$ cells (in red) and proliferating KI-67$^+$ cells (in green). **See Plate 9**.

cells were capable of giving rise to Olig2$^+$ oligodendrocyte precursors. These precursors were shown to migrate to the corpus callosum, striatum, and fimbria-fornix and to differentiate into mature oligodendrocytes. They also demonstrated that a single B cell could give rise to neurons and oligodendrocytes *in vitro* [48].

In Figure 4.2, we can see a graphic representation of the subventricular zone and a confocal microscopy picture of this same area showing proliferating cells (in green), PSA-NCAM-positive cells (in red), and GFAP-positive cells (in blue).

Human Subventricular Zone

The human subventricular zone is composed of four layers. Layer I contains the ependymal cells and layer II, called the hypocellular layer, contains extensions of astrocytes and virtually no cell bodies. Layer II is present in humans and cows, and its exact function is unknown. Layer III contains a range of three distinct populations of astrocyte cell bodies. Furthermore, in the third layer there are some oligodendrocytes and some displaced ependymal cells. Finally, layer IV is characterized by the presence of myelin. This layer is located in the transition zone between the ribbon of astrocytes and the brain parenchyma.

There is controversy in the literature regarding the existence of RMS in the human adult brain. Whereas one group has reported intense proliferation and neuroblast migration in the human subventricular zone and RMS, another group found evidence that this proliferation occurs only in infants, and in adults there are rare cells migrating toward the olfactory bulb [37,49,50]. In infants, it was found that not all neuroblasts migrate to the olfactory bulb, but instead there is another migratory chain toward the prefrontal cortex [49].

Subgranular Zone

The subgranular zone is located in the hippocampus, at the interface between the hilus and the granular layer of the hippocampus. It is estimated that approximately 100 to 150 neurons are generated per day in the subgranular zone of adult rodents. These neurons differentiate into granule neurons from the granular layer and form axonal projections that reach the CA3 layer, becoming indistinguishable from the local cells [51]. The newly born neurons that integrate into the local circuitry survive for at least 8 months in rodents and 2 years in humans [10,52].

The neural progenitor cells existing in the subgranular zone, similar to the subventricular zone stem cells, have astrocyte characteristics. These cells, called B cells or type I progenitors, express nestin and GFAP and have prominent radial processes [53,54].

The B cells give origin to the D cells, the intermediated progenitors (also called type II progenitors) that proliferate at a high rate and form aggregates between the B cells' radial processes [54]. They commonly form clusters of two to four cells and are positive to PSA-NCAM and doublecortin. Ultrastructure analysis showed three subpopulations of D cells: D1, D2, and D3 [55].

D cells generate the migratory neuroblasts—also called type III cells—that migrate to the inner granular cell layer and differentiate into granule neurons. Newly born neurons are located at the border of the subgranular zone–granular cell layer-and have a round soma, a well-developed dendritic arbor, and one axon that reaches the CA3 region. When these neurons mature, they become positive for NeuN and calbindin. The entire process of differentiation, maturation, and integration of the new granular cells occurs during 6 to 8 weeks after birth in mice [56,57]. Figure 4.3 shows the subgranular zone and its cytoarchitecture.

In the subgranular zone, it has been demonstrated that after treatment with antimitotic drugs, there is a transient interruption of neurogenesis, but eventually GFAP$^+$ cells or Sonic hedgehog-responsive cells give rise to granular neurons [58]. This demonstrates that type II cells are lost after the treatment, but slowly dividing radial cells (type I cells) remain alive and repopulate the missing cells (revised by [59]). Recently, it has been shown that only 10% of GFAP$^+$ neural progenitor cells divide asymmetrically, generating one GFAP$^+$ daughter cell and one neuron-committed progenitor. The remaining GFAP$^+$ neural progenitor cells divide symmetrically, and half of them generate two GFAP$^+$ cells whereas the other half gives rise to two neuron-committed cells [60].

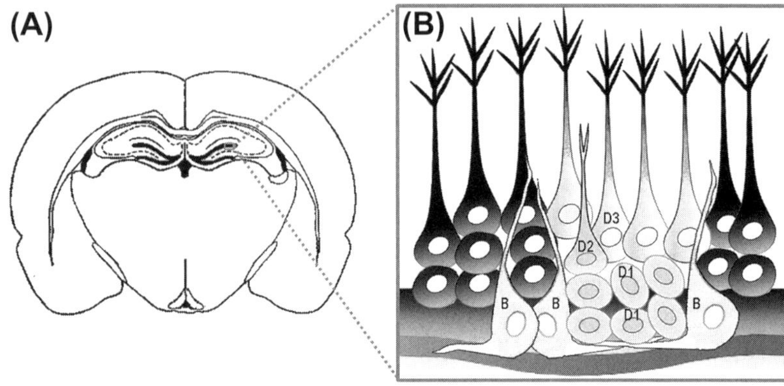

Figure 4.3 Subgranular zone of adult rodents. **(A)** Schematic drawing of a Coronal section of the adult mouse brain showing the hippocampus. The subgranular zone of the dentate gyrus is enlarged in (B). **(B)** Cytoarchitecture of the SGZ. The SGZ astrocytes —also called B cells—have radial processes and give rise to progenitor cells (D1). D1 cells divide and differentiate into new granule cells (D2) that integrate into the dentate gyrus granule cell layer (D3).

In an attempt to study single neural progenitor cells and to follow their progeny, two groups had different results. Whereas one group showed that GFAP$^+$ radial cells were the neural progenitor cells that have the capacity to generate neurons and astrocytes [55], the second group proposed that the Sox2+ nonradial cells could also self-renew and give rise to both neurons and astrocytes [61]. The data of the second group suggest there is a second population of neural progenitor cells in the subgranular zone. Whereas the radial cells remained quiescent, the Sox2+ nonradial cells self-renew actively. These cells express nestin but do not express GFAP, and the exact function of each population is still unclear.

The hippocampus is related to memory and learning. Therefore, it is logical to assume that an increased plasticity due to neurogenesis should be important for hippocampal function. Although some groups have demonstrated increased neurogenesis after learning stimulus [62], Gould and coworkers have shown that learning was not involved in the formation of new neurons, but instead it was related to the survival of neurons generated before the stimulus. These findings indicate that external factors could contribute to the survival of subgranular zone newly born neurons [63].

Physical activity and exposure of rodents to an enriched environment increase the number of functional neurons, and this increase is associated with a better performance of these animals in hippocampus-dependent tasks [62]. It has been demonstrated that physical activity is related to an increase in neural progenitor cells and cognitive stimulus is related to the survival of newly born neurons [64]. In general, it has been proposed that adult hippocampal neurogenesis is down-regulated by aging, stress, glucocorticoid hormones, and drug abuse, and it is up-regulated by an enriched environment, physical exercise, hippocampal-dependent learning, estrogens, and antidepressant drugs, among other factors (reviewed by [65]).

Neural Stem Cell Markers

A lot of progress has been made in the characterization of the neurogenic regions in the adult mammalian brain; however, as of now no specific marker for the neural stem cell is known. The identification of neural stem cell markers will allow researchers to study their functional properties and molecular regulation, and isolate these cells from more restricted progenitors.

In 2001, Rietze and co-workers were the first to suggest markers that could be used to purify a neural stem cell population based on negative selection. In their paper, they used peanut agglutinin (PNA) and heat-stable antigen (HSA) as exclusion markers to enrich neural stem cell populations. PNA has been used to fractionate murine hematopoietic stem cells and HSA, also known as CD24, is expressed in ependymal cells. They observed a higher frequency of neurosphere formation in the cells that have low expression of both markers [66].

In the following years, a neural stem cell marker was described—the cell surface carbohydrate Lex, also known as SSEA-1 or CD15 [33,67]. It was observed that the Lex-positive population was able to generate many neurospheres *in vitro*, whereas the negative population rarely generates any. Although Lex could be considered a positive marker, this antigen is not specific for neural stem cells, as it is expressed also in beta-III-tubulin-positive neuroblasts. It is interesting to note that astrocytes derived from non-neurogenic brain regions do not express Lex.

The neurosphere assay is, until now, the procedure more often used to demonstrate the presence of neural stem cells in a brain region. However, in case of the lack of a specific neural stem cell marker, it is necessary to use multiple markers to study these cells *in vivo*. It is especially important to distinguish the astrocytes with neural stem cell proprieties from other brain astrocytes. Pastrana and co-workers [68] used a multiple-marker approach to study subventricular zone neural stem cells. They isolated different populations based on the expression of three markers: GFAP, EGFR, and CD24. The capacity to generate neurospheres was tested in each population, and they observed that the population that expressed both GFAP and EGFR could generate one neurosphere for each three cells, and the population that expressed only EGFR could generate only one neurosphere for each 10 cells, whereas the other population did not generate any neurosphere. They proposed a time course of antigen expression during neurogenesis at the subventricular zone, which is illustrated in Figure 4.4.

Summarizing the different papers that tested neural stem cells markers in the subventricular zone, it can be concluded that astrocytes that do not form

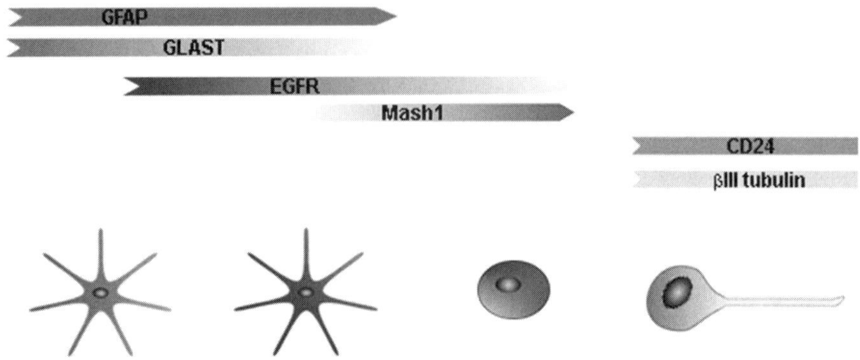

Figure 4.4 Schematic figure showing the subventricular zone cell lineage and the markers of each cell type. It has been demonstrated that GFAP$^+$ astrocytes give rise to transit amplifying cells that proliferate and differentiate into migrating neuroblasts. It is noticeable that some markers are not specific to one cell type and persist into the following cell lineage.

neurospheres are positive for GFAP and GLAST but negative for EGFR and Lex, whereas astrocytes with stem cell activity (B cells) express GFAP, GLAST, EGFR, and Lex. As the B cells differentiate into C cells, they maintain EGFR and Lex expression but lose GFAP and GLAST expression. The C cells are also positive for Dlx2 and Mash1. Although the C cells differentiate into neuroblasts (A cells), they lose Mash1 and EGFR expression, maintain the Dlx2 and Lex expression, and are positive for doublecortin, βIII-tubulin, and CD24^{low} [68].

Neurogenesis's Regulators

The neurogenic regions in the adult mammalian central nervous system consist of permissive niches for neural stem cell maintenance and neuronal differentiation. Once the neural stem cells from the subventricular zone and the subgranular zone are transplanted to non-neurogenic regions, such as the cerebral cortex, these cells differentiate into glial cells and have a limited survival time [69]. This data suggest that there is inhibitory signaling to neuronal differentiation present in the non-neurogenic regions.

One example of such inhibitory signaling is the bone morphogenic protein (BMP). It was demonstrated that BMP 2 and 4 are inhibitory to neurogenesis *in vivo* and *in vitro*. In the subventricular zone, the ependymal cells release noggin, a polypeptide that binds to BMP and prevents the activation of BMP receptors, creating a neurogenic environment [70]. In the subgranular zone, one of the first pro-neurogenic factors described is wnt-3, which is expressed by adult hippocampal astrocytes. Wnt-3 over-expression increases neurogenesis in the adult subgranular zone, and inhibition of wnt-3 decreases neurogenesis in this region [71].

As mentioned earlier, radial glia cells generate neurons. These neuronal precursor cells express cell-specific proteins and lack the active membrane properties seen in mature neurons [72]. Another feature of neuronal precursors is that they are coupled via gap junction intercellular channels [73]. One study proposed that the transfer of signaling molecules via gap junction channels plays a critical role in maintaining neuroblasts as nonmature neurons and maintaining their capacity to proliferate [74].

More recently, however, functions other than intercellular coupling have been proposed for the role played by connexins, the proteins forming gap junction channels during brain development. Connexins (Cx) have been proposed to be necessary for radial glial migration in the developing neocortex by acting as dynamic anchoring points for the migrating cells [75]. In addition to delayed migration, embryos of mice deficient in Cx43 show premature neuronal differentiation [76]. Similarly to *in vivo* conditions, neurospheres derived from Cx43-null embryos generated more neurons with morphologic and electrical properties similar to those of mature neurons than do those derived from wild-type mice [76]. Based on pharmacologic and mutation studies, it was proposed that neuronal differentiation was independent of gap junctional communication but dependent on the presence of the Cx43 carboxyl terminal domain that serves as a signaling molecule for neuronal differentiation [76].

An important component of the neurogenic niche is the vasculature. Neurogenic regions of both the subventricular zone and the subgranular zone are in close association with blood vessels [77]. In the subventricular zone, the majority of the B cells have a tangential process that contacts blood vessels [40]. In this region there is no astrocytic barrier in the blood vessels thus allowing for the exchange of signaling cues coming from the circulation that are probably important for the maintenance of the adult neurogenesis [78].

Many events can modulate neurogenesis at different moments, such as proliferation, differentiation, migration, integration, and survival in adult mammals. Age, for instance, seems to be an important factor, as there is a progressive decrease in neurogenesis during adulthood. The decrease in neurogenesis in the subgranular zone and the subventricular zone has a functional effect in memory and olfactory discrimination, respectively [79]. Adrenal steroids are probably related to this reduction during aging [80].

The signaling cues that regulates neurogenesis seems to respond to changes in local demand. Environmental enrichment and physical exercise, for example, stimulate neurogenesis only in the subgranular zone. In the subventricular zone, exposure to new smells or other olfactory stimuli promote neurogenesis and retention of the new neurons in the olfactory bulb (for revision see [79]).

Pathologic stimuli also modulate neurogenesis. Different models of cerebral ischemia induce an increase in the number of new neurons generated in both the subventricular zone and the subgranular zone. After focal ischemia, the new neurons born in the subventricular zone migrate to the lesion area at the striatum where they differentiate into striatal spiny neurons, although in a low percentage [81].

On the other hand, chronic stress, depression, and systemic inflammation decrease neurogenesis [82,83]. Neurodegenerative diseases such as Alzheimer's and Huntington's disease increase neurogenesis in the subgranular zone and subventricular zone. In Parkinson's disease there is a reduction in the number of the new neurons formed in the subventricular zone, in both rodents and humans [84].

The increase in neural stem cells proliferation rate in some types of injury may suggest an attempt of the nervous system to replace lost cells. In relation to this proposal, Sun and co-workers demonstrated that the neurologic improvement after traumatic brain injury is co-related with the increase in neurogenesis [85].

The molecules that regulate neurogenesis after physiologic and pathologic stimulus still have to be fully identified. To determine the influence of some of the growth factors or neurotrophins in neurogenesis, these factors are administrated locally or systemically and the proliferation and neuronal differentiation are analyzed. In this context, intraventricular or subcutaneous administration of FGF-2 was demonstrated to increase neurogenesis *in vivo* [86]. In accordance with these results, FGF-2 knockout mice showed a 50% reduction in neural stem cell proliferation in the subventricular zone [87]. Furthermore, FGF-2 seems to be one of the factors related to the observed increase in neurogenesis after injury, as this increase is not observed in FGF-2 knockout mice [88].

EGF also stimulates an increase in subventricular zone cell proliferation; however, EGF seems to affect mostly neural progenitor proliferation, as EGF receptors are extensively expressed in C cells [42]. Aguirre and coworkers demonstrated that signaling through EGFR expands the pool of neural progenitor cells and that down-regulation of Notch signaling results in a reduction in the number of neural stem cells in the subventricular zone [89].

Growth factors could have effect not only in proliferation but also in neural stem cell differentiation. PDGF, for example, seems to be important to regulate, through PDGFR alpha, the balance between neuron and oligodendrocyte production. The depletion of this receptor leads to a decrease in the number of new oligodendrocytes, whereas the number of new neurons remains similar [90].

Inflammation is also important, as it was proposed that the inflammatory cascade inhibits neuron maturation and survival [91]. However, depending on the type or amount of cytokines released by inflammatory cells, the effects in neurogenesis is different. For example, TNF-α and IL-6 seem to reduce neurogenesis, whereas IL-4, TGF-beta and low doses of IFN-γ promote neurogenesis [92,93,94].

Cell Therapy

Cell therapy with neural stem cells can be done using two different approaches. In one, endogenous neural stem cells are stimulated by using, for example, growth factors. The other approach is achieved by transplanting neural stem cells from a donor.

To make neural stem cell transplantation viable for a clinical approach, it is important to find a suitable donor source. In this regard, Palmer and coworkers isolated progenitor cells from human postmortem brain and surgical specimens. They used the brain of an 11-week-old newborn and the tissue surgically removed from a 27-year-old patient and were able to isolate neural progenitors. The cells from adult tissue could be passed in culture more than 30 times before reaching senescence. Viable progenitor cells could be isolated as long as 20 hours after death, and young tissues yielded more cells with greater proliferative capacity than older ones [95].

Because it is possible to isolate neural progenitors and even neural stem cells from postmortem brain or surgical resection, it is possible to think of neural stem cell transplantation as a potential clinical therapy. In this regard, several groups have tested cell therapy with neural stem cells in rodent models of neural injury or disease. Positive effects have been observed after neural stem cell transplantation in models of multiple sclerosis, Huntington's disease, and ischemia, among others [96-98]. Therefore, further investigations are needed in order to be able to translate the preclinical results to the clinical setting.

References

[1] Altman J. Autoradiographic study of degenerative and regenerative proliferation of neuroglia cells with tritiated thymidine. Exp Neurol 1962;5:302–18.

[2] Gross CG. Neurogenesis in the adult brain: death of a dogma. Nat Rev Neurosci 2000;1:67–73.

[3] Kaplan MS, Hinds JW. Neurogenesis in the adult rat: electron microscopic analysis of light radioautographs. Science 1977;197:1092–4.

[4] Rakic P. Limits of neurogenesis in primates. Science 1985;227:1054–6.

[5] Goldman SA, Nottebohm F. Neuronal production, migration, and differentiation in a vocal control nucleus of the adult female canary brain. Proc Natl Acad Sci USA 1983;80:2390–4.

[6] Kirn J, et al. Cell death and neuronal recruitment in the high vocal center of adult male canaries are temporally related to changes in song. Proc Natl Acad Sci USA 1994;91:7844–8.

[7] Richards LJ, Kilpatrick TJ, Bartlett PF. De novo generation of neuronal cells from the adult mouse brain. Proc Natl Acad Sci USA 1992;89:8591–5.

[8] Lois C, Alvarez-Buylla A. Proliferating subventricular zone cells in the adult mammalian forebrain can differentiate into neurons and glia. Proc Natl Acad Sci USA 1993;90:2074–7.

[9] Gould E, et al. Neurogenesis in the neocortex of adult primates. Science 1999;286:548–52.

[10] Eriksson PS, et al. Neurogenesis in the adult human hippocampus. Nat Med 1998;4:1313–7.

[11] Alvarez-Buylla A, Garcia-Verdugo JM, Tramontin AD. A unified hypothesis on the lineage of neural stem cells. Nat Rev Neurosci 2001;2:287–93.

[12] Noctor SC, et al. Neurons derived from radial glial cells establish radial units in neocortex. Nature 2001;409:714–20.

[13] Malatesta P, et al. Neuronal or glial progeny: regional differences in radial glia fate. Neuron 2003;37:751–64.

[14] Gotz M, Huttner WB. The cell biology of neurogenesis. Nat Rev Mol Cell Biol 2005;6:777–88.

[15] Rakic P. Neuronal migration and contact guidance in the primate telencephalon. Postgrad Med J 1978;54(Suppl. 1):25–40.

[16] Rakic P. Developmental and evolutionary adaptations of cortical radial glia. Cereb Cortex 2003;13:541–9.

[17] Noctor SC, et al. Dividing precursor cells of the embryonic cortical ventricular zone have morphological and molecular characteristics of radial glia. J Neurosci 2002;22:3161–73.

[18] Merkle FT, et al. Radial glia give rise to adult neural stem cells in the subventricular zone. Proc Natl Acad Sci USA 2004;101:17528–32.

[19] Spassky N, et al. Adult ependymal cells are postmitotic and are derived from radial glial cells during embryogenesis. J Neurosci 2005;25:10–8.

[20] Levitt P, Rakic P. Immunoperoxidase localization of glial fibrillary acidic protein in radial glial cells and astrocytes of the developing rhesus monkey brain. J Comp Neurol 1980;193:815–40.

[21] Pixley SK, de Vellis J. Transition between immature radial glia and mature astrocytes studied with a monoclonal antibody to vimentin. Brain Res 1984;317:201–9.

[22] Martinez-Cerdeno V, Noctor SC, Kriegstein AR. The role of intermediate progenitor cells in the evolutionary expansion of the cerebral cortex. Cereb Cortex 2006;16(Suppl. 1):i152–61.

[23] Bonfanti L, Peretto P. Radial glial origin of the adult neural stem cells in the subventricular zone. Prog Neurobiol 2007;83:24–36.

[24] Reynolds BA, Weiss S. Generation of neurons and astrocytes from isolated cells of the adult mammalian central nervous system. Science 1992;255:1707–10.

[25] Doetsch F, et al. Subventricular zone astrocytes are neural stem cells in the adult mammalian brain. Cell 1999;97:703–16.

[26] Morshead CM, et al. Neural stem cells in the adult mammalian forebrain: a relatively quiescent subpopulation of subependymal cells. Neuron 1994;13:1071–82.

[27] Seaberg RM, van der Kooy D. Adult rodent neurogenic regions: the ventricular subependyma contains neural stem cells, but the dentate gyrus contains restricted progenitors. J Neurosci 2002;22:1784–93.

[28] Koketsu D, et al. Nonrenewal of neurons in the cerebral neocortex of adult macaque monkeys. J Neurosci 2003;23:937–42.

[29] Rakic P. Adult neurogenesis in mammals: an identity crisis. J Neurosci 2002;22:614–8.

[30] Gould E, et al. Adult-generated hippocampal and neocortical neurons in macaques have a transient existence. Proc Natl Acad Sci USA 2001;98:10910–7.

[31] Gould E. How widespread is adult neurogenesis in mammals? Nat Rev Neurosci 2007;8:481–8.

[32] Laywell ED, et al. Identification of a multipotent astrocytic stem cell in the immature and adult mouse brain. Proc Natl Acad Sci USA 2000;97:13883–8.

[33] Imura T, et al. Phenotypic and functional heterogeneity of GFAP-expressing cells in vitro: differential expression of LeX/CD15 by GFAP-expressing multipotent neural stem cells and non-neurogenic astrocytes. Glia 2006;53:277–93.

[34] Luskin MB. Restricted proliferation and migration of postnatally generated neurons derived from the forebrain subventricular zone. Neuron 1993;11:173–89.

[35] Lois C, Alvarez-Buylla A. Long-distance neuronal migration in the adult mammalian brain. Science 1994;264:1145–8.

[36] Kornack DR, Rakic P. The generation, migration, and differentiation of olfactory neurons in the adult primate brain. Proc Natl Acad Sci USA 2001;98:4752–7.

[37] Sanai N, et al. Unique astrocyte ribbon in adult human brain contains neural stem cells but lacks chain migration. Nature 2004;427:740–4.
[38] Peterson DA. Stem cells in brain plasticity and repair. Curr Opin Pharmacol 2002;2:34–42.
[39] Doetsch F, Garcia-Verdugo JM, Alvarez-Buylla A. Cellular composition and three-dimensional organization of the subventricular germinal zone in the adult mammalian brain. J Neurosci 1997;17:5046–61.
[40] Mirzadeh Z, et al. Neural stem cells confer unique pinwheel architecture to the ventricular surface in neurogenic regions of the adult brain. Cell Stem Cell 2008;3:265–78.
[41] Lois C, Garcia-Verdugo JM, Alvarez-Buylla A. Chain migration of neuronal precursors. Science 1996;271:978–81.
[42] Doetsch F, et al. EGF converts transit-amplifying neurogenic precursors in the adult brain into multipotent stem cells. Neuron 2002;36:1021–34.
[43] Reya T, et al. Stem cells, cancer, and cancer stem cells. Nature 2001;414:105–11.
[44] Johansson CB, et al. Identification of a neural stem cell in the adult mammalian central nervous system. Cell 1999;96:25–34.
[45] Chiasson BJ, et al. Adult mammalian forebrain ependymal and subependymal cells demonstrate proliferative potential, but only subependymal cells have neural stem cell characteristics. J Neurosci 1999;19:4462–71.
[46] Coskun V, et al. CD133+ neural stem cells in the ependyma of mammalian postnatal forebrain. Proc Natl Acad Sci USA 2008;105:1026–31.
[47] Carlen M, et al. Forebrain ependymal cells are Notch-dependent and generate neuroblasts and astrocytes after stroke. Nat Neurosci 2009;12:259–67.
[48] Menn B, et al. Origin of oligodendrocytes in the subventricular zone of the adult brain. J Neurosci 2006;26:7907–18.
[49] Sanai N, et al. Corridors of migrating neurons in the human brain and their decline during infancy. Nature 2011;478:382–6.
[50] Curtis MA, et al. Human neuroblasts migrate to the olfactory bulb via a lateral ventricular extension. Science 2007;315:1243–9.
[51] Carlen M, et al. Functional integration of adult-born neurons. Curr Biol 2002;12:606–8.
[52] Dayer AG, et al. Short-term and long-term survival of new neurons in the rat dentate gyrus. J Comp Neurol 2003;460:563–72.
[53] Steiner B, et al. Type-2 cells as link between glial and neuronal lineage in adult hippocampal neurogenesis. Glia 2006;54:805–14.
[54] Seri B, et al. Astrocytes give rise to new neurons in the adult mammalian hippocampus. J Neurosci 2001;21:7153–60.
[55] Seri B, et al. Cell types, lineage, and architecture of the germinal zone in the adult dentate gyrus. J Comp Neurol 2004;478:359–78.
[56] van Praag H, et al. Functional neurogenesis in the adult hippocampus. Nature 2002;415:1030–4.
[57] Laplagne DA, et al. Functional convergence of neurons generated in the developing and adult hippocampus. PLoS Biol 2006;4:e409.
[58] Ahn S, Joyner AL. In vivo analysis of quiescent adult neural stem cells responding to Sonic hedgehog. Nature 2005;437:894–7.
[59] Suh H, Deng W, Gage FH. Signaling in adult neurogenesis. Annu Rev Cell Dev Biol 2009; 25:253–75.
[60] Namba, T., et al. Time-lapse imaging reveals symmetric neurogenic cell division of GFAP-expressing progenitors for expansion of postnatal dentate granule neurons. PLoS One 6:e25303.
[61] Suh H, et al. In vivo fate analysis reveals the multipotent and self-renewal capacities of Sox2+ neural stem cells in the adult hippocampus. Cell Stem Cell 2007;1:515–28.

[62] van Praag H, Kempermann G, Gage FH. Running increases cell proliferation and neurogenesis in the adult mouse dentate gyrus. Nat Neurosci 1999;2:266–70.

[63] Gould E, et al. Learning enhances adult neurogenesis in the hippocampal formation. Nat Neurosci 1999;2:260–5.

[64] Kempermann G. The neurogenic reserve hypothesis: what is adult hippocampal neurogenesis good for? Trends Neurosci 2008;31:163–9.

[65] Lee E, Son H. Adult hippocampal neurogenesis and related neurotrophic factors. BMB Rep 2009;42:239–44.

[66] Rietze RL, et al. Purification of a pluripotent neural stem cell from the adult mouse brain. Nature 2001;412:736–9.

[67] Capela A, Temple S. LeX/ssea-1 is expressed by adult mouse CNS stem cells, identifying them as nonependymal. Neuron 2002;35:865–75.

[68] Pastrana E, Cheng LC, Doetsch F. Simultaneous prospective purification of adult subventricular zone neural stem cells and their progeny. Proc Natl Acad Sci USA 2009;106:6387–92.

[69] Temple S. The development of neural stem cells. Nature 2001;414:112–7.

[70] Lim DA, et al. Noggin antagonizes BMP signaling to create a niche for adult neurogenesis. Neuron 2000;28:713–26.

[71] Lie DC, et al. Wnt signalling regulates adult hippocampal neurogenesis. Nature 2005;437: 1370–5.

[72] Hartfuss E, et al. Characterization of CNS precursor subtypes and radial glia. Dev Biol 2001;229:15–30.

[73] Lo Turco JJ, Kriegstein AR. Clusters of coupled neuroblasts in embryonic neocortex. Science 1991;252:563–6.

[74] Bruzzone R, Dermietzel R. Structure and function of gap junctions in the developing brain. Cell Tissue Res 2006;326:239–48.

[75] Elias LA, Wang DD, Kriegstein AR. Gap junction adhesion is necessary for radial migration in the neocortex. Nature 2007;448:901–7.

[76] Santiago MF, et al. The carboxyl-terminal domain of connexin43 is a negative modulator of neuronal differentiation. J Biol Chem 285:11836–45.

[77] Palmer TD, Willhoite AR, Gage FH. Vascular niche for adult hippocampal neurogenesis. J Comp Neurol 2000;425:479–94.

[78] Tavazoie M, et al. A specialized vascular niche for adult neural stem cells. Cell Stem Cell 2008;3:279–88.

[79] Taupin P. Adult neurogenesis in the mammalian central nervous system: functionality and potential clinical interest. Med Sci Monit 2005;11:RA247–252.

[80] Cameron HA, Gould E. Adult neurogenesis is regulated by adrenal steroids in the dentate gyrus. Neuroscience 1994;61:203–9.

[81] Arvidsson A, et al. Neuronal replacement from endogenous precursors in the adult brain after stroke. Nat Med 2002;8:963–70.

[82] Gould E, et al. Proliferation of granule cell precursors in the dentate gyrus of adult monkeys is diminished by stress. Proc Natl Acad Sci USA 1998;95:3168–71.

[83] Ekdahl CT, et al. Inflammation is detrimental for neurogenesis in adult brain. Proc Natl Acad Sci USA 2003;100:13632–7.

[84] Höglinger GU, et al. Dopamine depletion impairs precursor cell proliferation in Parkinson disease. Nat Neurosci 2004;7:726–35.

[85] Sun D, et al. Anatomical integration of newly generated dentate granule neurons following traumatic brain injury in adult rats and its association to cognitive recovery. Exp Neurol 2007;204:264–72.

[86] Kuhn HG, et al. Epidermal growth factor and fibroblast growth factor-2 have different effects on neural progenitors in the adult rat brain. J Neurosci 1997;17:5820–9.

[87] Zheng W, Nowakowski RS, Vaccarino FM. Fibroblast growth factor 2 is required for maintaining the neural stem cell pool in the mouse brain subventricular zone. Dev Neurosci 2004;26:181–96.

[88] Yoshimura S, et al. FGF-2 regulation of neurogenesis in adult hippocampus after brain injury. Proc Natl Acad Sci USA 2001;98:5874–9.

[89] Aguirre A, Rubio ME, Gallo V. Notch and EGFR pathway interaction regulates neural stem cell number and self-renewal. Nature 2010;467:323–7.

[90] Jackson EL, et al. PDGFR alpha-positive B cells are neural stem cells in the adult SVZ that form glioma-like growths in response to increased PDGF signaling. Neuron 2006;51:187–99.

[91] Carpentier PA, Palmer TD. Immune influence on adult neural stem cell regulation and function. Neuron 2009;64:79–92.

[92] Monje ML, Toda H, Palmer TD. Inflammatory blockade restores adult hippocampal neurogenesis. Science 2003;302:1760–5.

[93] Butovsky O, et al. Microglia activated by IL-4 or IFN-gamma differentially induce neurogenesis and oligodendrogenesis from adult stem/progenitor cells. Mol Cell Neurosci 2006;31:149–60.

[94] Mathieu P, et al. The more you have, the less you get: the functional role of inflammation on neuronal differentiation of endogenous and transplanted neural stem cells in the adult brain. J Neurochem 2010;112:1368–85.

[95] Palmer TD, et al. Cell culture. Progenitor cells from human brain after death. Nature 2001;411:42–3.

[96] Pluchino S, et al. Injection of adult neurospheres induces recovery in a chronic model of multiple sclerosis. Nature 2003;422:688–94.

[97] Seyed Jafari SS, et al. Investigating the effects of adult neural stem cell transplantation by lumbar puncture in transient cerebral ischemia. Neurosci Lett 2011;495:1–5.

[98] Vazey EM, et al. Transplanted adult neural progenitor cells survive, differentiate and reduce motor function impairment in a rodent model of Huntington's disease. Exp Neurol 2006;199:384–96.

Chapter 5

Resident Stem Cell in Skin

Radovan Borojevic
Instituto de Ciências Biomédicas Universidade Federal do Rio de Janeiro and Excellion Serviços Biomédicos SA, Petrópolis, Rio de Janeiro, Brazil

Introduction

Skin is our largest organ and it covers our body surface. It is our interface with the environment, responsible for protection from injury, dehydration, and infection. Therefore, the external surface of the body is covered by a water-impermeable and mechanically resistant squamous layer derived from skin epithelial cells. Our movements modify the relation of the skin with the underlying organs. Hence, the skin has to be elastic and only loosely attached to the subjacent tissues, and it has to be functionally integrated through blood and lymphatic vessels as well as by nerves. Skin has also sensory functions for touch and temperature that can lead to pleasure or pain. In human society, which depends largely on visual information, the aspect of our skin and hair can modulate social communications. Being in charge of our interaction with the external world, it has to be protective, adaptive, and informative throughout life.

Mammalian skin consists of two layers that are of different embryologic origins: epidermis and dermis, derived from ectoderm and mesoderm, respectively. They

Resident Stem Cells and Regenerative Therapy. http://dx.doi.org/10.1016/B978-0-12-416012-5.00005-0

interact at the dermo-epidermal junction in which they are separated by a thick basement membrane. This is a reference for the spatial organization of epidermis, which is essentially a two-dimensional structure. The underlying thicker dermis supplies nutrients to the epidermis and interacts with the rest of the body through a complex vascular tree and a nerve network. Because of permanent environmental stress, the epidermis is continuously renewed throughout the life. The turnover rate of the epidermal cell layer is on the order of days. Conversely, cells and extracellular matrix of dermis have a much slower renewal. Moreover, both layers have to respond promptly to any injury, and skin repair and regeneration have to be integrated within the two layers.

With the exception of the plantar and palmar regions, skin is provided by pilosebaceous follicles producing hair, whose shafts have a proper cycle of growth and renewal. Hair follicles are complex structures that include sebaceous glands, insertion of the *arrector pili* muscle, and nerves and blood vessels. Hair follicles are of epidermal origin, but they are deeply embedded in dermis or even in the subdermal adipose tissue. It is now clear that they are not simply "appendages" but have a central role in the kinetics of skin cell proliferation and turnover.

Dermis and the dermo-epidermal junction also harbor exogenous cells: sensorial cells and nerves, as well as Langerhans cells, mast cells, and migratory cells brought by blood, belonging to nervous and immune systems, respectively. They will not be discussed in this chapter. Cells associated with blood vessels, as well as melanocytes that actively participate in physiology of the skin, will be discussed when necessary.

Interfollicular Epidermis

Interfollicular epidermis, which is by far the largest part of the skin surface, has a relatively simple organization depending directly upon the dermo-epidermal junction. Only the cells that are in contact with the basement membrane maintain their ability to replicate. They are hierarchically ordered within the small, approximately hexagonal basal unit that corresponds to a single keratinized squame at the surface, named the epidermal proliferative unit (EPU) [1,2]. This structure is clearly visible in mice, but it can be only roughly recognized in human skin in the regions where skin is thin and flat [3]. Within an EPU, cells are arranged in a columnar pattern located over a group of a dozen cells lying on the basement membrane. These basal cells continuously produce the overlaying cells of the spinous and granular layers and finally transform into dead corneous squames of the *stratum corneum* that eventually slough off the skin surface. EPUs have a long life, and under normal conditions their distribution on the skin surface is stable. Each one has a cluster of a dozen proliferating committed progenitor cells. When they divide with their mitotic spindle parallel to the basement membrane, the two daughter cells remain in touch with the membrane and both can further divide. Conversely, when any one of them divides with the mitotic spindle perpendicular to the basement membrane, one of the daughter cells moves upward and loses contact with the basement membrane. These cells then engage into the terminal differentiation cascade and are bound to be lost from the skin surface (Figure 5.1).

In humans, particularly in the regions of thick papillary dermis, the EPUs are larger, containing up to four dozen cells. In the center, a group of slowly proliferating cells can be identified by bright staining with antibody that labels the integrin β1. They are surrounded by cells that lose this labeling, actively proliferate as indicated by their reactivity to Ki67, and acquire progressively higher expression of keratin-10 [4]. We understand that the former ones are local stem cells and the early committed epidermal progenitors, whereas the latter ones are cells engaged into the terminal keratinocyte differentiation.

Under a minor skin injury or in an in vitro culture, the basal stem cells can give origin to clonogenic regenerative cells. A single cell can generate in vitro a sheet of epidermis of more than 15 cm^2. These sheets can contain more than 150 evenly spaced patches of β1 integrin-bright cells with their cell number and spacing similar to those observed in vivo. They correspond roughly to EPUs. Each patch is centered by one or a few stem cells, selected among the progeny of the first clonogenic stem cell plated in culture. In a primary cell culture of human epidermis, the number of fully competent holoclones is independent of the proliferative characteristics and quantity of the plated cells, being thus autoregulated and independent of a putative

(A) (B)

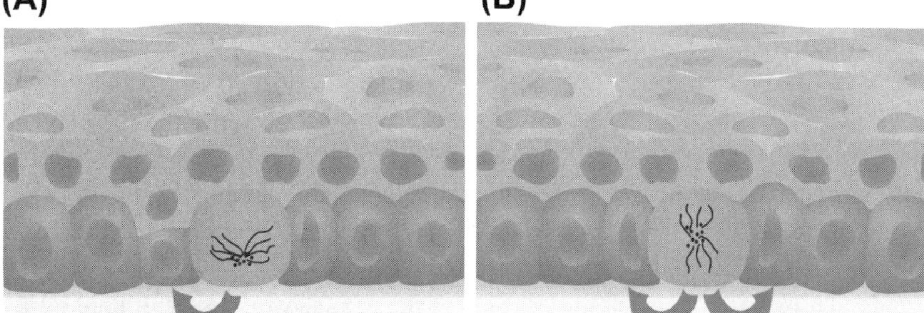

Figure 5.1 The position of the mitotic spindle is determinant of the destiny of epithelial cells after stem cell division. In (**A**), the spindle is perpendicular and the chromosome plate is parallel to the basement membrane, giving one cell that can further divide and one that is engaged in terminal differentiation. In (**B**), the spindle is parallel and the chromosome plate is perpendicular to the basement membrane, and both daughter cells can further divide. **See Plate 10.**

molecular "niche" in the basement region of the dermo-epidermal junction. The stem cell "niche" in the interfollicular epidermis is thus apparently dependent only on the environment provided by the surrounding keratinocytes, and it is established as a consequence of autonomous autoregulatory feedback signals [4]. Because the number of clones is independent of the quantity and quality of the plated cells able to replicate, the underlying hypothesis is that all the proliferating basal stem cells in epidermis are able to organize a new EPU when appropriately stimulated, returning thus from the transitory amplification cell pool to the central permanent stem cell pool of long life and capacity to maintain epidermal homeostasis. They can be understood as unipotent committed progenitors that can give rise by asymmetric divisions either to new progenitors or to suprabasal nonproliferating cells engaged in terminal differentiation [5].

In major superficial skin injuries that destroy epidermis but not the deep dermis, such as an extensive second-degree burn, lateral outgrowth of epidermal cells cannot supply the cells required for covering the denuded surface within a reasonably short period. However, a rapid reconstitution of the epidermal layer can be observed, and this clearly shows that a regularly widespread source of epidermal stem cells is available within the skin and that it is located deeply below the epidermis and the superficial dermis. We know now that they are resident within the hair follicles, which harbor an amazingly complex and functional source of stem cells [6].

Hair Follicle and Associated Structures

Each hair follicle has a proper life cycle. The follicle can be divided into two parts. The upper part is located between the skin surface and the sector that harbors the opening of the sebaceous gland duct, the point of fixation of the *arrector pili* muscle, and the thickened region named the *bulge*. This is the permanent part of the follicle maintained throughout the hair cycle. The lower part, ranging from the bulge to the *hair bulb,* is lengthened during hair growth and shortened during its regression. Hence, the distance between the bulge and the hair bulb is much larger in the former phase as compared to the latter one (Fig. 5.2, mature follicle).

The hair shaft grows in the *anagen* stage, when cells of the bulbar region at the base of the follicle rapidly proliferate and differentiate into keratinocytes that produce the hair shaft and the inner layer of the root sheath. This cell pool is sustained by a continuous inflow of progenitor cells coming from the bulge region. In the subsequent *catagen* stage, the hair growth ceases, the cellular mitotic activity slows down, and epithelial cells of the lower part of the follicle enter apoptosis. This leads to shortening of the hair shaft, raising the basal part of the follicle closer to the skin surface, and bringing the hair bulb close to the bulge. Removal of the hair shaft in the latent phase, named the *telogen* phase, leads eventually to a new hair cycle

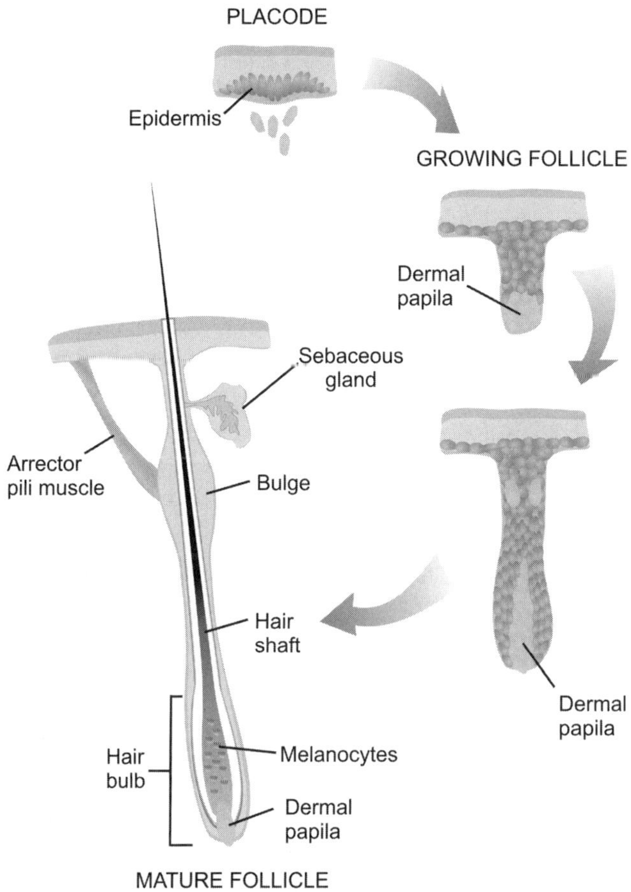

Figure 5.2 The placode initiates interaction between epithelial and mesenchymal cells, ending progressively in differentiation of the dermal papilla and the hair bulb. Thereafter, formation of a mature hair follicle is autonomous and self-sustained. **See Plate 11**.

beginning with the anagen stage. Renewal of the proliferating epithelium of the hair bulb required for hair shaft growth in the new cycle is a consequence of a direct interaction of epithelial cells with a specific group of mesenchymal cells, the *dermal papilla* cells, which are located in the dermis at the base of the follicle. Because in the telogen stage the hair bulb comes close to the bulge through the shortening of the lower part of the hair follicle, the direct interaction of epithelial and dermal papilla cells becomes feasible and the new shaft can be created [7].

This brief description of the hair cycle introduces two structures fundamental for hair growth: the *bulge* and the hair bulb with the underlying *dermal papilla*.

The bulge is a thickened region of the hair root sheet, marking the lowest part of the upper permanent region of the hair follicle. It is close to the attachment of the *arrector pili* muscle and the entrance of the sebaceous gland channel. It has a rich capillary network and terminal nerve fibers, granting its integration with adjacent tissues. A small group of undifferentiated epithelial cells that dwell in the bulge are cycling very slowly; albeit they can progressively produce numerous progeny that can be located both upward and downward along the root sheet [6]. Taken together, these characteristics are consistent with the concept of the "stem cell niche," able to protect and retrain stem cells from cycling as well as to respond to external signals and mobilize them when required [8].

In the anagen stage, the bulge epithelial cells can supply downward the highly proliferative population of epidermal cells resident in the hair bulb that are required for hair growth. This rapidly and transitory amplifying population is dependent on the input of early progenitor cells from the bulge. The same stem cell niche also supplies upward the cells that proliferate in response to superficial skin regeneration after trauma. This is the minor pathway of stem cell migration, as resident stems of the interfollicular area maintain the normal homeostasis of this tissue and respond efficiently to smaller demands of regeneration. The mobilization of bulge cells into the repair of epidermis normally occurs only following a major demand for lesion repair [9]. The finding that the stem cell reservoir resides in the bulge region helps to explain the general cycling nature with a repeated renewal of hair growth and the fact that hair bulbs are not required for hair reconstitution, but the stem cells resident in the bulge are. Further studies showed that bulge stem cells are pluripotent, not only in their capacity to generate both hair-producing and interfollicular epidermal keratinocytes but also in their ability to produce the cells of sebaceous glands [10].

In conclusion, the proposal that bulge stem cells are the ultimate totipotent epidermal stem cells is consistent with the fact that these resting stem cells of epidermis are very rare, even though both interfollicular epidermis renewal and hair growth require a high input of new cells. The bulge cells can supply the required proliferative clone-generating cells both to the hair root and to the epidermis. Because the quantity of cells generated by each clone follows the logarithmic progression, this hierarchical pattern of proliferation grants production of the required quantity of cells for a lifetime.

Bulge stem cells have several markers that are highly expressed but not exclusive, notably cytokeratin-15 (K15), which is also expressed at low levels in the hair bulb. Genetic manipulation using expression of enhanced green fluorescent protein (EGFP) under the control of K15 promoter generated mice with EGFP expression restricted to bulge cells [11]. This model enabled the isolation of living bulge cells by fluorescence-activated cell sorting (FACS). The sorted cells confirmed their multipotency by generating in vitro and in vivo holoclones that differentiate into epidermal, hair bulb, and sebaceous gland cells [10]. Moreover, the microarray analyses showed, as expected, a high expression of genes associated with cell arrest.

The analysis of cell surface markers now enables us to characterize the bulge stem cells as $CD200^{hi}$, $CD24^{lo}$, $CD34^{lo}$, and $CD146^{lo}$.

The sebaceous is a part of the upper region of the hair follicle. Its duct opens above the bulge, secreting sebum onto the skin surface. This is a holocrine gland, whose secretory cells reach death and are eliminated with the produced sebum. This is reminiscent of keratinocyte terminal differentiation and death followed by corneous scale elimination. The permanent loss of sebocytes requires a continuous cell renewal, which is dependent on the proliferative basal cell layer that is in contact with the basement membrane at the junction of the sebaceous gland the surrounding dermal tissue [12]. However, in postnatal skin development and growth in which formation of new sebaceous glands occurs, as well as following the increased local demand during skin regeneration, the multipotent stem cells of the bulge can differentiate into the cells of the sebaceous gland lineage, when they migrate up the hair shaft and enter the sebaceous gland duct. Bulge stem cells committed to this pathway initially substitute the expression of the general stem cell marker CD34 with the expression of Blimp1. This is a transcription factor originally known to participate in control of the terminal differentiation of B lymphocytes. In sebocyte progenitors, this factor controls the cell proliferation through the Myc pathway, and it induces the expression of PPARγ, the master gene that controls lipid metabolism in many cells lineages such as adipocytes, whose expression is also required for the production of lipids, which are major components of sebum. Moreover, when sebaceous gland progenitor proliferation is disturbed through the inhibition of Blimp1, enhanced activity of adjacent bulge stem cell is observed. This suggests that following the impaired or insufficient renewal of sebaceous gland cells, multipotent stem cells in the hair follicle bulge can be mobilized to reestablish normal gland function [13].

The bulge region is thus a typical stem cells niche in which physical contact with the local microenvironment and surrounding mesenchymal stroma protects epithelial stem cells and retains them in the resting state. A single stem cell of the bulge is sufficient to generate a new hair follicle [14]. When required, these cells are available for generation of proliferative cascades that lead to maintenance and renewal of all the epidermal and adnexal skin structures [11].

Dermal papilla is the second skin stem cell niche of the hair follicle that directly controls generation of hair. During embryogenesis, dermis is derived from the dermomyotome that underlies epidermis. The future hair bulbs are at first visible as small epidermal thickenings and invaginations into the underlying dermis, named placodes [15] (see Fig. 5.2). Their positioning and cell commitment are defined and controlled by mesenchyme. In conjunction with the signals from dermis, the ectodermal Wnt signals induce dermal cells to grow downward, forming the hair bud by condensation of the dermal papilla cells, which are thereafter associated with the tip of the growing hair follicle. Once formed, the dermal papilla mesenchymal cells establish the permanent physical association and molecular crosstalk with epidermal cells of the hair bud. The two cell populations are required and sufficient for inducing formation of the hair bulb and controlling the hair growth cycle.

The presence of neural crest–derived mesenchymal cells in the dermal papilla has been reported in embryo. They are maintained in young humans, but their number apparently decreases with age, as mobilization of resident mesenchymal cells of dermis also participates eventually in renewal of the papilla cell population [16]. These cells represent a common neural crest–derived pool of tissue pluripotent progenitors that can be induced in regenerative processes to differentiate into many cell lineages including neurons and glia [17].

During the catagen-anagen transition, when the lower part of the hair follicle is fully retracted and the papilla cells reach the vicinity of the bulge, they activate and mobilize production of a new wave of keratinocyte progenitors. Their association with dermal papilla grants the growth of the lower part of the follicle, at which point a new papilla surrounded by keratinocytes forms a new active hair bulb and sustains the new cycle of hair growth [18].

The association of papillary cells with proliferating keratinocytes is fully informative, required, and sufficient for generation of a new hair shaft. This has prompted proposals to generate new hair roots in the bald skin using dermal papilla cells, and this was shown to be feasible. In vitro culture of the two cell types supplemented with melanocytes is able to form a fully structured organoid with a proper extracellular matrix, morphologically and functionally equivalent to the hair bulb, which produces a hair fiber similar to the in vivo produced hair shaft [19]. Human hair bulb cells can be expanded in vitro for long periods without losing their capacity to generate new follicles in vivo [20], and extensive expansion was recently standardized under Good Medical Procedures (GMP) conditions that can be applied for human cell therapies [21]. When small aggregates of dermal papilla cells are implanted under keratinocytes, preferentially provided with a matrix that avoids their random dispersal after injection into the tissue, they can induce invagination of the adjacent epidermal layer, similar to the formation of a placode in embryogenesis, which can subsequently form a new hair root. This is feasible even in skin normally devoid of hair, such as the plantar or palmar skin [22]. The formation of organoids associating dermal papilla cells with proliferative keratinocytes is also sufficient to initiate the formation of a functional hair follicle when implanted into dermis, without direct contact with epidermis. We may conclude that the cells of dermal papilla, but not other mesenchymal cells, when associated with keratinocytes are competent and sufficient to form a new hair follicle.

Melanocytes

The color of skin and hair is genetically controlled and modulated by the environment through the activation of melanocytes. However, white skin can be associated with very dark hair and dark skin can be associated with brown or reddish hair, indicating that the controls of melanization that depend on the melanocyte metabolic activity are independent in the two sites. The fact that interfollicular epidermis, but not hair, responds to ultraviolet (UV) irradiation with a higher production of melanin and that maturing keratinocytes of the *stratum corneum* lose their color whereas the hair shaft retains it throughout its growth are only a few examples of these differences [23].

Melanocytes are normal components of skin and hair. They are migratory cells dwelling mainly at the dermo-epidermal junction. Their long extensions are inserted among keratinocytes of the basal and spinous layers of epidermis, as well as in the melanogenic zone above dermal papilla in the hair bulb. Keratinocytes receive melanin from melanocytes and accumulate it to different degrees. Melanocytes are not randomly distributed among keratinocytes but follow a pattern corresponding to "epidermal-melanin units" with a relatively constant ratio between the two cell types. The ratio varies between skin regions, but the skin color is due mainly to the quantity of melanin granules in keratinocytes and not to the quantity of melanocytes. Similar to keratinocytes, melanocyte niches at the dermo-epidermal junction can harbor small groups of cells with a high potential capacity to proliferate and differentiate into a new melanocyte population. In view of the longer life span of melanocytes, the kinetics of these populations is less clear than of the EPUs [24].

Both melanocytes of interfollicular epidermis and of hair follicles derive from uncolored melanoblasts. Early in embryogenesis, their progenitors exit from the neural crest, and some of them follow the cues that lead them to dermis. A fraction of them already starts differentiation into melanocytes, indicated by melanogenesis, in the embryo by the seventh week of gestation in humans, several weeks before the formation of hair follicles. Others remain as melanoblasts within the intensely proliferative cell population, and they are the transit-amplifying cells of the melanocytic lineage. The nascent hair follicles are progressively populated by a mixed population containing immature, proliferative, colorless cells and maturing melanogenic cells. Expression of the *kit* gene is required for their migration in response to stem cell factor (SCF)-producing epithelium of the follicle. Kit-positive melanocytes target the hair bulb, supplying melanin to the nascent hair shaft. Simultaneously, immature melanoblasts settle in the melanoblast stem cell niche, which is located in the bulge region proximally adjacent to the keratinocyte stem cell niche [25]. This melanocyte stem cell niche is spatially well delimited within the lowermost region of the permanent portion of hair follicle.

The pattern of melanocyte lineage stem cell interaction with their niche is relevant for understanding the transformation of the proliferating melanoblast population in the embryo into the nonproliferating stem cell population settled in their hair follicle niche. Studies by Nishikawa's group have proposed that Notch signaling is required, both for embryonic development of the lineage and for maintenance of stem cells in the hair follicle niche. Although Wnt signaling commits the differentiation into melanocytes, the less well understood signals induce the down-regulation of housekeeping genes consistent with the dormancy of stem cells, the suppression of the Wnt signal pathway, and the independence of Kit signaling. This dormant autonomous survival of stem cells is limited to a small number of ultimate stem cells of the melanocytic lineage [26]. In mice, the hair follicle niche can contain only a few or a single stem cell of this lineage. However, at the beginning of the anagen stage, a transitory amplifying stem cell population derived from stem cells generates colorless melanoblasts, which migrate through the external hair root sheet toward the bulb. Only in the bulb do these cells proliferate in contact with the dermal papilla. At its dome, they intensely produce melanin and transfer it to cortical keratinocytes incorporated into the hair cuticle. At the end of catagen, bulb melanocytes enter in apoptosis. The renewal of melanocytes required for the new hair shaft depends exclusively on the activation of stem cells in the melanocyte lineage niche located in the bulge, and there is cyclical control of hair renewal. Conversely, and similar to keratinocytes, the melanocyte production upward, in the direction of the interfollicular epidermis, can occur continuously or in response to demands of regeneration. The amplifying proliferative cells can give active melanocytes, as well as repopulate when required the vacant stem cell niches at the dermo-epidermal junction [25].

In conclusion, the two major epidermal cell systems, keratinocytes and melanocytes, are organized following similar patterns, notwithstanding their embryologic and functional differences. Whereas keratinocyte stem cells are pluripotent, those of the melanocytic cell lineage produce only melanocytes. Ultimate stem cells of both lineages dwell in the same region of the hair follicle and supply, in a cyclic manner, cells for hair growth, responding also, when required, to specific demands of regeneration. In the hair bulb, their activity depends on a secondary, well-defined niche sustained by mesenchymal cells of the dermal papilla. Conversely, in the interfollicular epidermis, both cell systems are organized in proliferative and functional units, centered on a single cell or a small stem cell group, which apparently has an intrinsic capacity to establish the hierarchical stem cell order.

Dermis

Dermis is essentially a thick layer of connective tissue. Its major components are fibroblasts, which have a long life and slow turnover. They are engaged in the production of elastic extracellular matrix whose quantity and orientation respond to the mechanical demand of the specific skin region inhabited by fibroblasts. Mechanotactic cell receptors organize spatially both the cells and the secreted matrix in order to homogenously distribute the elastic tension within the skin and grant its protective function. Besides its mechanic functions, the dermal extracellular matrix is hydrophilic, and it controls the water content within the skin as well as retains growth factors, cytokines, and intercellular mediators locally produced or brought into the skin by biologic fluids.

During skin formation in embryo, a vascular tree grows into dermis from the underlying blood and lymphatic vessels. It forms a delicate capillary network subjacent to the dermo-epidermal junction, which supplies nutrients to epidermis and hair follicles. Perpendicular small vessels link it to the central blood vessel network in the dermis, which is tributary of the larger vascular tree in subjacent tissues, mainly in the subdermal adipose tissue.

It has been known for decades that in the case of a surgical section of the skin, regeneration elicits local cell proliferation immediately. Resident fibroblasts adjacent to the cut can enter early into the cell cycle, as shown by thymidine incorporation. The subsequent and major proliferative responses involve perivascular cells: pericytes of the small blood vessels and adventitial cells of the larger ones, as well as cells surrounding skin appendages [27]. These cells are now known as resident perivascular mesenchymal stem cells, which are described in detail and discussed in Chapter 12.

Regenerative skin cell population of perivascular origin is characterized by expression of the smooth muscle α actin, which is absent in resident fibroblasts. They produce basement membrane, notably the laminin 511/521, which participates in tissue regeneration, and they have adhesive molecules, which facilitate cell migration [28]. They are now well known as myofibroblasts and represent the major component of reactive dermal tissue involved in repair and regeneration. They can differentiate into several connective-tissue cell types, including cells associated with the vascular wall, and they are considered to be resident mesenchymal cells of the skin. They are quite mobile and can remain associated with blood vessels or dwell in the interstitial environment.

The growth of new capillaries involves two events: the sprouting of endothelial cells and the subsequent stabilization of sprouts by pericytes, which can be mobilized from the surrounding interstitial cell pool or are brought by migration of pericytes already resident on growing vessels [29]. It is not clear whether the former source involves already differentiated fibroblasts or resident quiescent mesenchymal cells. Studies on formation of the granulation tissue indicate that myofibroblasts can be generated both from perivascular cells and from resident quiescent fibroblasts. The underlying proposal is that bona fide fibroblasts can return into the myofibroblast state, and they can thus be potentially considered to be connective tissue resident mesenchymal stem cells [30].

Simultaneous contractibility of the myofibroblast cytoskeleton and the tendency to increase adhesion onto the fibrous matrix generate tensile forces that close the wounds. When they retain their hyperactivity through permanent and intense inflammation and activation such as occurs in deep burns, they result in contracted hypertrophic cicatrices that can severely incapacitate movement and function of dermal and subdermal tissues. Scar hypertrophy is sustained by excessive blood vessel growth and stabilization, which is a counterpart of perivascular stem cell activation. Normally, a decrease of blood vessel density consequent to tissue hyperoxygenation also decreases myofibroblast activation, resulting in normal resolution of the scar in flat cicatrizes. However, the sustained mechanical load that is a consequence of cell contraction protects myofibroblasts from apoptosis, generating a loop that hampers scar regression [31]. Genetic background and sustained hyperactivity of the dermal connective tissue caused by chronic inflammation may thus result in permanent hypertrophic scars that are difficult to treat.

Cell Therapies and Clinical Application of Skin Stem Cells

Because of the frequency of skin lesion and the facility of surgical access to the external tissues, cell therapies have been applied to skin for a long time. Autologous transplantation of skin is relatively easy. Provided the receptor tissue is in a good physiologic state, the transplanted skin integrates functionally into the receptor site and gives satisfactory results. However, the availability of donor sites is often limiting, and ablation of the skin for transplantation can be associated with discomfort and the formation of permanent scars. In vitro culture of human skin cells has been known for decades [32]. Nowadays, cultured cells, whether or not they are associated with artificial or natural matrices, can be prepared for regenerative medicine, and several products are commercially available [33].

The culture of epithelial cells can be facilitated by feeder layers of irradiated connective tissue cells, such as murine 3T3 cells or human fibroblasts, but culture media that sustain keratinocyte growth without a feeder layer are now available from different sources. As already mentioned, epidermal stem cells have a high proliferative capacity, and under optimal conditions 1 cm^2 of normal skin from a donor site can generate a monolayer of up to 1 m^2 in 3 to 4 weeks, depending on the age and the biologic state of the patient. These monolayers are fragile, but can be carefully prepared and used for transplantation with good results, provided again the appropriate state of the receptor site. The clinical benefits of such therapies, particularly in the repair of severe and extensive skin lesions, are obvious, because successful treatments have been obtained in burn patients who have lost over 70% of their skin surface. This has raised expectations that could not always be attended, because of the technical difficulties of cell manipulation and transport, the high cost, and the relatively long time required for preparing an implant of sufficient size [34]. The possibilities of skin bioengineering are broad, and both technical and operational improvements are expected in the coming years.

Transplanted epidermal layers need the subjacent dermis and the normally organized and physiologically active dermo-epidermal junction. In deep skin wounds with loss of dermis, this can be substituted by bioengineered substrates, either synthetic, which are mimetic of dermal extracellular matrix, or obtained by treatment of donor skin by mild detergents to dissolve the cells. Because acellular skin is of low antigenicity, allogeneic donor skin can give good results. Both types of substrates have to be vascularized in order to grant the survival and integration of the implant. Skin is a flat tissue and the contact with the subjacent tissue is broad, so a full integration of the implant normally occurs within a short period. The therapy can be done thus in either a single or a sequential procedure. In the former option, the full-thickness implant is prepared with the dermis-equivalent layer covered by cultured epidermis. In the latter option, a dermis substitute is used to cover the lesion, protect it, and let it be colonized by blood vessels growing from the underlying tissue. When capillary tips reach the surface, cultured epidermis can be placed over the implanted area to regenerate the superficial tissue layer.

Melanocyte cultures can be also obtained from skin, either pure or associated with keratinocytes, depending on the culture conditions. When they are present in the epidermal culture used to cover a lesion, the resulting skin is melanized, and the degree of melanization is similar to the original skin. Melanocytes can also be seeded onto the dermis denuded by dermoabrasion or injected into the dermo-epidermal junction. Because they migrate well, are replicative, and have self-regulatory control of their distribution, a local melanization can be obtained with satisfactory results.

Dermis is responsible for physical properties of the skin. Aging and exposition to solar irradiation always lead to a progressive degeneration of dermis, which results in significant loss of vitality and elasticity. This results in the typical aspect of aged human skin. Homeostasis and regeneration of skin are maintained at the expense of the regular renewal of the major dermal cell population: the fibroblasts. As already stated, fibroblasts can be generated by self-renewal, but by far the major input of fibroblasts is via myofibroblasts, caused by the activation of perivascular mesenchymal stem cells. Intradermal introduction of fibroblasts or their progenitors has been used in reparative therapy both of aged skin and of local dysplasias, cicatrizes, radiotherapy, and solar lesions, with satisfactory results. Because the introduced fibroblasts are guided by tensergic information of the dermal matrix, their tendency is to physiologically integrate into the skin, homogenizing the distribution of the extracellular matrix and decreasing the existing irregularities. Phase III clinical studies have shown the efficacy and safety of the procedure [35].

An alternative approach is the use of perivascular mesenchymal cells, which produce in situ growth factors that stimulate tissue regeneration and differentiate into fibroblasts in the dermal environment, increasing the number of locally available cell progenitors. The use of adipose tissue–derived mesenchymal cells has been shown to be feasible and efficient [36]. Introduction of fibroblasts or mesenchymal cells can be associated with extracellular matrix that may function as temporary filler, providing an immediate aesthetic improvement, substrate for cell migration, and support for their proliferation [37].

Skin is permanently exposed to environmental stress, and occasionally to major trauma, and skin regeneration has deserved extensive studies. The participation of hierarchically ordered and precisely located cell populations with different degrees of stemness is now better understood. The use of cell therapies, as well as the development of more complex procedures involving bioengineering of complex skin surrogates, is promising and in the near future will probably offer relief and a solution to several pathologic situations that need rapid and efficient repair in order to maintain the integrity and protection of our bodies.

References

[1] Potten CS. The epidermal proliferative unit: the possible role of the central basal cell. Cell Tissue Kinet 1974;7:77–88.

[2] Potten CS, Morris RJ. Epithelial stem cells in vivo. J Cell Sci 1988;10(Suppl.):45–62.

[3] Jensen UB, Lowell S, Watt FM. The spatial relationship between stem cells and their progeny in the basal layer of human epidermis: a new view based on whole-mount labelling and lineage analysis. Development 1999;126:2409–18.

[4] Janes SM, Lowell S, Hutter C. Epidermal stem cells. J Pathol 2002;197:479–91.

[5] Jones P, Simons BD. Epidermal homeostasis: do committed progenitors work while stem cells sleep? Nat Rev Mol Cell Biol 2008;9:82–8.

[6] Taylor G, Lehrer MS, Jensen PJ, Sun TT, Lavker RM. Involvement of follicular stem cells in forming not only the follicle but also the epidermis. Cell 2000;102:451–61.

[7] Cotsarelis G. Epithelial stem cells: a folliculocentric view. J Invest Dermatol 2006;126:1459–68.

[8] Jones DL, Wagers AJ. No place like home: anatomy and function of the stem cell niche. Nature Rev Mol Cell Biol 2008:9e11.

[9] Ito M, Liu Y, Yang Z, Nguyen J, Liang F, Morris RJ, Cotsarelis G. Stem cells in the hair follicle bulge contribute to wound repair but not to homeostasis of the epidermis. Nat Med 2005;11: 1351–4.

[10] Blanpain C, Lowry WE, Geoghegan A, Polak L, Fuchs E. Self-renewal, multipotency, and the existence of two cell populations within an epithelial stem cell niche. Cell 2004;118:635–48.

[11] Morris RJ, Liu Y, Marles L, Yang Z, Trempus C, Li S, Lin JS, Sawicki JA, Cotsarelis G. Capturing and profiling adult hair follicle stem cells. Nat Biotechnol 2004;22:411–7.

[12] Ghazizadeh S, Taichman LB. Multiple classes of stem cells in cutaneous epithelium: a lineage analysis of adult mouse skin. EMBO J 2001;20:1215–22.

[13] Horsley V, O'Carroll D, Tooze R, et al. Blimp1 defines a progenitor population that governs cellular input to the sebaceous gland. Cell 2006;126:597–609.

[14] Claudinot S, Nicolas M, Oshima H, Rochat A, Barrandon Y. Long-term renewal of hair follicles from clonogenic multipotent stem cells. Proc Natl Acad Sci USA 2005;102:14677–82.

[15] Fuchs E. Scratching the surface of skin development. Nature 2007;445:834–42.

[16] Rendl M, Lewis L, Fuchs E. Molecular dissection of mesenchymal-epithelial interactions in the hair follicle. PLoS Biol 2005;3:e331.

[17] Hunt DPJ, Morris PN, Sterling J, Anderson JA, Joannides A, Jahoda C, Compston A, Chandran S. A highly enriched niche of precursor cells with neuronal and glial potential within the hair follicle dermal papilla of adult skin. Stem Cells 2008;26:163–72.
[18] Yang CC, Cotsarelis G. Review of hair follicle dermal cells. J Dermatol Sci 2010;57:2–11.
[19] Lindner G, Horland R, Wagner I, Ataç B, Lauster R. De novo formation and ultra-structural characterization of a fiber-producing human hair follicle equivalent *in vitro*. J Biotechnol 2011;152: 108–12.
[20] Qiao J, Zawadzka A, Philips E, Turetsky A, Batchelor S, Peacock J, Durrant S, Garlick D, Kemp P, Teumer J. Hair follicle neogenesis induced by cultured human scalp dermal papilla cells. Regen Med 2009;4:667–76.
[21] Marazzi M, Crovato F, Bucco M, Sironi MC, Tosca MC, Antonioli B, Chlapanidas T, Lucconi G, Scalise A, Vigo D, Faustini M, Torre ML. GMP-compliant culture of human hair follicle cells for encapsulation and transplantation. Cell Transplant 2012;21:373–8.
[22] Lin CM, Li Y, Ji YC, Keng H, Cai XN, Zhang JK. Microencapsulated human hair dermal papilla cells: a substitute for dermal papilla? Arch Dermatol Res 2008;300:531–5.
[23] Tobin DJ. Human hair pigmentation – biological aspects. Internat J Cosmetic Sci 2008;30:233–57.
[24] Lin JY, Fisher DE. Melanocyte biology and skin pigmentation. Nature 2007;445:843–50.
[25] Nishimura EK, Jordan SA, Oshima H, Yoshida H, Osawa M, Moriyama M, Jackson IJ, Barrandonk Y, Miyachi Y, Nishikawa SI. Dominant role of the niche in melanocyte stem-cell fate determination. Nature 2002;416:854–60.
[26] Nishikawa SI, Osawa M. Generating quiescent stem cells. Pigment Cell Res 2007;20:263–70.
[27] Oliver RF. An autoradiographic study of 3H-thymidine incorporation and distribution in the dermis of healing skin incisions in the pig. Br J Exp Pathol 1979;60:65–71.
[28] Paquet-Fifield S, Schlüter H, Li A, Aitken T, Gangatirkar P, Blashki D, Koelmeyer R, Pouliot N, Palatsides M, Ellis S, Brouard N, Zannettino A, Saunders N, Thompson N, Li J, Kaur P. A role for pericytes as microenvironmental regulators of human skin tissue regeneration. J Clin Invest 2009; 119:2795–806.
[29] Potente M, Gerhardt H, Carmeliet P. Basic and therapeutic aspects of angiogenesis. Cell 2011;146:873–87.
[30] Kis K, Liu X, Hagood JS. Myofibroblast differentiation and survival in fibrotic disease. Expert Rev Mol Med 2011;13:e27.
[31] Aarabi S, Bhatt KA, Shi Y, Paterno J, Chang EI, Loh SA, Holmes JW, Longaker MT, Yee H, Gurtner GC. Mechanical load initiates hypertrophic scar formation through decreased cellular apoptosis. FASEB J 2007;21:3250–61.
[32] Rheinwald JG, Green H. Serial cultivation of strains of human epidermal keratinocytes: the formation of keratinizing colonies from single cells. Cell 1975;6:331–44.
[33] Groeber F, Holeiter M, Hampel M, Hinderer S, Schenke-Layland K. Skin tissue engineering — In vivo and in vitro applications. Advanced Drug Deliv Rev 2011;128:352–66.
[34] Böttcher-Haberzeth S, Biedermann T, Reichmann E. Tissue engineering of skin. Burns 2010;36: 450–60.
[35] Weiss RA, Weiss MA, Beasley KL, Munivalli G. Autologous cultured fibroblast injection for facial contour deformities: A prospective, placebo-controlled, phase III clinical trial. Dermatol Surg 2007;33:263–8.
[36] Park BS, Jang KA, Sung JH, Park JS, Kwon YH, Kim KJ, Kim WS. Adipose-derived stem cells and their secretory factors as a promising therapy for skin aging. Dermatol Surg 2008;34:1323–6.
[37] Claudio-da-Silva C, Baptista LS, Carias RBV, Menezes-Neto HC, Borojevic R. Autologous mesenchymal stem cells culture from adipose tissue for treatment of facial rhytids. Rev Col Bras Cir 2009;36:288–91.

Chapter 6

Lung Resident Stem Cells

Mariana Alves Antunes[*,†,‡,§,**], Marcelo Marcos Morales[*,†,‡,§,**], Paolo Pelosi[††], Patricia Rieken Macêdo Rocco[*,†,‡,§,**]

[*] *Instituto de Biofísica Carlos Chagas Filho, Rio de Janeiro, Brazil*

[†] *Programa de Terapia Celular e Bioengenharia, Rio de Janeiro, Brazil*

[‡] *Laboratório de Investigação Pulmonar, Rio de Janeiro, Brazil*

[§] *Instituto Nacional de Ciência e Tecnologia de Biologia Estrutural e Bioimagem -INBEB, Rio de Janeiro, Brazil*

[**] *Universidade Federal do Rio de Janeiro, Rio de Janeiro, Brazil*

[††] *Department of Surgical Sciences and Integrated Diagnostics, University of Genoa, Genoa, Italy*

Lung Structure

The adult human lung consists of 23 generations of branching airways leading to the alveoli, the primary gas exchange units of the lung. The lung's gas exchange surface area, covering 70 m^2, is the largest surface of the body in direct contact with the environment. In contrast, gas diffusion occurs across a mere 1 μm, corresponding to the thickness of the alveolar-capillary barrier [1]. During normal respiration, the air transported through the airways is frequently contaminated with pollutants and infectious agents [2], leading to extreme exposure to both organic and inorganic insults. As a result, the lung has evolved multifaceted defense mechanisms that are critical for survival, including an important system to replace injured or damaged cells. In this system, damaged cells are repaired or die by necrosis or apoptosis [3,4]. Recent findings also suggest that a variety of cell types participate in this repair system. Although adult lung epithelial cells seem to be quiescent, with turnover times possibly higher than 100 days [5], consistent data have shown that differentiated airway and alveolar epithelial cell types may proliferate in response to epithelial injury. Several local epithelial cell types can act as both fully differentiated functional cells or transient-amplifying progenitors (cells that arise from stem cells

Resident Stem Cells and Regenerative Therapy. http://dx.doi.org/10.1016/B978-0-12-416012-5.00006-2

and divide a finite number of times until becoming differentiated). Alveolar epithelial type 2 cells are one example, rapidly giving rise to differentiated mature cell types, including alveolar epithelial type 1 cells [6].

In anatomic terms, the lung is characterized by complex branching of several transitional anatomic regions made up of unique cellular arrangements. The most proximal conducting airways, trachea and main bronchi (Fig. 6.1), display a columnar epithelium composed of ciliated cells, secretory cells, basal cells, and submucosal glandular epithelium. More distal conducting airways do not present basal cells and are populated by an epithelial surface with an increasing ratio of secretory (Clara) cells to ciliated epithelial cells with sparsely distributed neuro-epithelial cells that are either solitary or organized into clusters. At the bronchoalveolar duct junction, there is a transition from airway epithelium to a distal lung epithelium organized into functional alveoli. The alveolar epithelium consists of type 1 cells, which cover most of the pulmonary gas-exchange surface area, and cuboidal surfactant-expressing type 2 cells. Rare populations of multipotent resident epithelial stem cells have been identified in specific niches (Fig. 6.2), including the basal layer of the upper airways, within or near pulmonary neuroepithelial bodies, and at the bronchoalveolar junction [6-11]. This compartmentalization is logical because local stem cell niches are needed to repair local tissues in the context of daily attrition of cells. Classic lung resident stem/progenitor cells include (1) basal cells of the submucosal gland ducts and intercartilaginous zone in proximal airways [12-16], (2) Clara cells and variant-Clara cells [10,11], and (3) alveolar epithelial type 2 cells in the lung parenchyma [17-19]. These resident stem/progenitor cells are critical for lung development and growth during childhood; however, restricted residual pools of resident stem cells in adulthood are speculated as the cause of impaired regeneration

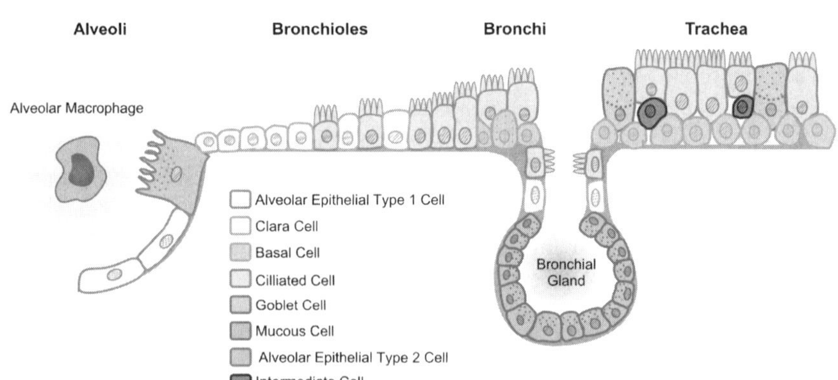

Figure 6.1 Schematic representation of pulmonary cell types. The lung is a multicellular organ, where some cells have been recognized as resident stem/progenitor cells, including basal cells, Clara cells, and alveolar epithelial type 2 cells. **See Plate 12.**

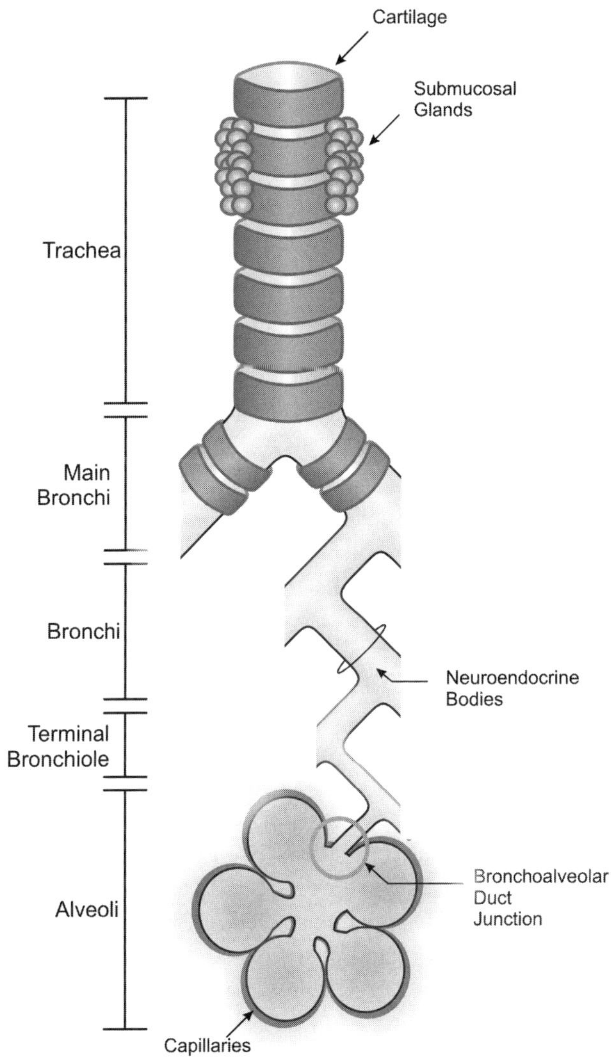

Figure 6.2 Putative stem cell niches in the lung. Stem/progenitor cells have been identified in certain regions of the tracheobronchial tree, including the submucosal glands and intercartilaginous zones in the proximal airways, neuroendocrine bodies, bronchoalveolar duct junction, and alveolar epithelium.

and repair capacity of the adult lung. Similar developmental signaling networks have been implicated in local progenitor cell activation, daughter cell proliferation, and migration during lung tissue repair after injury. This is an important research area, because it is likely that the type of airway injury is a major determinant of the type of progenitor cell to be activated.

Hierarchy of Lung Resident Stem Cells

Hierarchic organization of progenitor cells into what is commonly referred to as "stem cell hierarchy" has classically been described in tissues that display a high rate of cellular turnover, such as the small intestine and skin. The property of DNA label retention is still commonly used for identification and localization of resident tissue stem cells, but the high quiescence of adult lung epithelial cells and the large number of differentiated cell types present in lung tissue (Fig. 6.1) make the identification of stem cell hierarchy difficult in the lung. Additionally, because there is no specific protocol to identify stem/progenitor resident cells in lung, different techniques may produce different interpretations of the same cell population [11,20]; this underscores the urgency for specific markers to distinguish cell populations and methods of lineage tracing. Studies have indicated that the pseudostratified epithelium of tracheobronchial airways is maintained by multiple progenitor cell populations that are distinct from those that sustain the simple epithelium of bronchioles [21,22] and suggest that stem cell niches in the conducting airways must also be uniquely divided between the proximal and distal regions [12] (Fig. 6.2).

Resident Epithelial Progenitor Cells

Stem cell niches in the tracheobronchial tree can be easily recognized by the following characteristics: (1) stem cells undergo minimal expansion during regeneration and give rise to transient amplifying cells populations; (2) pulmonary cells feature a slow-cycling phenotype that can be used to mark potential stem cell compartments in vivo following incorporation of nucleotide analogs, such as bromodeoxyuridine (BrdU) and tritiated [H^3]-thymidine (these pulmonary cells are long-term label retaining cells [putative stem cell populations]); and (3) stem cell populations should be capable of regenerating all cellular phenotypes present in the airway. Because of the low basal rate of stem cell division and turnover of differentiated cells in the normal airway, injury models are required to study stem cell proliferation.

Basal Cells

In vivo models of tracheal epithelium injury following parenteral administration of naphthalene have shown that basal cells possess progenitor capacity [15]; in turn, following injury with nitrogen dioxide or ozone, secretory cells of rat airways appear to function as the main airway epithelial progenitor [23]. Using a murine model of SO_2 injury-induced regeneration, specific stem cell niches were identified in the submucosal gland ducts, in the proximal trachea (as indicated by the presence of label retaining cell expanding zones). Cell niches were also identified at cartilage-intercartilage junctions (Fig. 6.2), along the surface airway epithelium, in more distal trachea and bronchi [12]. To determine if airway-repopulating epithelial progenitors can be marked by surface antigens, studies with human fetal lung tissue transplanted into severe combined immunodeficiency (SCID) mice have screened an array of candidate markers, including lectin ligands, the CD44 and CD166 adhesion molecules, and aquaporin 3 (AQP3). AQP3, which acts as the membrane channel of water and other small solutes, plays a major role in fluid homeostasis and is selectively expressed on the surface of basal cells, allowing the separation by flow cytometry of AQP3^{+} basal cells and AQP3^{-} ciliated and secretory cells. Although AQP3^{+} basal cells and AQP3^{-} suprabasal cell subpopulations can restore in vivo a well differentiated, pseudostratified mucociliary surface epithelium and functional submucosal glands, faster engraftment of AQP3^{-} cells suggests their inclusion as more committed progenitors [24]. These data suggest that epithelial progenitors exist among both basal and suprabasal cell subsets within the human tracheobronchial mucosa and also that they may be positively or negatively selected with the AQP3^{+} surface marker.

Other evidence has demonstrated that basal cells are at least airway surface transient-amplifying epithelial cells. Human-specific basal cell markers cell-surface protein CD151 and tissue factor (TF) allow the differentiation and separation of positive basal cells from negative columnar cells through fluorescence-activated cell sorting (FACS). A basal cell population has been observed to proliferate on plastic and adhere on epithelium-denuded rat tracheae, as well as on collagen-coated porous membranes, where it was able to restore a fully differentiated mucociliary and functional airway epithelium, whereas viable CD151/TF-negative columnar cells did not. Telomerase activity was detected in CD151/TF-positive basal cell population, but not in CD151/TF-negative columnar cells. In the stem cell compartment of several tissues, heightened telomerase activity has been demonstrated, leading to retention of transient-amplifying cell properties [25]. Nevertheless, the relative contribution of suprabasal subsets versus basal progenitor cell populations to epithelial maintenance and regeneration following injury remains controversial [22,26,27].

A population of primitive adult precursor cells with increased colony-forming and proliferative potential in vitro, termed the side population, which can be found in many adult tissues including the lung [28], was recently isolated from normal and asthmatic human tracheobronchial epithelium [29]. These cells seem to maintain telomere length over passages in vitro, which is consistent with increased telomerase activity detected in the basal cells of adult nasal epithelium [30]. This side population is distinguished by its ability to efflux the DNA-binding dye Hoechst 33342 in FACS [28]. Also, it is negative for the hematopoietic marker (CD45) and presents positive immunohistochemistry staining for epithelial basal cell marker transcription factors p63 and cytokeratin-5, epithelial adherens junction protein E-cadherin, and tight junctional protein ZO-1. However, these cells were negative for the mesenchymal marker vimentin, suggesting that this isolated side population may represent putative proximal airway progenitors only for an epithelial phenotype [29].

Although real multipotent cells appear to exist in pulmonary tissue, the exact location of these cells is unclear. Mesenchymal stem cells were characterized in murine lung cells expressing stem cell antigen (Sca1) and negative for cell-surface proteins CD45 and CD31. These cells are clonogenic, adherent to plastic, and undergo adipogenic, osteogenic, chondrogenic, and musculogenic differentiation in vitro [31]. In addition, after being cultured in Matrigel and supplemented with basic fibroblast growth factor, hepatocyte growth factor, and epidermal growth factor, many cells lost Sca1 expression and strongly expressed differentiation lineage markers, including aquaporin subtype 5 (alveolar epithelial type 1 cell phenotype), surfactant protein-C (alveolar epithelial type 2 cell phenotype), cell-surface protein CD31 (endothelial cell phenotype), Clara cell secretory protein (Clara cell phenotype), cytokeratin 19 (pan-epithelial phenotype), smooth muscle actin, and vimentin (mesenchymal cell phenotype). Collectively, these findings suggest that these pulmonary multipotent cells may be pushed up or not to the "stemness axis" by changing the microenvironment in which they are immersed.

Clara and Variant-Clara Cells

Clara cell secretory protein (also known as CCSP, CC10, CC16, Clara cell antigen, secretoglobin, and uteroglobin) is the most abundant secretory protein found in airway surface fluid. There is evidence for a role of CCSP in homeostasis and repair, and as a potential biomarker of lung injury or disease. Clara cells, the most active producers of CCSP in lung, are found in bronchiolar epithelium and respiratory bronchioles (Fig. 6.1). Several subpopulations of CCSP-expressing cells have been suggested as true airway stem cells, including Clara cells [27,32,33] and variant-Clara cells, also known as octamer-binding transcription factor 4 (Oct-4)-expressing stem cells [34], pulmonary neuroendocrine cells [7,8,35,36], and bronchoalveolar stem cells [10].

Classical Clara cells present functional properties of differentiated cell types in their quiescent state, but they may change to an active proliferative state with properties similar to a transient-amplifying cell after injury. Clara cells express high levels of cytochrome P450 monooxygenases (specialized in metabolism of toxins), making them susceptible to toxicant-induced injury and death. Naphthalene, an environmental toxin found in cigarette smoke and mothballs, only becomes toxic after its metabolism by cytochrome P450, making it a common experimental system for studying Clara cell injury.

Neuroendocrine cells have also been speculated as resident stem cells. Nevertheless, evidence suggests that these cells may not be real airway stem cells but rather that they participate in the microenvironment necessary for stem cell expansion [12]. These cells are primordially organized into a discrete innervated zone inside neuroepithelial bodies (NEBs) in the airway (Fig. 6.2). Through the action of a number of neuropeptides, NEBs seem to be involved in regulating embryonic lung growth and maturation [37]. Two populations of label-retaining cells have been identified within the NEBs: one expressing the pulmonary neuroendocrine cell (PNEC)-specific marker calcitonin gene-related peptide (CGRP) and one subpopulation of Clara cell secretory protein (CCSP)-expressing cells. These cells have been associated with epithelial regeneration after naphthalene-induced classical Clara-cell ablation. Evidence demonstrates that after selective ablation of CCSP-expressing cells, there is an increase in proliferative rate and hyperplasia of PNECs; however, it is not sufficient to effectively regenerate a normal epithelium [36]. This indicates that PNECs are not themselves capable of renewing CCSP-expressing cell-depleted airway epithelium, and that naphthalene-resistant CCSP-expressing cells of the NEBs have a critical function in airway renewal after transient-amplifying cell ablation. However, previous findings suggest that, during repair, CCSP-expressing cells that are close to NEBs in adult lung have a higher proliferative rate [36], and also that during development airway epithelial cell proliferation decreases as a function of increasing distance from the NEBs [38,39]. These data suggest that PNEC-derived paracrine factors might play a role in regulating epithelial cell differentiation and proliferation during fetal lung development and possibly in the normal or injured adult lung.

The term *bronchoalveolar stem cell (BASC)* reflects the ability of cultured CCSP/SP-C (SP-C, surfactant protein C) dual positive cells to express markers of Clara cells (CCSP positive (cells) and alveolar epithelial type 2 cells (SP-C positive cells), and to originate alveolar type 1 cells (Aquaporin 5 positive cells). These are rare cells (1 to 2 cells per bronchoalveolar duct junction) with cuboidal morphology. BASC do not express the cytochrome P450 enzyme CYP2F2, which converts naphthalene into its toxic dihydrodiol metabolites, and for this reason, survive injury. BASC can be identified by their expression of cell-surface protein CD34 and Sca-1 (both markers are ordinarily expressed in hematopoietic stem cells) and deprived expression of mature hematopoietic cell CD45 and platelet-endothelial cell adhesion molecule (PECAM, for hematopoietic and endothelial lineages exclusion) by FACS. Immunophenotyping of FACS-sorted populations using the CCSP and SP-C markers has determined that Sca-1$^+$, CD45$^-$, and PECAM$^-$ populations contain BASC. Following distal airway injury using naphthalene and alveolar injury using bleomycin, resistant bronchoalveolar stem cells are stimulated to proliferate and originate daughter cells capable of renewing the airway and alveolar epithelium [11,20].

Several other stem cell markers have been identified in cultured lung, including Oct-4$^+$, a robust marker of pluripotency for both mouse and human embryonic stem cells [34]. Using serum-free selection media, neonatal and adult lung-derived Oct-4$^+$ epithelial colony cells may be obtained. In addition to Oct-4, neonatal lung cells also express other stem cell markers such as stage-specific embryonic antigen 1

(SSEA-1), Sca-1, and CCSP, but not c-Kit, CD34, and p63, suggesting an alternative subpopulation of Clara cells that have been implicated as lung stem/progenitor cells. In summary, the detection of Oct-4^+ long-term BrdU label-retaining cells at the bronchoalveolar duct junction of neonatal lungs and the in vitro differentiation of cultured cells into alveolar epithelial type-2- and type-1-like cells suggest these Oct-4^+ cells as putative neonatal lung stem/progenitor cells.

Alveolar Epithelial Type 2 Cells

Two distinct types of epithelial cells are found in the gas-exchange distal air sacs (alveoli): alveolar epithelial type 1 cells (AEC1) and alveolar epithelial type 2 cells (AEC2). AEC1 function by transporting oxygen and carbon dioxide across the alveolar space to the capillary beds that surround the alveoli (Fig. 6.2). These large, flat cells are presumed to be terminally differentiated and therefore incapable of cell division [17]. The smaller, cuboidal AEC2 contain prominent cytoplasmic lamellar bodies and perform the differentiated function of surfactant synthesis and secretion. AEC2 are thought to function as progenitors of alveolar epithelium, based on their capacity to replenish themselves and give rise to terminally differentiated AEC1 [40-42]. This has been demonstrated in vivo, through serial progression of [H^3]-thymidine labeling from AEC2 into AEC1 cell nuclei after lung injury [43-45], and in vitro with culture of primary AEC2 giving rise to an AEC1-like phenotype [41,46-49]. However, some reports suggest that different subsets of AEC2 may exist, including E-cadherin-positive and negative fractions of these cells [19] and some heterogeneity of transgene expression in AEC2 in genetically engineered mice [50,51]; therefore, it is not clear whether these different kinds of AEC2 present distinct progenitor or other functional capacities. AEC2 populations isolated from control and hyperoxia-treated animals may be segregated functionally based on the surface expression of E-cadherin. E-cadherin-positive and -negative populations physically separated by antibody-coated magnetic beads present distinct damage susceptibility, proliferation rate, and telomerase activity. E-cadherin-negative subpopulations are seemingly damage resistant and proliferative and exhibit high levels of telomerase activity compared to E-cadherin-positive subpopulations [19]. E-cadherin-negative subpopulations of AEC2 appear to be responsible for renewing and repairing injured alveolar epithelium and may be the transient amplifying progenitor subpopulation of the alveolar compartment.

Resident Endothelium Progenitor Cells

Less information is available concerning potential progenitor cell populations contributing to other cell populations such as interstitial, smooth muscle, or endothelial cells in the lung. Nevertheless, recent research has shown that the pulmonary microvasculature is a rich endothelial progenitor niche [52,53]. Resident microvascular endothelial progenitor cells isolated from lung are highly proliferative cells with vasculogenic capacity in vitro (using Matrigel assays) and in vivo, forming ultrastructurally normal de novo vessels. These cells may apparently be identified by immunophenotyping through their expression of endothelial cell markers (CD31, CD144, endothelial nitric oxide synthase, and von Willebrand factor), progenitor cell markers (CD34 and vascular endothelial growth factor receptor-2), and lack of leukocyte marker CD45 [52]. However, there is in vitro evidence of at least two different subsets of endothelial progenitor cells in murine lung microvasculature: one that has the intrinsic capacity to differentiate into blood endothelial cells and another that generates lymphatic endothelial cells. This latter subpopulation seems to express lymph endothelial cell markers, including Lyve1, VEGFR-3, and podoplanin, whereas a blood endothelium progenitor subpopulation was identified through LA5 and MECA-32/PV-1, markers very recently described for mouse blood endothelial cells [53]. These data are compatible with the rich blood supply and vascularization of the lung, whose remodeling is triggered during inflammatory events, such as *Mycoplasma pulmonis* infection [54]. In addition, lung side population cells, which are thought to be resident lung precursor cells because of their epithelial and mesenchymal potential, apparently also have endothelial potential. Lung side populations can be sorted on the basis of their capacity to efflux Hoechst 33342 and through the expression of cell-surface proteins Flk-1 (a marker of both hemangioblast and mesenchymal lung cell precursors during development) and CD45 [55]. $CD45^{+}$/$Flk\text{-}1^{+}$ and $CD45^{-}$/$Flk\text{-}1^{+}$–expressing cells are able to bind isolectin B4 (a property specific to endothelial cells) and form networks in a Matrigel tube forming assay, indicating endothelial cell characteristics. However, this phenomenon is temporarily affected/impaired by hyperoxic conditions, resulting in a reduction in the number of lung side population cells, as well as diminished pulmonary alveolar and microvessel density [55]. Despite advances in detecting real lung resident endothelial progenitor cells, the exact location of these cell niches is still unknown, and reliable identification markers in the intact lung are still missing.

Resident Mesenchymal Progenitor Cells

Several differentiated mesenchymal cell types contained in the adult lung (cartilage, smooth muscle, and myofibroblasts) provide structural support for airways and vessels. The pathogenesis of a variety of lung diseases includes alterations in the number and phenotype of these cells. Although such cells appear to be renewed locally, progenitors have yet to be purified [56,57]. Different studies have tried to isolate mesenchymal progenitor cell populations from lung tissue employing immunohistochemistry, flow cytometry-based procedures, and in vitro differentiation potential [58-61]. One of the first studies to isolate a putative mesenchymal progenitor cell population in lung used Hoechst-effluxing cells separated by a flow cytometry-based procedure. These Hoechst-effluxing cells are negative for pan-hematopoietic ($CD45^-$) and endothelial ($CD31^-$) cell-surface markers and positive for ordinary mesenchymal progenitor cell population markers (Sca-1, CD106, and CD44). Additionally, they are able to differentiate in vitro (co-culture with mesenchymal stem cell) into smooth muscle, cartilage, bone, and fat [59]. $CD31^-/CD45^-/Sca\text{-}1^+$ cells have been pointed to as markers for identification of bronchoalveolar stem cell candidates at the bronchoalveolar duct junction that also co-express surfactant protein C and the Clara cell specific protein [11]. However, a recent report has demonstrated that $CD45^-/CD31^-/Sca\text{-}1^+/CD34^+$ cells also co-express immunophenotypic markers that define lung fibroblastic rather than epithelial cells, such as thymus cell antigen 1 (Thy-1) and platelet-derived growth factor receptor α [60]. Indeed, the culture of $CD45^-/CD31^-/Sca\text{-}1^+$ cells in serum-supplemented (20%) modified Eagle medium may induce an increased proliferation rate of fibroblast-like cells (up to 20%), whereas epithelial progenitors growth from $CD45^-/CD31^-/Sca\text{-}1^+$ cells is very low in either serum-supplemented or serum-free culture [60]. Although $CD45^-/CD31^-/Sca\text{-}1^+/CD34^+$ represent a good pool of markers for obtaining endogenous clonogenic progenitor cells in the adult lung, they seem to exhibit more similarity for mesenchymal than epithelial cell lineages.

Interestingly, recent reports suggest that progenitor cells from human adult lung tissue isolated and expanded with multicellular spheroid methodology, which already present mixed phenotype of cells with alveolar type 2 and Clara cell features, also present mesenchymal features like cytoplasmic collections of vimentin-like intermediate filaments, well-developed tight junctions, and expression of mesenchymal stem cell antigens CD90 and CD105 [62]. When the *Slug* gene is silenced, which has a pivotal role in the process that governs transition from mesenchyme to epithelium during embryogenesis, mRNA of several epithelial markers (cytokeratin 8, cytokeratin 18, thyroid transcription factor-1, surfactant protein-A) is up-regulated, whereas stemness markers (OCT 3/4, CD133, lymphoid enhancer-binding factor-1) are down-regulated, indicating that *Slug* down-regulation accompanies the differentiation process of lung stem/progenitor cells.

Detecting exact locations of resident stem cell niches in solid organs, such as lung, is not a simple task. A recent study has suggested a putative niche for mesenchymal progenitor cells in perivascular cells of the lung parenchyma, demonstrated by immunohistology showing co-expression of CD31 and Sca-1 antigens in the adult lung [60]. Nevertheless, more studies are needed in order to confirm this and other possible resident mesenchymal stem cell niches.

Lung Resident Stem Cells in Respiratory Diseases

Impaired airway epithelial repair has been suggested as a precocious event at the onset of chronic pulmonary diseases. The perpetuation of defects related to regenerative capacity and cellular composition may contribute to disease progression and exacerbation [63]. This implies that normal functions of facultative progenitor and stem cells may be compromised.

Reports have suggested that PNECs and their amine/peptide mediators could be involved in the pathogenesis of cystic fibrosis (CF) [64,65], because expression of the cystic fibrosis transmembrane conductance regulator (CFTR), the defective protein in CF, has been identified in NEB cells of rabbit lung [64] and in human tumor cell lines of PNEC in normal lungs [65]. Although no conclusive data have been generated correlating PNEC and CF development, CFTR-null mice do present widespread alterations in airway NEB cells, suggesting that endogenous airway progenitor cell pathways in CF lungs may be altered, contributing to lung dysfunction in CF patients.

Chronic diseases like chronic obstructive pulmonary disease (COPD) and asthma are characterized by remarkable epithelial and mesenchymal remodeling, which also involves decreased numbers of Clara cells [66]. There is evidence that epithelial repair can be severely affected by environment. In vitro studies have demonstrated that cigarette smoke compromises epithelial wound repair by decreasing the rate of cell proliferation and migration and altering extracellular matrix remodeling [67,68]. This can also be observed in in vivo models: prolonged cigarette smoke exposure in mice has been shown to diminish epithelial repair after pulmonary injury induced by naphthalene. Recently, lung-derived mesenchymal stem cells have been shown to be an option for treatment of chronic lung injuries such as emphysema [61]. Autologous transplantation of multipotent pulmonary cells was able of increase extracellular matrix content (collagen and elastin) and lung perfusion, suggesting tissue healing and regeneration after elastase-induced damage [61]. However, more research is necessary in order to identify the real benefits of lung-derived stem/progenitor cells in health and disease.

Main Challenges for the Identification of Lung Resident Stem Cells

In the absence of markers that are uniquely expressed by lung stem cells (Table 6.1), researchers have developed multiparameter cell separation strategies for the isolation and identification of lung stem cells. For that, phenotypic features that typify known stem cell populations have been used, such as antibodies against various cell surface antigens, including Sca-1, CD34, and E-cadherin [28,69-72], or exploration of

TABLE 6.1 Surface Markers Expressed in Putative Pulmonary Resident Stem Cells

	Basal	Clara	PNEC	BASC	Other Variant-Clara	AEC2	EPC	MPC	Multipotent
AQP3	+								
AQP5									
CCSP		+	+	+	+				
CD31								−	−
CD34				+	−		+	+	
CD44								+	
CD45	−			−			−	−	−
CD90								+	
C105								+	
CD106								+	
CD144							+		
CD151	+								
CGRP			+						
CK5	+								
CK19									
c-Kit					−				

E-cadherin	+			+/−			
eNOS					+		
LA5					+		
Lyvel					+		
MECA32/PV1					+		
Oct-4			+				
PDGFRα						+	
PECAM		−					
Sca1		+	+			+	+
SP-C		+		+			
SSEA-1			+				
Thy-1						+	
VEGFR2					+		
VEGFR3					+		
ZO-1	+						

PENC, pulmonary neuroendocrine cell; BASC, bronchoalveolar stem cell; AEC2, alveolar epithelial type 2 cells; EPC, endothelial progenitor cell; MPC, mesenchymal progenitor cell; AQ, aquaporin; CGRP, calcitonin gene-related peptide; CK, cytokeratin; E-cadherin, epithelial adherens junction protein; c-Kit, proto-oncogene highly expressed by stem cells; eNOS, endothelial nitric oxide synthase; LA5, monoclonal antibody for blood vessels; LYEL, specific receptor for hyaluronic acid expressed by lymphatic vessel cells; MECA32/PV1, monoclonal antibody for blood vessels; Oct-4, octamer-binding transcription factor; PDGFRα, platelet-derived growth factor receptor α; PECAM, platelet-endothelial cell adhesion molecule; Sca1, stem cell antigen 1; SP-C, surfactant protein C; SSEA-1, stage-specific embryonic antigen 1; Thy-1, thymus cells antigen 1; VEGFR, vascular endothelial growth factor receptor; ZO-1, tight junction protein.

activity of highly conserved ATP-dependent cell membrane transport proteins that confer the ability to efflux vital dyes, like Hoechst 33342, to characterize pulmonary side population phenotypes [59,73]. These approaches have proved limited in complex organs such as the lung because of the variable expression of markers by cells, depending on their developmental potential, activation state, and maturational age within and among different cell lineages (Sca-1 may be expressed on rare bronchoalveolar stem cells, endothelium of pulmonary arteries, veins, and capillaries, whereas c-kit and CD34 are similarly expressed in subtypes of leukocytes and endothelial cells). Another strategy, adapted from bone marrow–derived stem cell identification, is to isolate multipotent populations of lung cells for in vitro assays designed to test multipotency [74]. Well-defined culture conditions that have been developed to display the potential of bone marrow mesenchymal stem cells to self-renew or differentiate into bone, cartilage, and fat were directly transposed for putative lung stem cell identification [61,62]. So far, however, no standard protocol for the isolation and characterization of such cells has been developed.

References

[1] Warburton D, Lee MK. Current concepts on lung development. Curr Opin Pediatr 1999; 11(3):188–92.

[2] Hogg JC, Chu F, Utokaparch S, Woods R, Elliott WM, Buzatu L, et al. The nature of small-airway obstruction in chronic obstructive pulmonary disease. N Engl J Med 2004; 350(26):2645–53.

[3] Uhal BD, Joshi I, True AL, Mundle S, Raza A, Pardo A, et al. Fibroblasts isolated after fibrotic lung injury induce apoptosis of alveolar epithelial cells in vitro. Am J Physiol 1995; 269(6 Pt 1):L819–28.

[4] Yokohori N, Aoshiba K, Nagai A. Increased levels of cell death and proliferation in alveolar wall cells in patients with pulmonary emphysema. Chest 2004;125(2):626–32.

[5] Blenkinsopp WK. Proliferation of respiratory tract epithelium in the rat. Exp Cell Res 1967;46(1):144–54.

[6] Rawlins EL, Hogan BL. Epithelial stem cells of the lung: privileged few or opportunities for many? Development 2006;133(13):2455–65.

[7] Reynolds SD, Giangreco A, Power JH, Stripp BR. Neuroepithelial bodies of pulmonary airways serve as a reservoir of progenitor cells capable of epithelial regeneration. Am J Pathol 2000;156(1):269–78.

[8] Reynolds SD, Hong KU, Giangreco A, Mango GW, Guron C, Morimoto Y, et al. Conditional Clara cell ablation reveals a self-renewing progenitor function of pulmonary neuroendocrine cells. Am J Physiol Lung Cell Mol Physiol 2000;278(6):L1256–63.

[9] Engelhardt JF. Stem cell niches in the mouse airway. Am J Respir Cell Mol Biol 2001;24(6):649–52.

[10] Giangreco A, Reynolds SD, Stripp BR. Terminal bronchioles harbor a unique airway stem cell population that localizes to the bronchoalveolar duct junction. Am J Pathol 2002;161(1): 173–82.

[11] Kim CF, Jackson EL, Woolfenden AE, Lawrence S, Babar I, Vogel S, et al. Identification of bronchioalveolar stem cells in normal lung and lung cancer. Cell 2005;121(6):823–35.

[12] Borthwick DW, Shahbazian M, Krantz QT, Dorin JR, Randell SH. Evidence for stem-cell niches in the tracheal epithelium. Am J Respir Cell Mol Biol 2001;24(6):662–70.

[13] Otto WR. Lung epithelial stem cells. J Pathol 2002;197(4):527–35.

[14] Hong KU, Reynolds SD, Watkins S, Fuchs E, Stripp BR. Basal cells are a multipotent progenitor capable of renewing the bronchial epithelium. Am J Pathol 2004;164(2):577–88.

[15] Hong KU, Reynolds SD, Watkins S, Fuchs E, Stripp BR. In vivo differentiation potential of tracheal basal cells: evidence for multipotent and unipotent subpopulations. Am J Physiol Lung Cell Mol Physiol 2004;286(4):L643–9.

[16] Schoch KG, Lori A, Burns KA, Eldred T, Olsen JC, Randell SH. A subset of mouse tracheal epithelial basal cells generates large colonies in vitro. Am J Physiol Lung Cell Mol Physiol 2004;286(4):L631–42.

[17] Adamson IY, Bowden DH. The type 2 cell as progenitor of alveolar epithelial regeneration. A cytodynamic study in mice after exposure to oxygen. Lab Invest 1974;30(1):35–42.

[18] Adamson IY, Bowden DH. Derivation of type 1 epithelium from type 2 cells in the developing rat lung. Lab Invest 1975;32(6):736–45.

[19] Reddy R, Buckley S, Doerken M, Barsky L, Weinberg K, Anderson KD, et al. Isolation of a putative progenitor subpopulation of alveolar epithelial type 2 cells. Am J Physiol Lung Cell Mol Physiol 2004;286(4):L658–67.

[20] Nolen-Walston RD, Kim CF, Mazan MR, Ingenito EP, Gruntman AM, Tsai L, et al. Cellular kinetics and modeling of bronchioalveolar stem cell response during lung regeneration. Am J Physiol Lung Cell Mol Physiol 2008;294(6):L1158–65.

[21] Evans MJ, Cabral-Anderson LJ, Freeman G. Role of the Clara cell in renewal of the bronchiolar epithelium. Lab Invest 1978;38(6):648–53.

[22] Evans MJ, Shami SG, Cabral-Anderson LJ, Dekker NP. Role of nonciliated cells in renewal of the bronchial epithelium of rats exposed to NO2. Am J Pathol 1986;123(1):126–33.

[23] Evans MJ, Johnson LV, Stephens RJ, Freeman G. Renewal of the terminal bronchiolar epithelium in the rat following exposure to NO2 or O3. Lab Invest 1976;35(3):246–57.

[24] Avril-Delplanque A, Casal I, Castillon N, Hinnrasky J, Puchelle E, Peault B. Aquaporin-3 expression in human fetal airway epithelial progenitor cells. Stem Cells 2005;23(7):992–1001.

[25] Chan SW, Blackburn EH. New ways not to make ends meet: telomerase, DNA damage proteins and heterochromatin. Oncogene 2002;21(4):553–63.

[26] Randell SH, Comment CE, Ramaekers FC, Nettesheim P. Properties of rat tracheal epithelial cells separated based on expression of cell surface alpha-galactosyl end groups. Am J Respir Cell Mol Biol 1991;4(6):544–54.

[27] Liu JY, Nettesheim P, Randell SH. Growth and differentiation of tracheal epithelial progenitor cells. Am J Physiol 1994;266(3 Pt 1):L296–307.

[28] Asakura A, Rudnicki MA. Side population cells from diverse adult tissues are capable of in vitro hematopoietic differentiation. Exp Hematol 2002;30(11):1339–45.

[29] Hackett TL, Shaheen F, Johnson A, Wadsworth S, Pechkovsky DV, Jacoby DB, et al. Characterization of side population cells from human airway epithelium. Stem Cells 2008; 26(10):2576–85.

[30] Hajj R, Baranek T, Le Naour R, Lesimple P, Puchelle E, Coraux C. Basal cells of the human adult airway surface epithelium retain transit-amplifying cell properties. Stem Cells 2007; 25(1):139–48.

[31] Hegab AE, Kubo H, Fujino N, Suzuki T, He M, Kato H, et al. Isolation and characterization of murine multipotent lung stem cells. Stem Cells Dev 2010;19(4):523–36.

[32] Johnson NF, Hubbs AF. Epithelial progenitor cells in the rat trachea. Am J Respir Cell Mol Biol 1990;3(6):579–85.

[33] Ford JR, Terzaghi-Howe M. Basal cells are the progenitors of primary tracheal epithelial cell cultures. Exp Cell Res 1992;198(1):69–77.
[34] Ling TY, Kuo MD, Li CL, Yu AL, Huang YH, Wu TJ, et al. Identification of pulmonary Oct-4+ stem/progenitor cells and demonstration of their susceptibility to SARS coronavirus (SARS-CoV) infection in vitro. Proc Natl Acad Sci USA 2006;103(25):9530–5.
[35] Peake JL, Reynolds SD, Stripp BR, Stephens KE, Pinkerton KE. Alteration of pulmonary neuroendocrine cells during epithelial repair of naphthalene-induced airway injury. Am J Pathol 2000;156(1):279–86.
[36] Hong KU, Reynolds SD, Giangreco A, Hurley CM, Stripp BR. Clara cell secretory protein-expressing cells of the airway neuroepithelial body microenvironment include a label-retaining subset and are critical for epithelial renewal after progenitor cell depletion. Am J Respir Cell Mol Biol 2001;24(6):671–81.
[37] Van Lommel A, Bolle T, Fannes W, Lauweryns JM. The pulmonary neuroendocrine system: the past decade. Arch Histol Cytol 1999;62(1):1–16.
[38] Hoyt Jr RF, McNelly NA, McDowell EM, Sorokin SP. Neuroepithelial bodies stimulate proliferation of airway epithelium in fetal hamster lung. Am J Physiol 1991;260(4 Pt 1):L234–40.
[39] Hoyt Jr RF, Sorokin SP, McDowell EM, McNelly NA. Neuroepithelial bodies and growth of the airway epithelium in developing hamster lung. Anat Rec 1993;236(1):15–22; discussion 22–14.
[40] Mason RJ, Williams MC. Type II alveolar cell. Defender of the alveolus. Am Rev Respir Dis 1977;115(6 Pt 2):81–91.
[41] Brody JS, Williams MC. Pulmonary alveolar epithelial cell differentiation. Annu Rev Physiol 1992;54:351–71.
[42] Rawlins EL. Lung epithelial progenitor cells: lessons from development. Proc Am Thorac Soc 2008;5(6):675–81.
[43] Evans MJ, Cabral LJ, Stephens RJ, Freeman G. Renewal of alveolar epithelium in the rat following exposure to NO2. Am J Pathol 1973;70(2):175–98.
[44] Evans MJ, Cabral LJ, Stephens RJ, Freeman G. Transformation of alveolar type 2 cells to type 1 cells following exposure to NO2. Exp Mol Pathol 1975;22(1):142–50.
[45] Adamson IY, Bowden DH. Bleomycin-induced injury and metaplasia of alveolar type 2 cells. Relationship of cellular responses to drug presence in the lung. Am J Pathol 1979;96(2):531–44.
[46] Mason R, Williams MC, Clements JA. Isolation and identification of type 2 alveolar epithelial cells. Chest 1975;67(Suppl. 2):36S–7S.
[47] Dobbs LG, Williams MC, Brandt AE. Changes in biochemical characteristics and pattern of lectin binding of alveolar type II cells with time in culture. Biochim Biophys Acta 1985; 846(1):155–66.
[48] Danto SI, Zabski SM, Crandall ED. Reactivity of alveolar epithelial cells in primary culture with type I cell monoclonal antibodies. Am J Respir Cell Mol Biol 1992;6(3):296–306.
[49] Danto SI, Shannon JM, Borok Z, Zabski SM, Crandall ED. Reversible transdifferentiation of alveolar epithelial cells. Am J Respir Cell Mol Biol 1995;12(5):497–502.
[50] Roper JM, Staversky RJ, Finkelstein JN, Keng PC, O'Reilly MA. Identification and isolation of mouse type II cells on the basis of intrinsic expression of enhanced green fluorescent protein. Am J Physiol Lung Cell Mol Physiol 2003;285(3):L691–700.
[51] Perl AK, Wert SE, Loudy DE, Shan Z, Blair PA, Whitsett JA. Conditional recombination reveals distinct subsets of epithelial cells in trachea, bronchi, and alveoli. Am J Respir Cell Mol Biol 2005;33(5):455–62.
[52] Alvarez DF, Huang L, King JA, ElZarrad MK, Yoder MC, Stevens T. Lung microvascular endothelium is enriched with progenitor cells that exhibit vasculogenic capacity. Am J Physiol Lung Cell Mol Physiol 2008;294(3):L419–30.

[53] Schniedermann J, Rennecke M, Buttler K, Richter G, Stadtler AM, Norgall S, et al. Mouse lung contains endothelial progenitors with high capacity to form blood and lymphatic vessels. BMC Cell Biol 2010;11:50.
[54] Baluk P, Tammela T, Ator E, Lyubynska N, Achen MG, Hicklin DJ, et al. Pathogenesis of persistent lymphatic vessel hyperplasia in chronic airway inflammation. J Clin Invest 2005; 115(2):247–57.
[55] Irwin D, Helm K, Campbell N, Imamura M, Fagan K, Harral J, et al. Neonatal lung side population cells demonstrate endothelial potential and are altered in response to hyperoxia-induced lung simplification. Am J Physiol Lung Cell Mol Physiol 2007;293(4):L941–51.
[56] Sabatini F, Petecchia L, Tavian M, Jodon de Villeroche V, Rossi GA, Brouty-Boye D. Human bronchial fibroblasts exhibit a mesenchymal stem cell phenotype and multilineage differentiating potentialities. Lab Invest 2005;85(8):962–71.
[57] da Silva Meirelles L, Chagastelles PC, Nardi NB. Mesenchymal stem cells reside in virtually all post-natal organs and tissues. J Cell Sci 2006;119(Pt 11):2204–13.
[58] Lama VN, Smith L, Badri L, Flint A, Andrei AC, Murray S, et al. Evidence for tissue-resident mesenchymal stem cells in human adult lung from studies of transplanted allografts. J Clin Invest 2007;117(4):989–96.
[59] Summer R, Fitzsimmons K, Dwyer D, Murphy J, Fine A. Isolation of an adult mouse lung mesenchymal progenitor cell population. Am J Respir Cell Mol Biol 2007;37(2):152–9.
[60] McQualter JL, Brouard N, Williams B, Baird BN, Sims-Lucas S, Yuen K, et al. Endogenous fibroblastic progenitor cells in the adult mouse lung are highly enriched in the sca-1 positive cell fraction. Stem Cells 2009;27(3):623–33.
[61] Ingenito EP, Tsai L, Murthy S, Tyagi S, Mazan M, Hoffman A. Autologous lung-derived mesenchymal stem cell transplantation in experimental emphysema. Cell Transplant 2011.
[62] Tesei A, Zoli W, Arienti C, Storci G, Granato AM, Pasquinelli G, et al. Isolation of stem/progenitor cells from normal lung tissue of adult humans. Cell Prolif 2009;42(3):298–308.
[63] Holgate ST, Holloway J, Wilson S, Bucchieri F, Puddicombe S, Davies DE. Epithelial-mesenchymal communication in the pathogenesis of chronic asthma. Proc Am Thorac Soc 2004; 1(2):93–8.
[64] Yeger H, Pan J, Fu XW, Bear C, Cutz E. Expression of CFTR and Cl(-) conductances in cells of pulmonary neuroepithelial bodies. Am J Physiol Lung Cell Mol Physiol 2001;281(3): L713–21.
[65] Pan J, Bear C, Farragher S, Cutz E, Yeger H. Cystic fibrosis transmembrane conductance regulator modulates neurosecretory function in pulmonary neuroendocrine cell-related tumor cell line models. Am J Respir Cell Mol Biol 2002;27(5):553–60.
[66] Lumsden AB, McLean A, Lamb D. Goblet and Clara cells of human distal airways: evidence for smoking induced changes in their numbers. Thorax 1984;39(11):844–9.
[67] Wang H, Liu X, Umino T, Skold CM, Zhu Y, Kohyama T, et al. Cigarette smoke inhibits human bronchial epithelial cell repair processes. Am J Respir Cell Mol Biol 2001; 25(6):772–9.
[68] Schweitzer KS, Johnstone BH, Garrison J, Rush NI, Cooper S, Traktuev DO, et al. Adipose stem cell treatment in mice attenuates lung and systemic injury induced by cigarette smoking. Am J Respir Crit Care Med 2011;183(2):215–25.
[69] Quaini F, Urbanek K, Beltrami AP, Finato N, Beltrami CA, Nadal-Ginard B, et al. Chimerism of the transplanted heart. N Engl J Med 2002;346(1):5–15.
[70] Tamaki T, Akatsuka A, Ando K, Nakamura Y, Matsuzawa H, Hotta T, et al. Identification of myogenic-endothelial progenitor cells in the interstitial spaces of skeletal muscle. J Cell Biol 2002;157(4):571–7.

[71] Welm BE, Tepera SB, Venezia T, Graubert TA, Rosen JM, Goodell MA. Sca-1(pos) cells in the mouse mammary gland represent an enriched progenitor cell population. Dev Biol 2002; 245(1):42–56.

[72] Petersen BE, Grossbard B, Hatch H, Pi L, Deng J, Scott EW. Mouse A6-positive hepatic oval cells also express several hematopoietic stem cell markers. Hepatology 2003;37(3):632–40.

[73] Summer R, Kotton DN, Sun X, Fitzsimmons K, Fine A. Translational physiology: origin and phenotype of lung side population cells. Am J Physiol Lung Cell Mol Physiol 2004; 287(3):L477–83.

[74] Kajstura J, Rota M, Hall SR, Hosoda T, D'Amario D, Sanada F, et al. Evidence for human lung stem cells. N Engl J Med 2011;364(19):1795–806.

Chapter 7

Skeletal Resident Stem Cells

Alex Balduino*, Maria Eugenia Leite Duarte[†,‡], Russell S. Taichman[§]

*Coordenação de Pesquisa, Universidade Veiga de Almeida (UVA), Rio de Janeiro, Brazil

[†] Centro de Pesquisa em Terapia Celular e Bioengenharia Ortopédica (CTCel), Instituto Nacional de Traumatologia e Ortopedia (INTO), Rio de Janeiro, Brazil

[‡] Instituto de Ciências Biomédicas (ICB), Universidade Federal do Rio de Janeiro (UFRJ), Rio de Janeiro, Brazil

[§] Department of Periodontics and Oral Medicine, University of Michigan, Ann Arbor, Michigan, USA

Bone Tissue

Bone is a specialized type of connective tissue, whose extracellular matrix is embedded with chemical variations of hydroxyapatite crystals deposited through the process of mineralization. The presence of hydroxyapatite crystals gives bone the characteristic of a hard tissue, with limited flexibility, and the ability to play proper roles in sustaining and protecting organs and tissues. Besides working as a real body scaffold, bone tissue plays a significant role in calcium and phosphate metabolism, serving as a major reservoir for both minerals. Macroscopically, two types of bone tissue can be identified: compact bone and spongy, or trabecular bone. Compact bone is thick and has a solid appearance, is located in the outer portion of the bone, and plays an important role in the mechanical support ability of the tissue. Trabecular bone, in turn, is formed by a network of bone trabeculae or spicules, which defines and sustains the bone marrow. Although they are anatomically distinct and occupy different regions in the bones, trabecular and compact bone tissues have the same extracellular matrix components.

Resident Stem Cells and Regenerative Therapy. http://dx.doi.org/10.1016/B978-0-12-416012-5.00007-4

The bone matrix is comprised of an organic fraction and an inorganic fraction. The inorganic fraction corresponds to approximately 65% of the dry mass of bone tissue and mainly consists of hydroxyapatite crystals of different compositions and chemical structures, in which the OH^- ion can be replaced by fluoride, chloride, or carbonate, giving rise to fluorapatite, chlorapatite, and carboapatite. These crystals are deposited on the lacunar regions of collagen fibrils and overlapping regions. The organic fraction is represented by approximately 35% of the dry mass of bone tissue, of which 80% to 90% consists of collagen type I. The remaining 10% to 20% is composed of noncollagen proteins such as proteoglycans, particularly aggrecan, and the glycoproteins osteocalcin, osteopontin, and bone sialoprotein. The limited flexibility and supportive function depends on the molecular structure of type I collagen and its distribution and organization in the bone matrix.

Bone tissue can also be classified as woven bone, or primary bone tissue, and lamellar bone, or secondary bone tissue, depending on the matrix organization. Woven bone is usually found during development or in new formed nonremodeled bones, and its main feature is a disorganized matrix. Lamellar bone matrix, in turn, is well organized into lamellae, which guide osteocyte distribution in the tissue. Most of the compact bone tissue is formed by haversian systems (or osteons), each of which is composed of cylinders of lamellae concentrically arranged around a channel through which small-caliber blood vessels and nerves nurture the tissue. The haversian systems communicate with each other through the canals of Volkmann arranged perpendicular to the axis of the channel. On the outer surface, bones are covered by the periosteum, a structure consisting of several cell layers, rich in blood vessels and nerves, in which osteoblasts compose the most basal layer and maintain direct contact to the bone surface. Bone inner surface (marrow cavity and haversian canals) is covered by the endosteum, which, different from periosteum, is composed by a single layer of active or dormant (quiescent) osteoblasts (Fig. 7.1).

The three main cell types that participate in bone formation and maintenance are the osteoblasts, osteocytes, and osteoclasts. Osteocytes are considered the mature end point cell of the osteogenic lineage cascade. They are trapped in the mineralized matrix in a space called a lacuna, and they appear to have low activity on producing and secreting bone matrix components. Osteocytes communicate to each other via long cytoplasmic extensions through tiny canals termed canaliculi, in order to exchange nutrients, communicate, and eliminate waste products. Although they are not thought to undergo mitosis, there is evidence that they are actively involved in bone turnover and remodeling because of their role as mechanosensory cells.

Osteoblasts are bone-producing cells. These cells synthesize and secrete collagen fibers and other noncollagen proteins and begin the process of mineralization of bone matrix. They are found primarily in the endosteal and periosteal surfaces. During bone formation, about 10% of osteoblasts are trapped in the extracellular matrix and

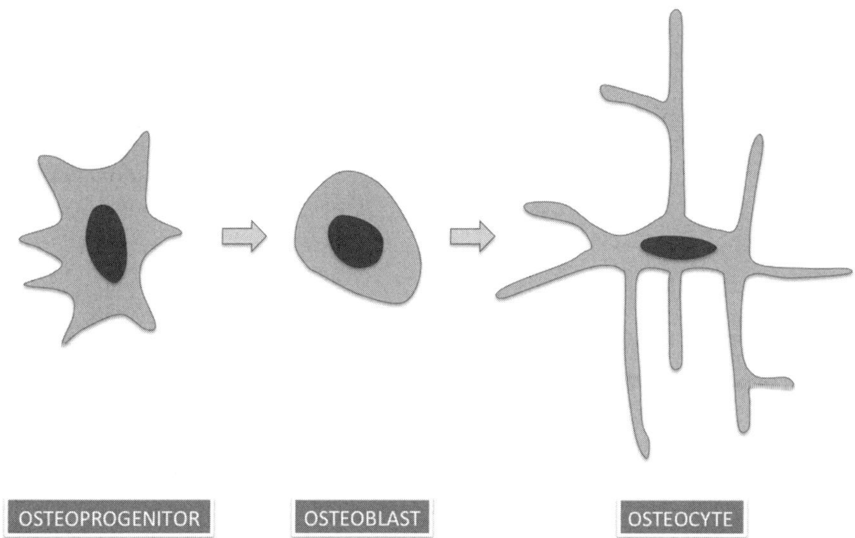

Figure 7.1 Osteogenic differentiation cascade.

give rise to osteocytes. It is not known exactly how this selection occurs or which markers these cells may express in order to be driven to the terminal differentiation. When fully engaged in matrix synthesis, osteoblasts appear histologically as having a cuboidal morphology. When in an inactive state, osteoblasts present a flat or fusiform shape until they are re-recruited to activity.

Osteoclasts are the third specialized cell type found in bone tissue and are responsible for bone resorption. These cells have the ability to dissolve hydroxyapatite crystals and to digest the organic matrix. A crucial balance in activities between osteoblasts and osteoclasts is responsible for maintaining bone homeostasis, but when lost or disrupted it results in conditions or diseases including osteoporosis and osteopetrosis.

Osteoblasts and osteocytes derive from a common precursor, identified as a mesenchymal stem cell, also known as a stromal stem cell or skeletal stem cell. Induced by the activity of the bone morphogenic proteins (BMP) and other agents, MSCs commit to the osteogenic lineage by first differentiating into osteoprogenitors, which ultimatately give rise to osteoblasts and osteocytes (Fig. 7.2). Each step of the osteogenic differentiation cascade can be identified by the expression of markers such as alkaline phosphatase (ALP), bone sialoprotein (BSP), and osteocalcin (OTC), and transcription factors, such as Osterix and CBFA-1.

Osteoclasts, are derived from the hematopoietic lineage (Fig. 7.3), as a result of the fusion of monocytes stimulated by M-CSF and RANKL factors. In some cases it has been reported that up to 50 monocytes may merge and give rise to a single

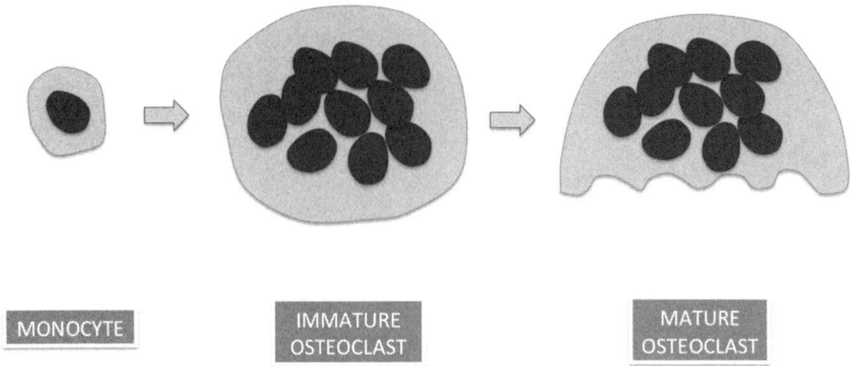

Figure 7.2 Osteoclast differentiation cascade.

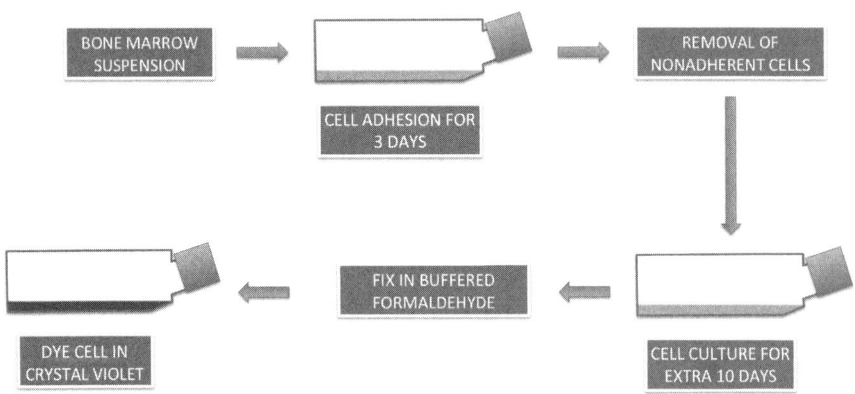

Figure 7.3 CFU-F assay used to quantify mesenchymal stem/progenitor cells.

osteoclast. However, in most cases, osteoclasts are formed by the fusion of approximately 10 monocytes.

Bone Marrow as a Source of Mesenchymal Stem Cells–Stem Cells for Bones

During vertebrate development, the hematopoietic system specifically colonizes different anatomic sites, which provide conditions for expansion, maintenance, and differentiation of blood cells [1,2]. Near birth, the system is established definitively in the bone marrow, which is the main site of blood cell production and maintenance of hematopoietic stem cells in adulthood [3-5]. In the postnatal bone marrow microenvironment, hematopoietic cells coexist and interact with stromal cells, responsible for the physical and biologic support of blood cell production in the marrow cavity. Stromal cells—osteoblasts, adipocytes, reticular cells, and endothelial cells—are distributed in an organized manner within the marrow cavity and participate in the formation of distinct niches, which will harbor hematopoietic cells at different stages of differentiation and contribute to the homeostasis of the hematopoietic system [6-9].

Osteoblasts, adipocytes, reticular cells chondrocytes, and smooth muscle cells [10-13] derive from a common progenitor, the mesenchymal stem cell. Friedenstein and Owen [14-16] performed the first studies in vitro and in vivo to characterize mesenchymal stem/progenitor cells from adult bone marrow. Using low-density bone marrow cell cultures, they demonstrated that stromal cells quickly adhered to the culture flask and were easily separated from hematopoietic cells (nonadherent fraction). Under proper conditions, some of these cells formed colonies of fibroblastoid cells, each colony derived from a single cell clone, a single precursor, termed a *colony forming unit–fibroblast (CFU-F)*. Linked to the observation of the potential to differentiate into cells of the musculoskeletal tissues, the term CFU-F is often used to define bone marrow derived mesenchymal stem cells [17].

Although stem cells are recognized by their differentiation potential, what really defines them is the ability for self-renewal. Unlike other systems of adult stem cells, the description of the in vitro and in vivo self-renewal process undergone by mesenchymal stem cells isolated from bone marrow was not clear until recently. Sacchetti and colleagues [18] have shown by in vitro assays and in vivo grafts that some mesenchymal cells from adult human bone marrow, capable of forming fibroblastoid colonies and expressing MCAM or CD146 (a surface marker present in mesenchymal stem cells with high differentiation potential), are pluripotent and capable of self-renewal, indicating the existence and characterization of a legitimate mesenchymal stem cell. Recently, several authors have suggested that mesenchymal stem cells could also follow a nonorthodox differentiation cascade and that these cells are capable of differentiating into cardiac muscle cells [19], skeletal muscle [20-22], hepatocytes, glial cells, and neurons [23].

Unlike the hematopoietic system, the mesenchymal stem cells and their committed progenitors do not have their function and phenotype clearly defined. According to the criteria proposed by the International Society for Cellular Therapy [24], the mesenchymal stem cells should be adherent, capable of differentiating into osteoblasts, chondrocytes and adipocytes, and must be positive for the markers CD73, CD90, and CD105 and negative for the markers CD14, CD19, CD34, CD45, and HLA-DR.

Simmons and Torok-Storb [25] developed an antibody to STRO-1, which identifies all colony-forming units (CFU-F) from the human bone marrow. It has also been used to identify a perivascular niche as a preferential niche of the mesenchymal stem/progenitor cells [26]. Nonetheless, STRO-1 is not a specific marker of mesenchymal stem/progenitor cells, and its expression is lost during the expansion phase in vitro [27]. Other markers have also been used to identify mesenchymal progenitors, such as CD13, CD73, CD90, and DC105 [28-30]. However, similar to STRO-1 expression, all mesenchymal stem/progenitor cells express the adhesion molecule associated with melanoma (MCAM/CD146) [26]. Its expression in vivo is also detected in cells located in the perivascular region of the sinusoids. However, unlike STRO-1, CD146 expression is maintained in vitro in cells that do not differentiate, even after expansion [18].

The CFU-F assay (Fig. 7.4) is the most widely used to quantify mesenchymal stem/progenitor cells from bone marrow and several other tissues; it is important to note that CFU-F colonies are very heterogenous [31,32]. Each colony has a distinct potential, both in vitro and in vivo, which suggests each "forming colony unit" is generated by a progenitor in a different stage of the differentiation cascade hierarchy [31-33], suggesting that not all of them are uncommitted primitive stem cells.

Muraglia and colleagues [32] observed that the pluripotent stem cells capable of generating osteoblasts, chondrocytes, and adipocytes give rise to preadipocytes and bipotent progenitors, which will give rise to osteoblasts and chondrocytes. Osteo-adipo and chondro-adipo bipotent progenitors were not observed.

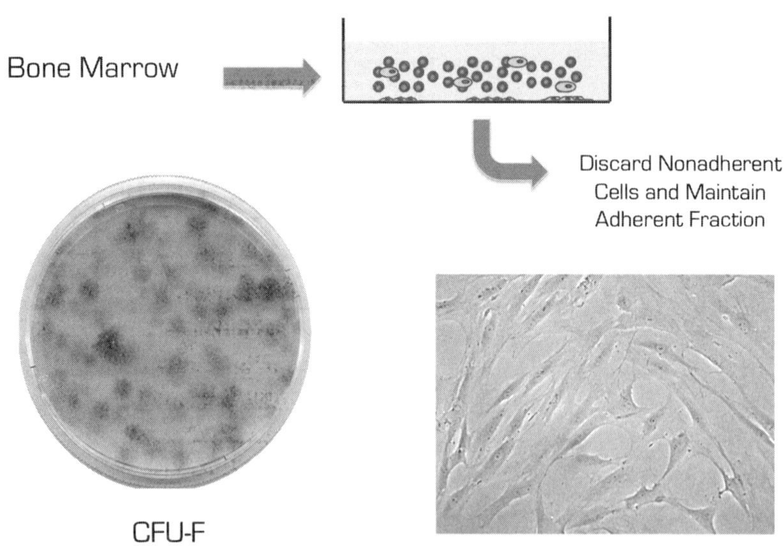

Figure 7.4 CFU-F colonies (Crystal violet stain). **See Plate 13.**

Notwithstanding, these tests were performed only in vitro and with cells expanded for long periods, which may truly not reflect normal physiology. In this regard, it is important to consider that the potential observed in vitro does not always correspond to what is observed in vivo [31].

Satomura and colleagues [31] showed that, regardless of the multipotency of mesenchymal cells derived from bone marrow, they all constitutively expressed CBFA-1, which is a transcription factor considered a master gene for the osteogenic lineage [34], suggesting that many if not most of mesenchymal stem/progenitor cells are programmed to become cells associated with the osteogenic lineage. Following the same idea, Sacchetti and colleagues [18] demonstrated that, on average, half of the colonies CD146+ in a CFU-F assay are precommitted to the osteogenic lineage, and that, when implanted in vivo, they give rise only to compact ossicles with no bone marrow. The heterogeneity of the populations obtained by current isolation methods may obscure what the identity of the true stem cell is and therefore at present probably needs to be defined functionally. Yet when used in tissue regeneration, the heterogeneity of the precursor populations may not be a disadvantage; tissue recombination in postfetal life requires participation of numerous cell types, some responsible for depositing new bone matrix, whereas others will undergo proliferation before reaching full differentiation and others are still required to remain as stem cells. Yet it is likely that some fidelity to the desired cell lineage is a desirable trait, particularly when attempts are made to use cell therapy for the generation of other musculoskeletal tissues including cartilage, muscle, and tendon. This is due to the notion that inappropriate cell proliferation can result in undesirable outcomes. For this, it is necessary to improve the existing protocols for isolating progenitors from adult bone marrow to obtain homogeneous populations of multipotent mesenchymal stem cells.

If the goal is to expand bone marrow progenitor cells that may regenerate skeletal tissues in vivo other than bone, then, to move forward with therapies, specific multipotent mesenchymal cells must be identified and characterized among the heterogeneous stromal population. Wang and colleagues [35] found that low-density in vivo 5-fluorouracil resistant mesenchymal stem cell-like cells are capable of osseous tissue formation. Freshly isolated Lin−Sca-1+CD45− (VSEL) cells, when used in an in vivo model, demonstrated that as few as 500 cells are able to generate bonelike tissues. Importantly, when transplanted to a bone marrow environment, the cells are able to differentiate into multiple mesenchymal lineages. Our group has also described that multipotent mesenchymal cells may reside preferentially in the same niche of the hematopoietic stem cells [36].

The ability of mesenchymal stem/progenitor cells to adhere to culture flask is an important feature applied in most isolation procedures [37-39]. However, some authors have shown that mesenchymal stem cells, in the presence or lack of specific factors, can be grown and expanded in nonadherent culture systems and still maintain their clonal ability as well as their differentiation potential [40,41]. Recently, Dellavalle and colleagues [42] and Crisan and colleagues [43] demonstrated that pericytes isolated from skeletal muscle take more than 5 days to adhere to the culture flask bottom. Although the relationship between mesenchymal stem cells and pericytes is not yet well defined, both are capable of differentiating into osteogenic, adipogenic, and chondrogenic lineages [44,45] and share a set of surface markers [43,46]. Just like mesenchymal stem cells, pericytes express CD146 and are negative for hematopoietic cell (CD45), endothelial (CD31), and myogenic (CD56) markers. However, unlike mesenchymal stem cells, pericytes express NG2 and PDGF-Rb, markers that allow their isolation [47].

Mesenchymal Cells in Orthopedic Bioengineering and Bone Repair

The characteristics of mesenchymal pluripotent cells—especially the ease with which they can be isolated from bone marrow [39], the possibility of obtaining, by ex vivo expansion, a large number of cells from relatively small volumes of bone marrow, their versatility in differentiating into mesenchymal tissues [48], the possibility of expanding a significant number of times without showing signs of senescence [49,50], besides the fact that, in normal culture, the osteogenic lineage is the default differentiation of this population [51]—make mesenchymal stem cells a very attractive tool for applications in orthopedic bioengineering.

The biologic potential of mesenchymal stem cells has been explored in the repair or regeneration of injured mesenchymal tissues such as bone [52], tendons [53], and cartilage [54,55]. For bone the location and the biologic and biomechanical characteristics of the affected site continue to make the reconstruction a challenging process. Large losses in areas with poor vascularization, such as the diaphysis of long bones, remain a significant therapeutic challenge for the orthopedic surgeon. An attractive strategy for treating this type of injury is the association of resorbable biomaterials with mesenchymal stem/progenitor cells, which would maintain the cells at the injury site, and the utilization of growth factors that become bioactive (osteoinductors), which helps and stimulates cell proliferation and differentiation.

Hydroxyapatite and its chemical variations are classically indicated as a biomaterial with high-potential applications in orthopedics. Using a diaphyseal defect model in sheep, Kon and colleagues [56] observed that a cylinder of synthetic porous hydroxyapatite combined with autologous mesenchymal stem cells expanded in vitro promoted the consolidation of the graft and bone regeneration when placed in a tibial gap. Despite the widespread use of hydroxyapatite and its variants, particularly in dental surgery, some aspects related to the time of remodeling should be considered. Although hydroxyapatite is a resorbable biomaterial, once integrated into the bone, its overall remodeling (replacement by natural bone tissue) can be very slow. Physically and chemically distinct from normal bone tissue, its ability to support mechanical pressure differs substantially from the normal tissue, making the site more susceptible to fractures. One of many strategies used to reduce the period of remodeling of the hydroxyapatite is the incorporation of other compounds such as calcium phosphate crystals (tricalcium phosphate alpha and beta). However, the reduced mechanical properties of these materials limit their utility in large and weight-bearing defects. Despite these advances, biomaterials alone are often not sufficient to promote complete tissue regeneration for larger defects. The addition of autologous mesenchymal cells isolated from bone marrow and cultivated by the classical model described by Friedenstein [14-16] can prove to be an excellent strategy to increase locally the potential for biologic repair.

Skeletal Stem Cells and the Vascular System

Bone tissue is highly vascularized, and its relationship with the vascular system is essential for the maintenance of metabolism and the remodeling process. The vascular system also plays crucial roles during all stages of repair and regeneration of bone tissue in cases of injury.

The intimate relationship of bone tissue with the vascular system is established early in embryogenesis. Bone marrow development is closely linked to the process of bone formation (ossification). In mammals, bones are formed by two processes of ossification that are apparently distinct: intramembranous ossification and endochondral ossification. For intramembranous bone development, mesenchymal cells derived from the neural crest and mesoderm cells migrate to the sites where the future bone will be formed, and they differentiate directly into osteoblasts [49,57]. Facial bones, several bones of the skull, and part of the clavicle are formed by an intramembranous ossification process. The majority of bones in the body are formed through a process termed endochondral bone formation. In this process, mesodermal cells migrate to the sites where the future bones will be formed, grouping. Unlike the process of intramembranous ossification, the cells do not differentiate directly into osteoblasts, but first into chondrocytes, the main cell type in cartilage [58]. Cells from the edge of the cell aggregate form the perichondrium, rich in chondroblasts. The chondrocytes of the cartilaginous mold center are characterized by the expression of type II collagen and aggrecan. The cartilage analog therefore serves as an initial scaffold to model the future bone. Chondrocytes in the center of the cartilage bone analog soon reduce their proliferative rate and begin to hypertrophy while at the same time initiating the synthesis of collagen type X. Soon thereafter, hypertrophic chondrocytes begin to calcify their extracellular matrix and secrete factors that are chemotactic for endothelial cells (VEGF) and for chondroclasts (similar to osteoclasts, responsible for the resorption of the cartilage matrix), which guide and orchestrate vascularization of the cartilage mold. The hypertrophic chondrocytes also induce the differentiation of the perichondrium into periosteum (rich in osteoblasts) by secreting factors such as Indian hedgehog. With the invasion of blood vessels, stromal and hematopoietic systems migrate and colonize the newly formed bone marrow. It is suggested that hypertrophic chondrocytes undergo apoptosis and are eventually colonized by osteoprogenitor cells.

At least two major theories have been used to explain the origin of osteoprogenitor cells that colonize the newly formed marrow cavity [10,12,13]. The first theory suggests that cells of the perichondrium/periosteum migrate into the new tissue using the abluminal side of blood vessels that invade the matrix of cartilage during endochondral ossification possible mechanisms. One possibility is that mesenchymal cells derived from the periosteum and already committed to the skeletal tissue could serve as the source of mesenchymal stem/progenitor cells. Alternatively, cells in the perichondrium/periosteum that are known to include a few mesenchymal stem cells, including ALCAM+ cells, migrate, colonize and give rise to the population of mesenchymal stem and osteoprogenitor cells [59]. There is evidence for both mechanisms, and neither is mutually exclusive. The second theory suggests that mesenchymal stem cells and the entire population of mesenchymal bone marrow progenitors are descendants of perivascular pericytes located around small vessels, which migrate and colonize the marrow cavity, regardless of the perichondrium/periosteum. It is noteworthy that the mesenchymal stem cells, or at least the cells that exhibit the same phenotype and functional properties of adult mesenchymal stem cells during embryonic development, co-localize with the main sites from which hematopoietic stem cells are derived [60].

From a global perspective when a fracture occurs, blood vessels from the bone marrow and the periosteum are disrupted, originating a blood clot at the fracture site. Some time later the clot is invaded by osteoprogenitor and endothelial cells (in the form of small blood vessel), both derived from bone marrow and periosteum. Osteoprogenitor cells proliferate and differentiate into chondrocytes and osteoblasts, through a process similar to endochondral ossification, giving rise to the callus. The callus is essential in the consolidation stage of the fracture. With the mechanical stabilization of the fracture, the callus is then remodeled and bone tissue returns to its original morphology.

Here we wish to draw attention to some of the molecular and cellular events occurring at the fracture site during the repair process of bone tissue. Platelets in the clot are known to release several growth factors and chemokines, among which we can highlight the transforming growth factor-ß (TGF-ß), fibroblast growth factor (FGF), endothelial growth factor (VEGF), and platelet-derived growth factor (PDGF). Many of these agents are known to stimulate osteoprogenitor and endothelial cell proliferation and migration into the fracture sites.

In the bone marrow microenvironment, skeletal stem cells are not randomly distributed. They are located mainly around the precapillary arteriole, a site identified as a perivascular niche. Although blood vessel growth during the process of bone tissue regeneration is of extreme importance for the supply of nutrients and oxygen, these newly formed vessels also serve as a guide for the distribution of osteoprogenitor cells at the site of injury. Osteoprogenitor cells derived from the periosteum and the bone marrow cells are recruited from the perivascular niche, as most of the bone marrow blood vessels and mesenchymal progenitors are distributed in the subendosteal region [36,61]. Cells with similar behavior and phenotype of the bone marrow mesenchymal progenitors were isolated in large numbers from other tissues and organs, from mice and humans. They all have in common the perivascular niche. We know now that cells with the potential to differentiate into musculoskeletal tissues can be isolated virtually from all tissues and organs, suggesting that other tissues besides the bone marrow may be used as a source of mesenchymal cells.

Mesenchymal progenitors have been isolated from many tissues other than the bone marrow, including adipose tissue [62], synovial membrane [63], umbilical cord blood [64], peripheral blood [65], and periosteum [66,67]. Regardless of the source, cells derived from different tissues share several similarities, expressing similar membrane markers (CD73, CD90, and CD105) and presenting similar ability to differentiate into chondrocytes, osteoblasts, and adipocytes in vitro. However, recent published data indicate that progenitor biology depends on tissue origin, in which local factors would induce a so-called precommitment (bias) to a specific lineage. For example, bone marrow and adipose tissue–derived progenitors would preferentially differentiate into osteoblasts and adipocytes, respectively. Considering the heterogenous population isolated from each different source, it is conceivable that not all cells are real stem cells. Suggesting that committed progenitors (intermediate progenitors) would differentiate faster and would take over the culture, masking the result.

As previously mentioned, mesenchymal progenitors and blood vessels are intimately and functionally associated. This means that widespread distribution of mesenchymal progenitors in the whole organism is dependent on tissue vascularization. Da Silva Meirelles et al. [68] noted that "mesenchymal stem cells" reside virtually in all organs and tissues, and that this distribution is related to their existence in a perivascular niche. They also showed that, in adipose tissue, CFU-F numbers and vascular density present a strong correlation, corroborating their previous observation [69]. Although mesenchymal stem cells and pericytes present different phenotypes, several authors suggest that the cells are the same or that, pericytes would represent a less differentiated category.

Combined Action: Therapeutic Use of Mesenchymal Progenitors and Angiogenic Progenitors

Recent data indicate that in addition to hematopoietic stem cells and mesenchymal stem cells, angiogenic progenitors also reside in the bone marrow microenvironment. These cells are capable of contributing to new blood vessel growth. Current data suggest that blood vessels can be formed through two processes: angiogenesis, in which new blood vessels are generated from a vessel already formed, and vasculogenesis, in which vessels are formed from the differentiation of progenitors. For many years it was believed that vasculogenesis would occur only during embryogenesis. Data emerging during the early 2000s suggest that the bone marrow may also serve as a source of angiogenic progenitors. The presence of angiogenic precursors in bone marrow has spurred the development of vasculogenic or angiogenic therapies. Unlike mesenchymal progenitors, endothelial progenitors, also termed angioblasts, are expensive to culture at present for they require supplementation with recombinant factors, and the initial isolation procedures are technically more difficult and complex.

The presence of angiogenic progenitors in the bone marrow has increased the interest in the utility of these cells for the treatment of diseases and injuries to musculoskeletal tissues. For bone formation is known to be highly dependent on the presence of not only mesenchymal stem/progenitor cells but on collaborative efforts with the cells of the vascular system. Based on this principle, we believe that the use of angiogenic progenitors in association with mesenchymal progenitors in orthopedic bioengineering protocols could enhance the ability of bone tissue regeneration. But how can these cells be used therapeutically in a combined manner?

The search for isolating endothelial progenitors began in the 1970s, when different research groups [70-72], in order to establish a protocol to purify monocytes and lymphocytes from peripheral blood and bone marrow, developed cell separation methods that were based on cell density using known solutions such as Percoll and Ficoll-Hypaque. This procedure allows the separation of cells into two fractions: a fraction with density higher than 1.077 g/cm3, which mostly comprises erythrocytes and granulocytes, and a fraction with density lower than 1.077 g/cm3, which mostly comprises monocytes, lymphocytes, and hematopoietic progenitors. The less dense fraction is enriched for mononuclear cells, as it contains fewer granulocytes. With the improvement of techniques used to analyze the phenotypic characteristics of different cell populations, researchers identified within the mononuclear population mesenchymal and angiogenic progenitors, using different markers and flow cytometry analyses, as well as in vitro progenitor quantification assays. In orthopedic bioengineering, the mononuclear fraction therefore represents a great promise of a combined-action cell therapy protocol, simultaneously accelerating the processes of osteogenesis and vasculogenesis. The combination of these two cell populations has become a highly attractive option for the treatment of nonunion fractures (nonunion, delayed union, and pseudarthrosis).

Cell Therapy in Pseudarthrosis

Pseudarthrosis occurs when the process of biologic repair of a fracture site is disrupted, resulting in a nonunion (gap). Since the 1990s, several techniques have been proposed for the treatment of nonunion fractures, including invasive procedures (internal fixation and bone grafting with or without bone substitutes) and noninvasive procedures (ultrasound and the application of electromagnetic pulses) [73,74]. However, the initial step to be taken in the treatment of a nonunion is the evaluation of biologic and mechanical conditions involved in the lesions. Several reports point to the use of bone marrow mononuclear cells as a new option for the treatment of a nonunion. The basis for this type of therapy is the combined use of osteoprogenitor cells and angiogenic progenitors. Hernigou and colleagues [75,76] showed that in 60 patients with nonunion fractures treated with percutaneous injection of bone marrow mononuclear cells, 53 (~88%) presented fracture healing, confirming the efficacy of treatment. An additional advantage of using mononuclear cells as an option for the treatment of nonunion of fractures is the fact that the cell pool injection is performed in a closed and minimally invasive procedure (percutaneous).

Cell Therapy in Avascular Necrosis of the Femoral Head (Osteonecrosis)

Cell therapy with mononuclear cells also has emerged as a promising therapy for the treatment of avascular necrosis of the femoral head. Osteonecrosis occurs when there is a disruption of local blood flow and necrosis of bone tissue resulting from sickle cell disease, the use of steroids and alcohol abuse. The bone matrix often becomes devoid of cellular elements (devitalized) yet remains in its place but without proper biologic and mechanical functions, resulting in the breakdown of bone tissue and impairment of joint anatomy.

The femoral head is one of the sites most affected by osteonecrosis. With ischemic necrosis of bone tissue, cells are eliminated, but the matrix still gives it temporary physical support. With the evolution of the disease, the matrix is degraded followed by the collapse of the articular cartilage. Once the joint collapse occurs, it becomes necessary to replace the joint with a prosthesis. However, in the early stages of the disease (necrosis without articular collapse), injection of autologous mononuclear cells can result in significant clinical improvement. The rationale of the treatment is that cells injected into the damaged area of the femoral head are able to repopulate the devitalized matrix by revascularization. The findings of Hernigou and colleagues [77-79] show promising results in treating osteonecrosis of the femoral head with mononuclear cells, particularly in lesions secondary to sickle cell anemia. It is noteworthy that the main objective of cell therapy in osteonecrosis is to delay joint replacement by a metal prosthesis. In the pseudoarthrosis situation, cell therapy has a curative role through the final biologic repair of the lesion.

Although the main focus of using mononuclear cells in the regeneration and reconstruction of bone tissue is the combined action of osteoprogenitor cells and angioblasts, several groups described a likely paracrine activity by lymphocytes and monocytes present in this population. These are the main cells responsible for the secretion of cytokines such as interleukin-6 and TGFb, which stimulate the proliferation and differentiation of cells from the bone and the vascular system. However, these studies are very recent, and real paracrine effects still need to be clarified.

Acknowledgments

We are indebted to Wallace de Mello for the critical review of the present text. We also thank our "cell expert team," which always does excellent work and helps us keep the train on the rails. Way to go!

References

[1] Godin I, Cumano A. The hare and the tortoise: an embryonic haematopoietic race. Nat Rev Immunol 2002;2:593–604.

[2] Ikuta K, Weissman IL. Evidence that hematopoietic stem cells express mouse c-kit but do not depend on steel factor for their generation. Proc Nat Acad Sci USA 1992;89:1502–6.

[3] Christensen JL, Wright DE, Wagers AJ, Weissman IL. Circulation and chemotaxis of fetal hematopoietic stem cells. PLoS Biol 2004;2:E75.

[4] Medvinsky A, Dzierzak E. Definitive hematopoiesis is autonomously initiated by the AGM region. Cell 1996;86:897–906.

[5] Morrison SJ, Uchida N, Weissman IL. The biology of hematopoietic stem cells. Annu Rev Cell Dev Biol 1995;11:35–71.

[6] Wolf NS, Trentin JJ. Differential proliferation of erythroid and granuloid spleen colonies following sublethal irradiation of the bone marrow donor. J Cell Physiol 1970;75:225–9.

[7] Charbord P, Tavian M, Humeau L, Péault B. Early ontogeny of the human marrow from long bones: an immunohistochemical study of hematopoiesis and its microenvironment. Blood 1996; 87:4109–19.

[8] Kuznetsov SA, Mankani MH, Gronthos S, et al. Circulating skeletal stem cells. J Cell Biol 2001;153:1133–40.

[9] Deans RJ, Moseley AB. Mesenchymal stem cells: biology and potential clinical uses. Exp Hematol 2000;28:875–84.

[10] Bianco P, Robey PG, Simmons PJ. Mesenchymal stem cells: revisiting history, concepts, and assays. Cell Stem Cell 2008;2:313–9.

[11] Charbord P, Tamayo E, Deschaseaux F, et al. The hematopoietic microenvironment: phenotypic and functional characterization of human marrow vascular stromal cells. Hematology (Amsterdam, Netherlands) 1999;4:257–82.

[12] Bianco P, Kuznetsov SA, Riminucci M, Gehron Robey P. Postnatal skeletal stem cells. Methods Enzymol 2006;419:117–48.

[13] Bianco P, Gehron Robey P. Marrow stromal stem cells. J Clin Invest 2000;105:1663–8.

[14] Owen M, Friedenstein AJ. Stromal stem cells: marrow-derived osteogenic precursors. Ciba Found Symp 1988;136:42–60.

[15] Friedenstein AJ, Chailakhjan RK, Lalykina KS. The development of fibroblast colonies in monolayer cultures of guinea-pig bone marrow and spleen cells. Cell Tissue Kinet 1970;3:393–403.

[16] Friedenstein AJ, Piatetzky-Shapiro II , Petrakova KV. Osteogenesis in transplants of bone marrow cells. J Embryol Exp Morphol 1966;16:381–90.

[17] Caplan AI. Mesenchymal stem cells. J Orthop Res 1991;9:641–50.

[18] Sacchetti B, Funari A, Michienzi S, et al. Self-renewing osteoprogenitors in bone marrow sinusoids can organize a hematopoietic microenvironment. Cell 2007;131:324–36.

[19] Toma C, Pittenger MF, Cahill KS, Byrne BJ, Kessler PD. Human mesenchymal stem cells differentiate to a cardiomyocyte phenotype in the adult murine heart. Circulation 2002;105:93–8.

[20] Lee J-H, Kosinski PA, Kemp DM. Contribution of human bone marrow stem cells to individual skeletal myotubes followed by myogenic gene activation. Exp Cell Res 2005;307:174–82.

[21] Dezawa M, Ishikawa H, Itokazu Y, et al. Bone marrow stromal cells generate muscle cells and repair muscle degeneration. Science 2005;309:314–7. New York, NY.

[22] Corti S, Strazzer S, Del Bo R, et al. A subpopulation of murine bone marrow cells fully differentiates along the myogenic pathway and participates in muscle repair in the mdx dystrophic mouse. Exp Cell Res 2002;277:74–85.

[23] Verfaillie CM, Schwartz R, Reyes M, Jiang Y. Unexpected potential of adult stem cells. Ann NY Acad Sci 2003;996:231–4.

[24] Dominici M, Le Blanc K, Mueller I, et al. Minimal criteria for defining multipotent mesenchymal stromal cells. The International Society for Cellular Therapy position statement. Cytotherapy 2006;8:315–7.

[25] Simmons PJ, Torok-Storb B. Identification of stromal cell precursors in human bone marrow by a novel monoclonal antibody, stro-1. Blood 1991;78:55–62.

[26] Shi S, Gronthos S. Perivascular niche of postnatal mesenchymal stem cells in human bone marrow and dental pulp. J Bone Miner Res 2003;18:696–704.

[27] Kolf CM, Cho E, Tuan RS. Mesenchymal stromal cells. biology of adult mesenchymal stem cells: regulation of niche, self-renewal and differentiation. Arthritis Res Ther 2007;9:204.

[28] Horwitz EM, Le Blanc K, Dominici M, et al. Clarification of the nomenclature for MSC: the International Society for Cellular Therapy position statement. Cytotherapy 2005;7:393–5.

[29] Gang EJ, Bosnakovski D, Figueiredo CA, Visser JW, Perlingeiro RCR. Ssea-4 identifies mesenchymal stem cells from bone marrow. Blood 2007;109:1743–51.

[30] Anjos-Afonso F, Bonnet D. Nonhematopoietic/endothelial ssea-1+ cells define the most primitive progenitors in the adult murine bone marrow mesenchymal compartment. Blood 2007;109:1298–306.

[31] Satomura K, Krebsbach P, Bianco P, Gehron Robey P. Osteogenic imprinting upstream of marrow stromal cell differentiation. J Cell Biochem 2000;78:391–403.

[32] Muraglia A, Cancedda R, Quarto R. Clonal mesenchymal progenitors from human bone marrow differentiate in vitro according to a hierarchical model. J Cell Sci 2000;113 (Pt 7):1161–6.

[33] Dennis JE, Charbord P. Origin and differentiation of human and murine stroma. Stem Cells 2002;20:205–14 (Dayton, Ohio).

[34] Ducy P, Zhang R, Geoffroy V, Ridall AL, Karsenty G. Osf2/cbfa1: a transcriptional activator of osteoblast differentiation. Cell 1997;89:747–54.

[35] Wang Z, Song J, Taichman RS, Krebsbach PH. Ablation of proliferating marrow with 5-fluorouracil allows partial purification of mesenchymal stem cells. Stem Cells 2006;24:1573–82.

[36] Taichman RS, Wang Z, Shiozawa Y, et al. Prospective identification and skeletal localization of cells capable of multilineage differentiation in vivo. Stem Cells Dev 2010;19:1557–70.

[37] Jaiswal N, Haynesworth SE, Caplan AI, Bruder SP. Osteogenic differentiation of purified, culture-expanded human mesenchymal stem cells in vitro. J Cell Biochem 1997;64:295–312.

[38] Bruder SP, Jaiswal N, Haynesworth SE. Growth kinetics, self-renewal, and the osteogenic potential of purified human mesenchymal stem cells during extensive subcultivation and following cryopreservation. J Cell Biochem 1997;64:278–94.

[39] Pittenger MF, Mackay AM, Beck SC, et al. Multilineage potential of adult human mesenchymal stem cells. Science 1999;284:143–7. New York, NY.

[40] Baksh D, Davies JE. Stem Cell Culture: Elsevier California, USA; 2008.

[41] Baksh D, Davies JE, Zandstra PW. Adult human bone marrow-derived mesenchymal progenitor cells are capable of adhesion-independent survival and expansion. Exp Hematol 2003;31:723–32.

[42] Dellavalle A, Sampaolesi M, Tonlorenzi R, et al. Pericytes of human skeletal muscle are myogenic precursors distinct from satellite cells. Nat Cell Biol 2007;9:255–67.

[43] Crisan M, Yap S, Casteilla L, et al. A perivascular origin for mesenchymal stem cells in multiple human organs. Cell Stem Cell 2008;3:301–13.

[44] Doherty MJ, Ashton BA, Walsh S, et al. Vascular pericytes express osteogenic potential in vitro and in vivo. J Bone Miner Res 1998;13:828–38.

[45] Farrington-Rock C, Crofts NJ, Doherty MJ, et al. Chondrogenic and adipogenic potential of microvascular pericytes. Circulation 2004;110:2226–32.

[46] Covas DT, Panepucci RA, Fontes AM, et al. Multipotent mesenchymal stromal cells obtained from diverse human tissues share functional properties and gene-expression profile with CD146+ perivascular cells and fibroblasts. Exp Hematol 2008;36:642–54.

[47] Crisan M, Huard J, Zheng B, et al. Purification and culture of human blood vessel-associated progenitor cells. Curr Protoc Stem Cell Biol 2008. Chapter 2:Unit 2B.2.1–2B.2.13.

[48] Seshi B, Kumar S, Sellers D. Human bone marrow stromal cell: coexpression of markers specific for multiple mesenchymal cell lineages. Blood Cells Mol Dis 2000;26:234–46.

[49] Jiang Y, Jahagirdar BN, Reinhardt RL, et al. Pluripotency of mesenchymal stem cells derived from adult marrow. Nature 2002;418:41–9.

[50] Jiang Y, Vaessen B, Lenvik T, et al. Multipotent progenitor cells can be isolated from postnatal murine bone marrow, muscle, and brain. Exp Hematol 2002;30:896–904.

[51] Banfi A, Bianchi G, Notaro R, et al. Replicative aging and gene expression in long-term cultures of human bone marrow stromal cells. Tissue Eng 2002;8:901–10.

[52] Moutsatsos IK, Turgeman G, Zhou S, et al. Exogenously regulated stem cell-mediated gene therapy for bone regeneration. Mol Ther 2001;3:449–61.

[53] Awad HA, Butler DL, Boivin GP, et al. Autologous mesenchymal stem cell-mediated repair of tendon. Tissue Eng 1999;5:267–77.

[54] Quintavalla J, Uziel-Fusi S, Yin J, et al. Fluorescently labeled mesenchymal stem cells (MSCs) maintain multilineage potential and can be detected following implantation into articular cartilage defects. Biomaterials 2002;23:109–19.

[55] Kadiyala S, Young RG, Thiede MA, Bruder SP. Culture expanded canine mesenchymal stem cells possess osteochondrogenic potential in vivo and in vitro. Cell Transplant 6:125-34.

[56] Kon E, Muraglia A, Corsi A, et al. Autologous bone marrow stromal cells loaded onto porous hydroxyapatite ceramic accelerate bone repair in critical-size defects of sheep long bones. Journal Biomed Mater Res 2000;49:328–37.

[57] Cobourne MT. Construction for the modern head: current concepts in craniofacial development. J Orthod 2000;27:307–14.

[58] Kronenberg HM. Developmental regulation of the growth plate. Nature 2003;423:332–6.

[59] Arai F, Ohneda O, Miyamoto T, Zhang XQ, Suda T. Mesenchymal stem cells in perichondrium express activated leukocyte cell adhesion molecule and participate in bone marrow formation. J Exp Med 2002;195:1549–63.

[60] Mendes SC, Robin C, Dzierzak E. Mesenchymal progenitor cells localize within hematopoietic sites throughout ontogeny. Development 2005;132:1127–36 (Cambridge, England).

[61] Balduino A, Hurtado SP, Frazão P, et al. Bone marrow subendosteal microenvironment harbours functionally distinct haemosupportive stromal cell populations. Cell Tissue Res 2005;319:255–66.

[62] Zuk PA, Zhu M, Mizuno H, et al. Multilineage cells from human adipose tissue: implications for cell-based therapies. Tissue Eng 2001;7:211–28.

[63] De Bari C, Dell'Accio F, Tylzanowski P, Luyten FP. Multipotent mesenchymal stem cells from adult human synovial membrane. Arthritis Rheum 2001;44:1928–42.

[64] Erices A, Conget P, Minguell JJ. Mesenchymal progenitor cells in human umbilical cord blood. Brit J Haematol 2000;109:235–42.

[65] Zvaifler NJ, Marinova-Mutafchieva L, Adams G, et al. Mesenchymal precursor cells in the blood of normal individuals. Arthritis Res 2000;2:477–88.

[66] Nakahara H, Dennis JE, Bruder SP, et al. In vitro differentiation of bone and hypertrophic cartilage from periosteal-derived cells. Exp Cell Res 1991;195:492–503.

[67] De Bari C, Dell'Accio F, Luyten FP. Human periosteum-derived cells maintain phenotypic stability and chondrogenic potential throughout expansion regardless of donor age. Arthritis Rheum 2001;44:85–95.

[68] Meirelles L da Silva, Chagastelles PC, Nardi NB. Mesenchymal stem cells reside in virtually all post-natal organs and tissues. J Cell Sci 2006;119:2204–13.

[69] Meirelles L da Silva, Sand TT, Harman RJ, Lennon DP, Caplan AI. MSC frequency correlates with blood vessel density in equine adipose tissue. Tissue Eng Part A 2009;15: 221–9.

[70] Ali FM, May A, McLaren GD, Jacobs A. A two-step procedure for obtaining normal peripheral blood T-lymphocytes using continuous equilibrium density gradient centrifugation on percoll. J Immunol Methods 1982;49:185–91.

[71] Gehrz RC, Knorr SO. Characterization of the role of mononuclear cell subpopulations in the in vitro lymphocyte proliferation assay. Clin Exp Immunol 1979;37:551–7.

[72] Messner HA, Fauser AA, Lepine J, Martin M. Properties of human pluripotent hemopoietic progenitors. Blood Cells 1980;6:595–607.

[73] Lerner A, Fodor L, Soudry M. Is staged external fixation a valuable strategy for war injuries to the limbs? Clin Orthop Relat Res 2006;448:217–24.

[74] Harris AM, Patterson BM, Sontich JK, Vallier HA. Results and outcomes after operative treatment of high-energy tibial plafond fractures. Foot & ankle international / American Orthopaedic Foot and Ankle Society [and] Swiss Foot and Ankle Society 2006;27:256–65.

[75] Hernigou P, Poignard A, Beaujean F, Rouard H. Percutaneous autologous bone-marrow grafting for nonunions influence of the number and concentration of progenitor cells. J Bone Joint Surg Am 2005;87:1430–7.

[76] Hernigou P, Mathieu G, Poignard A, et al. Percutaneous autologous bone-marrow grafting for nonunions. Surgical technique. J Bone Joint Surg Am 2006;88(Suppl. 1):322–7.

[77] Flouzat-Lachaniete CH, Roussignol X, Poignard A, et al. Multifocal joint osteonecrosis in sickle cell disease. Open Orthop J 2009;3:32–5.

[78] Hernigou P, Poignard A, Zilber S, Rouard H. Cell therapy of hip osteonecrosis with autologous bone marrow grafting. Indian J Orthop 2009;43:40–5.

[79] Hernigou P, Daltro G, Filippini P, Mukasa MM, Manicom O. Percutaneous implantation of autologous bone marrow osteoprogenitor cells as treatment of bone avascular necrosis related to sickle cell disease. Open Orthop J 2008;2:62–5.

Chapter 8

Cardiac Stem Cells

Adriana B. Carvalho*,†,‡,§,**, Bernd K. Fleischmann††, Antonio Carlos Campos de Carvalho*,†,‡,§,**,§§

*Instituto de Biofísica Carlos Chagas Filho, Rio de Janeiro, Brazil

†Programa de Terapia Celular e Bioengenharia, Rio de Janeiro, Brazil

‡Laboratório de Cardiologia Celular e Molecular, Rio de Janeiro, Brazil

§Instituto Nacional de Ciência e Tecnologia de Biologia Estrutural e Bioimagem -INBEB

**Universidade Federal do Rio de Janeiro, Rio de Janeiro, Brazil

††Institute of Physiology I, Life and Brain Center, University of Bonn, Bonn, Germany

§§Instituto Nacional de Cardiologia, Rede de Terapia Celular, Brazil

Not many years ago, the existence of cardiac stem cells would have been considered highly unlikely, because the heart has been classically viewed as an organ incapable of self-renewal [1]. The basis for this assumption started to be constructed more than eight decades ago, and many still consider it a definitive characteristic of the heart [2,3]. In 1925, Karsner and coworkers demonstrated that macroscopic cardiac hypertrophy occurred by an increase in cell size instead of an increase in cell number and stated that no mitotic figures could be found in the heart [2]. These observations laid the grounds for envisaging the heart as a postmitotic organ, a view that turned into common knowledge and remained widely accepted for most of the previous century.

Additional arguments for the postmitotic state of the adult heart were that the organ was unable to recover from myocardial infarctions and that primary heart tumors were rarely observed in adults [4,5]. However, as Anversa and coworkers pointed out, regardless of the proliferative capacity of their parenchymal cells, the outcome of infarction is identical in several organs, including the testis, skin, kidney, brain, and intestine [6]. Additionally, using the rarity of primary heart tumors as an argument is also faulty: despite the fact that neurons do not usually proliferate, several tumors arise from the interstitial/supporting cells in the central nervous

Resident Stem Cells and Regenerative Therapy. http://dx.doi.org/10.1016/B978-0-12-416012-5.00008-6

system. On the other hand, although the heart has a vast number of interstitial/supporting cells, tumors originating from these cells are almost as rare as the ones originating from cardiomyocytes. This could possibly indicate that the infrequency of primary heart tumors has more to do with the structural, mechanical, and functional characteristics of the organ than with the rate of cardiomyocyte proliferation.

Definitive evidence that the human heart is capable of some degree of self-renewal came in 2009. Bergmann and coworkers [7] published an important study in which they used the integration of carbon-14, generated by nuclear bomb tests during the Cold War, into DNA to establish the age of cardiomyocytes composing the human heart. They reported that 1% of human cardiomyocytes are renewed annually at the age of 25 and that this rate is reduced to 0.45% at the age of 75. Moreover, total cell renewal over the entire human life span corresponds to approximately 50% of the cardiomyocytes.

Nevertheless, this work did not prove that the heart had an actual subset of progenitor cells since renewal could be attributed to the proliferation of resident cardiomyocytes. Indication that heart regeneration was due to the presence of progenitor cells came from the work of Hsieh and colleagues [8]. Using an elegant α-myosin heavy chain *Cre-Lox* transgenic mouse model, these authors demonstrated that up to 15% of cardiomyocytes could be regenerated in adult hearts after myocardial infarction. In this model, the activation of Cre recombinase by tamoxifen removed the β-galactosidase gene from cardiomyocytes through the cleavage of loxP sites, leaving green fluorescent protein (GFP) under the control of β-actin (Fig. 8.1). This strategy had 80% efficiency, resulting in 80% GFP-positive and 20% β-galactosidase-positive cardiomyocytes.

Therefore, after the tamoxifen pulse, if heart regeneration were the result of myocyte proliferation, the proportion between GFP and β-galactosidase cardiomyocytes would be expected to remain unchanged. Although the authors were not able to detect new myocyte formation with aging, after myocardial infarction they detected a 15% increase in the fraction of β-galactosidase-positive cardiomyocytes, indicating that new cells, which had never been exposed to Cre recombinase, were being generated [8]. The origin of the stem/progenitor cells giving rise to the newly formed muscle was not described (whether endogenous or exogenous to the heart).

Evidence that the newborn mammalian heart is capable of complete regeneration and that this occurs primarily via cardiomyocyte division was recently provided. Porrello and coworkers showed that one-day-postnatal mouse hearts were able to fully regenerate a portion of the apex after surgical removal through cardiomyocyte mitosis of resident cardiomyocytes, but this capacity was lost after one week of postnatal life [9]; it is important to note that progenitor/stem cell contribution cannot be ruled out. These data are reminiscent of the zebra fish heart, where myocyte regeneration and proliferation were found to underlie the prominent regenerative

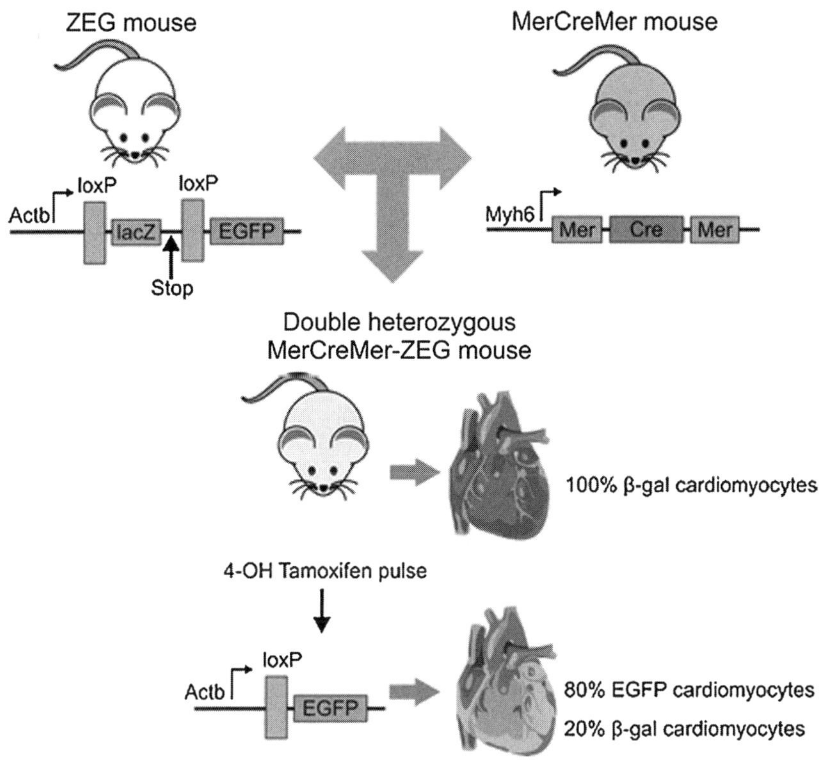

Figure 8.1 The Mer-CreMer-ZEG mouse after treatment with tamoxifen turns 80% of its cardiomyocytes (originally blue) into green cells. The remaining 20% cardiomyocytes remain blue. Any alteration in this proportion of green to blue cardiomyocytes indicates that new cells were formed by a process that does not involve cardiomyocyte mitosis. **See Plate 14.**

potential upon myocardial lesions even in adult animals [10,11], in contrast to the mammalian heart.

In this context, several different cardiac stem cell candidates have been described [12-20]. As pointed out by Laflamme and Murry [21], it is unlikely that the heart would harbor multiple nonoverlapping sets of cardiomyocyte progenitors. Thus, it remains to be defined which of these candidates, if any, is the true cardiac progenitor or stem cell, what is the degree of overlap between the different markers, and what is the specific function of each one of these subsets [22].

In this chapter, we will examine the four most studied subsets of cardiac stem cells. These cells can be isolated according to (1) presence of the cell-surface marker c-kit, (2) tissue culture of cardiac explants with spontaneous shedding of cardiac stem cells that generate cardiospheres, (3) ability to efflux Hoechst dye, which defines a side population (SP), and (4) expression of the islet-1 gene.

c-Kit Positive Cells

One of the most extensively investigated markers potentially labeling cardiac progenitor/stem cells (CPCs) is c-kit (also know as CD117), a member of the receptor tyrosine kinase family. Originally, c-kit was used as a marker for hematopoietic stem cells, but it is also known to be expressed at different stages of lineage commitment in germ, mast, endothelial, and smooth muscle cells [23-25]. Conclusive evidence for c-kit expression during mesodermal development has come from studies in which the early cardiac transcription factor Nkx2.5 was used to drive the expression of the live reporter enhanced green fluorescent protein (EGFP) in transgenic embryonic stem (ES) cells [26]. The c-kit$^+$ cell fraction of Nkx2.5$^+$ cells proved to be less differentiated, more proliferative, and occasionally yield, upon single cell isolation and expansion, cardiomyocytes and smooth muscle cells. Similarly, in E8.5 hearts of transgenic Nkx2.5-EGFP mouse embryos, c-kit$^+$ cells could be identified within the Nkx2.5$^+$ cell fraction and c-kit expression was found to decline during further embryonic development. Although the developmental origin of these c-kit$^+$ cells in the heart is not clear, these data suggest that Nkx2.5 and c-kit mark a population of bipotent CPCs. Having said this, the functional role of c-kit for heart development is still unknown, because spontaneous mutations of the W locus encoding c-kit have not been reported to result in abnormalities of hematopoietic and cardiovascular development [27]. Complete knockout models as well as some of the mutations of c-kit yielded different types of defects and the mice died at the perinatal stage [28]. Interestingly, studies using the transcription factor Isl1 in ES cells allowed the identification of precursor cells from the second heart field, which displayed tripotential developmental plasticity into cardiomyocytes, and smooth muscle and endothelial cells [29]. These precursor cells were reported to be c-kit$^-$, indicating that, in the developing heart, different precursor populations with potentially diverse plasticities could coexist. In order to identify directly c-kit$^+$ CPCs in the heart during development and at the adult stage and to explore their cardiomyogenic potential in vitro, Tallini et al. [30] have generated BAC transgenic mice, where EGFP was placed under control of the c-kit locus. This approach also circumvents known caveats for the labeling of c-kit$^+$ cells in heart using commercially available antibodies. The cardiac c-kit$^+$ cells were characterized at different time points during embryonic development, at the postnatal and adult stages. At E14.5, c-kit$^+$ cells could be identified in the atria and in the ventricles and their numbers increased, peaking shortly after birth. Thereafter, a rapid decline of c-kit$^+$ cells was observed and, in the adult stage, only very few mostly noncardiomyocytes were found. At the time of the peak of c-kit$^+$ cells within the heart, these were still below 1% and preferentially located in vicinity to the atrioventricular valves and to a lesser degree in the atria and ventricles. Most the c-kit$^+$ cells were of the CD45$^-$

type, as the $CD45^+$ subpopulation was preferentially located at the epicardial side; most of the c-kit^+ $CD45^-$ fraction was formed either by endothelial cells ($PECAM^+$) or cardiomyocytes (α-$actinin^+$). FACS-sorting and plating of strong $EGFP^+$ cells revealed cell proliferation and formation of clusters of $EGFP^+$ cells, while immunostaining showed that in these cultures almost no $CD45^+$ and $PECAM^+$ cells could be found. When culturing the cells in bFGF-supplemented medium, clusters of spontaneously contracting cardiomyocytes appeared. Their single cell analysis with patch clamp yielded typical electrophysiologic properties of developing cardiomyocytes, namely prolonged action potential duration and diastolic depolarizations. In analogy to these reported experiments in ES cells, single c-kit^+ cells from neonatal mouse hearts were FACS-sorted and plated into individual wells and their developmental plasticity explored at different time points after the isolation. In some of the wells, colony formation and differentiation into cardiomyocytes and endothelial and smooth muscle cells were detected, proving the presence of tripotential c-kit^+ CPCs in the neonatal heart [30]. Taken together, these findings suggest that, during cardiac development, a subpopulation of CPCs is characterized by the expression of cardiac-specific transcription factors in conjunction with c-kit and that CPCs can still be found at the early postnatal stages.

Similar efforts to identify CPCs in the adult heart were undertaken, and Beltrami et al. [12] reported the isolation of clonogenic and self-renewing c-kit^+, but lineage marker negative (Lin^-) stem cells from adult rat hearts. The cells were reported to be negative for cardiomyocyte, smooth muscle, endothelial cell, and fibroblast markers, and only a small fraction was found to express early cardiomyocyte transcription factors. These c-kit^+ cells were reported to be multipotent stem cells and to differentiate in vitro into cardiomyocytes as well as smooth muscle and endothelial cells. The cells only reached an immature stage in vitro, whereas upon injection into infarcted rat hearts differentiating myocardial structures were described. The existence of similar c-kit^+ progenitor/stem cells has in the meanwhile also been reported in dog [31] and human hearts [18]. However, in the mouse, low cardiomyogenic plasticity of c-kit^+ cells harvested from adult hearts has been reported after co-culturing with fetal cardiomyocytes or injecting into the border zone of infarcted hearts [32]. Interestingly, in the same study, the cardiomyogenic potential of MACS-sorted c-kit^+ cells from neonatal murine hearts was also probed in co-culture conditions, and 2% to 3% of these cells were found to adopt a cardiomyocyte fate within 7 days in culture. These data are consistent with the previously reported presence of multipotent c-kit^+ CPCs in embryonic and neonatal hearts. In fact, when inducing cardiac injury in adult transgenic c-kit-EGFP mice, c-kit^+ endothelial and smooth muscle cells were found in the border zone of the injury, whereas only very few differentiated cardiomyocytes became $EGFP^+$ [30]. These observations are in agreement with the augmented expression of c-kit in cardiomyocytes upon pressure overload [33]. Further characterization of the c-kit-EGFP$^+$ cardiomyocytes suggested that c-kit is reexpressed in resident, differentiated cardiomyocytes. The reported role of c-kit$^+$ cells in the heart postinjury was also explored by Fazel et al. [34], who revealed that bone marrow–derived c-kit$^+$ cells in the heart were responsible for establishing a "pro-angiogenic milieu" in the border zone of the infarct. When using the W-mutant mice, the c-kit$^+$ cells were no longer recruited to the heart resulting in defective angiogenesis, qualitative changes of scar tissue formation, and cardiac failure. In regard to the potential impact of marrow-derived c-kit$^+$ cells on heart regeneration, an intriguing study has been published [35]. The authors have performed cell therapy using bone marrow–derived c-kit$^+$ cells upon induction of a myocardial lesion in the above-mentioned tamoxifen inducible mouse model [8], and this treatment caused lowering of the EGFP$^+$ cardiomyocyte population, an increase of the beta-gal$^+$ population, and improved function suggesting activation of a heart-resident progenitor population. But once again the molecular identity and origin of the stem/progenitor cell pool was not investigated.

Taken together, these results suggest that resident c-kit$^+$ cells are found also in the adult heart; however, their developmental origin, lineage commitment, and physiologic relevance for endogenous myocardial repair postinjury or more generally during cardiovascular disease require further investigations. It is well possible that the different experimental results are related to differences in the harvesting of the cells, their manipulation in vitro, species, or lesion models. Future studies should be therefore directed at establishing novel mouse models, which would enable the performance of fate mapping studies to determine the origin of c-kit$^+$ cells in the heart, their molecular identity, and their potential contribution to heart regeneration. In this regard, a recent clinical trial, using autologous c-kit$^+$ cells isolated from atrial appendages during cardiac surgery and expanded in vitro, reported significant improvement in ejection fraction and decrease in infarct area by magnetic resonance imaging (MRI) after injection of 1 million cells into the coronary arteries of patients [36]. These results although encouraging have yet to be confirmed by a trial designed specifically to address the efficacy of this type of cell therapy.

Cardiosphere-Derived Cells

The formation of cardiospheres from human and murine heart tissue was first described in 2004 by Messina and coworkers [16]. It was demonstrated that, when placed in adherent plates, heart explants generated a layer of fibroblast-like cells over which small, phase-bright cells migrated. These phase-bright cells were collected and transferred to nonadherent plates where they originated three-dimensional structures named cardiospheres. Cardiospheres were clonogenic and when cocultured with rat neonatal cardiomyocytes expressed troponin I and connexin 43.

Additionally, there was visual evidence that cardiospheres showed synchronous contractions with cardiomyocytes. When transplanted into infarcted hearts, these cells started to express myosin heavy chain as well as α-smooth muscle actin and platelet endothelial cell adhesion molecule, which resulted in functional improvement. Curiously, when the expression of surface molecules was analyzed by flow cytometry, cardiospheres showed a 25% expression of c-kit. Thus, it is possible that c-kit^{+} cells contribute to the characteristics observed in cardiospheres, explaining the similar findings between these two cardiac progenitor/stem cell types.

However, it was only in 2007 that Marbán's group described cardiosphere-derived cells (CDCs) [19]. They slightly changed Messina's protocol by placing cardiospheres in adherent plates where cells were grown in monolayers instead of three-dimensional structures. The advantage of this step was that cell expansion in monolayers was easier and faster, which would facilitate future clinical use. Flow cytometry showed that c-kit expression was still present in similar levels to those described by Messina and coworkers. Additionally, high expression levels of CD105 and CD90 were found, indicating a mesenchymal phenotype. When co-cultured with rat neonatal cardiomyocytes, CDCs presented spontaneous intracellular calcium transients and action potentials, as well as I_{Na}, I_{K1} and $I_{Ca,L}$ currents. In vivo, injection of CDCs in acute myocardial infarctions (MI) prevented further ejection fraction deterioration 3 weeks after MI when compared to placebo and fibroblast injected mice. In 2009, Marbán's group also reported functional benefit and reduction of infarct size in a porcine animal model after CDC injection [37], a preclinical model that prompted a phase I clinical trial (CADUCEUS, ClinicalTrials.gov, Identifier NCT00893360).

Nonetheless, the usage of cardiospheres as a source of cardiac stem cells has been refuted. Andersen and coworkers showed that even though cardiospheres can be produced from heart specimens, they do not hold cardiomyogenic potential and simply represent aggregated fibroblasts [38]. This group also found cells that expressed cardiac contractile proteins, such as myosin heavy chain and troponin T, in cardiospheres. However, the findings were attributed to the presence of contaminating heart tissue fragments in the explant-derived cell suspension. By adding a filtration step in which explant-derived cells were passed through cell strainers prior to cardiosphere formation, the presence of cells expressing cardiac contractile proteins was eliminated. In addition, this group showed that phase-bright cells were of hematopoietic origin and did not organize into spheric structures, a characteristic attributed to the fibroblast-like cells.

In response to Andersen's findings, Marbán's group published a revalidation of the CDC isolation method [39]. Using a strategy identical to the one described by Hsieh and colleagues [8], cardiomyocytes were irreversibly labeled with GFP after a tamoxifen pulse (see Fig. 8.1). Isolation of CDCs from these transgenic mouse hearts did not reveal the presence of GFP^{+} cells, refuting the possibility that cardiac differentiation of CDCs was due to the presence of contaminating myocardial tissue fragments. Additionally, they reported that cardiospheres were consistently negative for CD45, indicating that CDCs do not contain cells of hematopoietic origin. The authors also emphasized that Andersen and coworkers used different isolation protocols, which could justify the discrepancies found in results.

Even though they demonstrated that CDCs expressed myosin heavy chain after transplantation into myocardial infarctions in mice [17], indicating that cardiomyogenic differentiation was possible in vivo, an additional mechanism was proposed to explain the improvement in cardiac function. Chimenti and colleagues studied the relative roles of direct regeneration versus paracrine effects promoted by human CDCs in a mouse infarction model [40]. The paracrine hypothesis has been used frequently to explain the beneficial effects observed with several types of adult stem cells or bone marrow–derived cells used in cell therapy experiments. According to this hypothesis, stem cells could act secreting signaling molecules, which may influence cardiomyocyte survival and angiogenesis and could also recruit endogenous cardiac stem cells. Chimenti and coworkers demonstrated that CDCs secrete high levels of insulin growth factor-1 (IGF-1), hepatocyte growth factor (HGF), and vascular endothelial growth factor (VEGF). Moreover, using in vivo bioluminescence assays, the authors showed that no cells could be found in the heart 1 week after injection, even though functional improvement persisted until 3 weeks post-MI. Therefore, it seems that cell persistence is not important for functional improvement, strengthening the paracrine hypothesis. To address this issue, the authors quantified capillary density and viable myocardium analyzing the contributions of human (injected) and mouse (endogenous) cells to each of these variables 1-week post-MI [40]. They found that, for both variables, the contribution of endogenous cells was more prominent than that of injected cells. Hence, the release of factors seems to be more important than direct regeneration in the improvement of cardiac function after cell therapy with CDCs.

Recently, results of a phase I clinical trial using CDCs were published [41]. The safety of autologous intracoronary delivery of CDCs to patients 1.5 to 3 months after MI was evaluated. Cells were obtained from endomyocardial biopsies and cultured according to the protocols previously established by Eduardo Marbán's group. Patients with a recent MI (less than 4 weeks) and left ventricular ejection fraction ranging from 25% to 45% were eligible for inclusion. Twelve months after cell therapy, patients treated with CDCs had a 12.3% decrease in scar size, whereas the control group had a 2.2% reduction, as measured by late enhancement after gadolinium MRI. However, no differences were detected in ejection fraction between cell-treated and control groups. It is important to note that this was a safety study; therefore, phase II double-blinded placebo-controlled clinical trials still need to be performed to access efficacy of therapy with CDCs in humans. Additionally, a more thorough cell biologic characterization of CDCs is required to understand provenience, molecular identity, and mechanism of action of these cells as potential cardioprotective agents.

Side Population

Side population (SP) cells are characterized by their intrinsic capacity to efflux Hoechst dye through an ATP-binding cassette transporter. Upon Hoechst 33342 exposure, SP cells efficiently clear the fluorescent dye thereby becoming a "Hoechst-low" population [42]. Because the cells characteristically appear to the side of Hoechst dye retaining cells in flow cytometry analysis, they were termed side population. These cells were first isolated from the bone marrow and proved to be enriched in long-term repopulating hematopoietic stem cells [43]. A few years later, SP cells started to be isolated from other organs, such as lung, testis, kidney, skin, and skeletal muscle [44].

In 2002, SP cells were first isolated from the heart: myocardial tissue was enzymatically digested and the presence of a side population comprising 1% of total myocardial cells was demonstrated [13]. Later on, in 2005, Pfister and coworkers showed that a specific subset of mouse cardiac SP cells, which were negative for the endothelial marker CD31 and positive for stem cell antigen-1 (Sca-1), exhibited functional cardiomyogenic differentiation capacity [43]. CD31-/Sca-1$^+$ SP cells expressed the cardiac transcription factors GATA4 and MEF2C and the cardiac contractile proteins α-actinin and troponin I, even though mature sarcomeric organization was not present. Additionally, upon co-culture with adult cardiomyocytes, SP cells expressed connexin 43 and 10% of them exhibited ordered myofibrils with striation pattern and spontaneous contractions at low frequency (< 1Hz) [42].

The presence of Sca-1$^+$ cells in the mouse heart had been previously demonstrated by Schneider's group [13,45]. The authors claimed that 14% to 17% of the nonmyocyte cells in the heart were positive for Sca-1, which is a very high frequency for progenitor/stem cells in any organ. These cells were negative for CD45, CD34, Flt-1, Flk-1, and VE-cadherin, whereas 11% to 14% of them were positive for CD31. Cardiac SP cells constitute only 0.03% of heart-derived cells, but 93% of SP cells are positive for Sca-1 [45]. Therefore, in the mouse heart Sca-1 expression seems to be widely distributed through different subsets of nonmyocyte cells and will need to be combined with other markers in order to isolate specific progenitor subpopulations.

Regarding the expression of c-kit, both groups claim that no c-kit$^+$ cells were found in the heart [42,45]. However, Pfister and coworkers argued that c-kit is cleaved by the enzymes used to digest the heart and, thus, its expression cannot be assessed in this type of experiment. Another important point is that Sca-1 is an antigen found in mouse cells and no correspondent molecule exists in humans, which is a major setback for clinical translation. Hence, when isolating human cells, only the SP flow cytometry strategy could be used.

In another work by Pfister and coworkers, the identity of the ATP-binding cassette transporter in cardiac SP cells was analyzed [46]. Among the various transporters in this family, *Abcg2* and *Mdr1* are both capable of extruding Hoechst. However, the SP phenotype in bone marrow cells was attributed exclusively to *Abcg2*, as *Mdr1* knockout mice did not lose the capacity to extrude Hoechst. In the adult heart, the authors demonstrated that $Abcg2^{-/-}$ mice retained their SP phenotype, whereas $Mdr1^{-/-}$ mice completely lost their capacity to extrude Hoechst [46]. Nevertheless, even though *Abcg2* did not participate in the cardiac SP phenotype, knockout mice showed decreased proliferation and increased apoptosis of SP cells, indicating that this molecule might be an important regulator of cardiac SP cell progenitor function [46].

Engraftment and differentiation of Sca-1+/CD31- cardiac SP cells were studied in a myocardial infarction mouse model [47]. PKH26-labeled cells were injected in the peri-infarct area and, after 14 days, were found to co-localize with α-actinin and troponin T, suggesting cardiac differentiation. However, these findings corresponded to less than 4% of the PKH26-labeled cells. Moreover, no functional data were presented in this work or in any other study using cardiac SP cells. Thus, the utility of these cells for cell therapy remains to be defined.

Finally, until very recently, the existence of cardiac SP cells in human hearts had never been demonstrated. Sandstedt and coworkers [48] studied the presence of SP cells in human atria myocardial biopsies. They stated that the left atrium, but not the right, possesses a subset of cells capable of extruding Hoechst, although these results still need to be confirmed.

Islet-1 Positive Cells

Islet-1 (Isl-1) is a LIM homeodomain transcription factor expressed in embryonic cells of the secondary heart field that drives expression towards cardiac lineage cells [49]. The contribution of Isl-1^+ cells to the formation of the atria, the right ventricle, the outflow tract, and specific regions of the left ventricle is substantial during cardiopoiesis [49]. Using transgenic mice and lineage tracing, Laugwitz and coworkers [17] showed that Isl-1^+ cells could be isolated and expanded in culture from neonatal hearts and, after co-culture with wild-type neonatal cardiomyocytes, the Isl-1^+ precursors were found to differentiate into myocytes that fired action potentials and displayed calcium transients; both proved to be sensitive to beat rate and adrenergic stimulation. Importantly, the authors were able to find Isl-1^+ cells in human hearts from 2- to 8-day-old patients.

One year later, the same group, again using transgenic mice and lineage tracing, was able to show that postnatal Isl-1^+ cells gave rise not only to cardiomyocytes, but also to endothelial and smooth muscle cells [29]. This multipotentiality, coupled to the maintenance of the Isl-1$^+$ undifferentiated state when cultured in a cardiac mesenchymal feeder layer, led to the proposal that Isl-1$^+$ cells represented true stem/progenitor cardiac cells [29]. However, a major problem for the use of these cells was their rarity in the adult myocardium. As mentioned previously, in human samples, the cells could only be detected at 2 to 8 days after birth, and even then in extremely small numbers. Therefore, Moretti and coworkers proposed a strategy for derivation and expansion of Isl-1 cells from embryonic stem cells [29]. Using a transgenic ES cell line expressing β-galactosidase under control of the Isl-1 promoter, they were able to isolate the beta-gal$^+$ cells and, although Isl-1 is expressed in many cell types during embryogenesis, they showed that culturing in a cardiac mesenchymal feeder layer led to expansion of the cardiac Isl-1$^+$ precursors. This ES cell derived cardiac precursor expressing Isl-1 was shown to give rise to cardiomyocytes and endothelial and smooth muscle cells, but lineage specification seemed to depend on the expression of other markers besides Isl-1. In fact, based on these results, Moretti et al. [29] proposed a model in which Isl-1 and Nkx2.5$^+$ precursors give rise to cardiomyocytes and smooth muscle cells, whereas Isl-1 and Flk-1$^+$ cells give rise to smooth muscle and endothelial cells.

Multipotent Isl-1$^+$ cells were also found in human fetal hearts. Bu et al. [50] were able to identify Isl-1$^+$ cells in human fetal hearts at 11 and 18 weeks of gestation. They further showed that some Isl-1$^+$ cells were also positive for cardiac (cardiac troponin T), smooth muscle (smooth muscle myosin heavy chain), and endothelial (platelet/endothelial cell adhesion molecule) markers. Using human ES cells and the same strategy of culturing the cells on cardiac mesenchymal feeders, the authors were able to show expansion and differentiation of the Isl-1$^+$ ES-derived cells in vitro, into all three cell types.

In an elegant experiment using transgenic mice that had first and secondary heart field cells differentially labeled by green and red fluorescence, Domian et al. [51] were able to illustrate the existence of distinct progenitor populations in the heart, based on genomic-wide profiling. This work suggests that more than one stem/progenitor cell population may be present in the heart, even though this organ has a limited regenerative capacity. It further points to the possibility of using distinct progenitors, depending on the region of the heart that has to be regenerated.

Although the existence and multipotentiality of Isl-1$^+$ cells in fetal and neonatal hearts is well established, these cells are extremely rare in the adult heart, precluding their use for regenerative therapies [52,53]. Furthermore, we still lack evidence for their regenerative capacity when injected in vivo in murine models of cardiovascular diseases. In fact, a recent work shows that Isl-1$^+$ cells in the adult heart are mainly found in the sinus-atrial node and in the outflow tract, and that their numbers do not increase with aging or after a myocardial infarction [54]. In addition, the authors have been unable to find Isl-1$^+$ cells in the infarct or peri-infarct zone, a necessary although not sufficient criteria to propose a role for these cells in regeneration of the adult heart.

Conclusions and Future Directions

The adult heart has limited regenerative potential and thus it seems counterintuitive that it would be endowed with multiple endogenous stem/progenitor cell types. The distinct stem/progenitor cells described in the literature may represent different developmental stages of a primordial cardiac stem cell, although the distinct developmental origins of the various regions of the hearts may give rise to more than one cardiac stem cell. In this context, it is important to mention that besides progenitors/stem cells, the plasticity of the heart could also, similar to the zebra fish, be based on the redivision of existing cardiomyocytes.

It is obvious that the putative identification of the true cardiac stem cell—presuming that such cells do exist in the heart—using a single marker has yielded many conflicting results and that identifying a set of markers would provide a more robust method for progenitor/stem cell isolation, characterization, and expansion. On the other hand, given the dependence of stem cell properties on the niche, it is arguable that for cell therapy, administering a mixed cell population may yield better results for regenerative purposes. The amazing progress in the field with the discovery of induced pluripotency [55] and the recently proposed direct differentiation of fibroblasts into cardiomyocytes in vitro [56-58] and in vivo [57-59] opens new and exciting possibilities for cell-based therapies in cardiac diseases, although robustness [60] and mechanisms underlying these procedures need to be further investigated.

References

[1] Anversa P, Leri A, Rota M, Hosoda T, Bearzi C, Urbanek K, et al. Stem cells, myocardial regeneration, and methodological artifacts. Stem Cells 2007;25:589–601.

[2] Karsner HT, Saphir O, Todd TW. The state of the cardiac muscle in hypertrophy and atrophy. Am J Pathol 1925;1:351–72.

[3] Anversa P, Kajstura J. Ventricular myocytes are not terminally differentiated in the adult mammalian heart. Circ Res 1998;83:1–14.

[4] Soonpaa MH, Field LJ. Survey of studies examining mammalian cardiomyocytes DNA synthesis. Circ Res 1998;83:15–26.

[5] Von Harsdorf R. Can cardiomyocytes divide? Heart 2001;86:481–2.

[6] Anversa P, Leri A, Kajstura J, Nadal-Ginard B. Myocyte growth and cardiac repair. J Mol Cell Cardiol 2002;34:91–105.

[7] Bergmann O, Bhardwaj RD, Bernard S, Zdunek S, Barnabé-Heider F, Walsh S, et al. Evidence for cardiomyocyte renewal in humans. Science 2009;324:98–102.

[8] Hsieh PC, Segers VF, Davis ME, MacGillivray C, Gannon J, Molkentin JD, et al. Evidence from a genetic fate-mapping study that stem cells refresh adult mammalian cardiomyocytes after injury. Nat Med 2007;13:970–4.

[9] Porrello ER, Mahmoud AI, Simpson E, Hill JA, Richardson JA, Olson EN, et al. Transient regenerative potential of neonatal mouse heart. Science 2011;331:1078–80.

[10] Jopling C, Sleep E, Raya M, Martí M, Raya A, Izpisúa Belmonte JC. Zebrafish heart regeneration occurs by cardiomyocytes dedifferentiation and proliferation. Nature 2010;464:606–9.

[11] Kikuchi K, Holdway JE, Werdich AA, Anderson RM, Fang Y, Egnaczyk GF, et al. Primary contribution to zebrafish heart regeneration by gata4(+) cardiomyocytes. Nature 2010;464:601–5.

[12] Beltrami AP, Barlucchi L, Torella D, Baker M, Limana F, Chimenti S, et al. Adult cardiac stem cells are multipotent and support myocardial regeneration. Cell 2003;114:763–76.

[13] Hierlihy AM, Seale P, Corrinne GL, Rudnicki MA, Megeney LA. The post-natal heart contains a myocardial stem cell population. FEBS Letters 2002;530:239–43.

[14] Oh H, Bradfute SB, Gallardo TD, Nakamura T, Gaussin V, Mishina Y, et al. Cardiac progenitor cells from the adult myocardium: homing, differentiation and fusion after infarction. Proc Natl Acad Sci USA 2003;100:12313–8.

[15] Martin CM, Meeson AP, Robertson SM, Hawke TJ, Richardson JA, Bates S, et al. Persistent expression of the ATP-binding cassette transporter, Abcg2, identifies cardiac SP cells in the developing and adult heart. Dev Biol 2004;265:262–75.

[16] Messina E, De Angelis L, Frati G, Morrone S, Chimenti S, Fiordaliso F, et al. Isolation and expansion of adult cardiac stem cells from human and murine heart. Circ Res 2004;95:911–21.

[17] Laugwitz KL, Moretti A, Lam J, Gruber P, Chen Y, Woodard S, et al. Postnatal isl1+ cardioblasts enter fully differentiated cardiomyocyte lineages. Nature 2005;433:647–53.

[18] Bearzi C, Rota M, Hosoda T, Tillmanns J, Nascimbene A, De Angelis A, et al. Human cardiac stem cells. Proc Natl Acad Sci USA 2007;104:14068–73.

[19] Smith RR, Barile L, Cho HC, Leppo MK, Hare JM, Messina E, et al. Regenerative potential of cardiosphere-derived cells expanded from percutaneous endomyocardial biopsy specimens. Circulation 2007;115:896–908.

[20] Bu L, Jiang X, Martin-Puig S, Caron L, Zhu S, Shao Y, et al. Human ISL1 heart progenitors generate diverse multipotent cardiovascular cell lineages. Nature 2009;460:113–7.

[21] Laflamme MA, Murry CE. Regenerating the heart. Nat Biotechnol 2005;23:845–56.

[22] Barile L, Messina E, Giacomello A, Marbán E. Endogenous cardiac stem cells. Prog Cardiovasc Dis 2007;50:31–48.

[23] Yasuda H, Galli SJ, Geissler EN. Cloning and functional analysis of the mouse c-kit promoter. Biochem Biophys Res Commun 1993;191:893–901.

[24] Reber L, Da Silva CA, Frossard N. Stem cell factor and its receptor c-Kit as targets for inflammatory diseases. Eur J Pharmacol 2006;533:327–40.

[25] Alexeev V, Yoon K. Distinctive role of the cKit receptor tyrosine kinase signaling in mammalian melanocytes. J Invest Dermatol 2006;126:1102–10.

[26] Wu SM, Fujiwara Y, Cibulsky SM, Clapham DE, Lien CL, Schultheiss TM, et al. Developmental origin of a bipotential myocardial and smooth muscle precursor in the mammalian heart. Cell 2006;127:1137–50.

[27] Reith AD, Rottapel R, Giddens E, Brady C, Forrester L, Bernstein A. W mutant mice with mild or severe developmental defects contain distinct point mutations in the kinase domain of the c-kit receptor. Genes Dev 1990;4:390–400.

[28] Mauduit C, Hamamah S, Benahmed M. Stem cell factor/c-kit system in spermatogenesis. Hum Reprod Update 1999;5:535–45.

[29] Moretti A, Caron L, Nakano A, Lam JT, Bernshausen A, Chen Y, et al. Multipotent embryonic isl1+ progenitor cells lead to cardiac, smooth muscle, and endothelial cell diversification. Cell 2006;127:1151–65.

[30] Tallini YN, Greene KS, Craven M, Spealman A, Breitbach M, Smith J, et al. c-kit expression identifies cardiovascular precursors in the neonatal heart. Proc Natl Acad Sci USA 2009;106:1808–13.

[31] Linke A, Müller P, Nurzynska D, Casarsa C, Torella D, Nascimbene A, et al. Stem cells in the dog heart are self-renewing, clonogenic, and multipotent and regenerate infarcted myocardium, improving cardiac function. Proc Natl Acad Sci USA 2005;102:8966–71.

[32] Zaruba MM, Soonpaa M, Reuter S, Field LJ. Cardiomyogenic potential of C-kit(+)-expressing cells derived from neonatal and adult mouse hearts. Circulation 2010;121:1992–2000.

[33] Li M, Naqvi N, Yahiro E, Liu K, Powell PC, Bradley WE, et al. c-kit is required for cardiomyocyte terminal differentiation. Circ Res 2008;102:677–85.

[34] Fazel S, Cimini M, Chen L, Li S, Angoulvant D, Fedak P, et al. Cardioprotective c-kit+ cells are from the bone marrow and regulate the myocardial balance of angiogenic cytokines. J Clin Invest 2006;116:1865–77.

[35] Loffredo FS, Steinhauser ML, Gannon J, Lee RT. Bone marrow-derived therapy stimulates endogenous cardiomyocyte progenitors and promotes cardiac repair. Cell Stem Cell 2011;8:389–98.

[36] Bolli R, Chugh AR, D'Amario D, Loughran JH, Stoddard MF, Ikram S, et al. Cardiac stem cells in patients with ischaemic cardiomyopathy (SCIPIO): initial results of a randomized phase 1 trial. Lancet 2011;378:1847–57.

[37] Johnston PV, Sasano T, Mills K, Evens R, Lee S, Smith RR, et al. Engraftment, differentiation, and functional benefits of autologous cardiosphere-derived cells in porcine ischemic cardiomyopathy. Circulation 2009;120:1075–83.

[38] Andersen DC, Andersen P, Schneider M, Jensen HB, Sheikh SP. Murine "cardiospheres" are not a source of stem cells with cardiomyogenic potential. Stem Cells 2009;27:1571–81.

[39] Davis DR, Zhang Y, Smith RR, Cheng K, Terrovitis J, Malliaras K, et al. Validation of the cardiosphere method to culture cardiac progenitor cells from myocardial tissue. PLoS ONE 2009;4:e7195.

[40] Chimenti I, Smith RR, Li T, Gerstenblith G, Messina E, Giacomello A, et al. Relative roles of direct regeneration versus paracrine effects of human cardiosphere-derived cells transplanted into infracted mice. Circ Res 2010;106:971–80.

[41] Makkar RR, Smith RR, Cheng K, Malliaras K, Thomson LEJ, Berman D, et al. Intracoronary cardiosphere-derived cells for heart regeneration after myocardial infarction (CADUCEUS): a prospective, randomized phase 1 trial. Lancet 2012;379:895–904.

[42] Pfister O, Mouquet F, Jain M, Summer R, Helmes M, Fine A, et al. CD31- but not CD31+ cardiac side population cells exhibit functional cardiomyogenic differentiation. Circ Res 2005;97:52–61.

[43] Goodell MA, Brose K, Paradis G, Conner AS, Mulligan RC. Isolation and functional properties of murine hematopoietic stem cells that are replicating in vivo. J Exp Med 1996;183:1797–806.

[44] Asakura A, Rudnicki MA. Side population cells from diverse adult tissues are capable of in vitro hematopoietic differentiation. Exp Hematol 2002;30:1339–45.

[45] Oh H, Chi X, Bradfute SB, Mishina Y, Pocius J, Michael LH, et al. Cardiac muscle plasticity in adult and embryo by heart-derived progenitor cells. Ann N Y Acad Sci 2004;1015:182–9.

[46] Pfister O, Oikonomopoulos A, Sereti K, Sohn RL, Culler D, Fine GC, et al. Role of the ATP-binding cassette transporter Abcg2 in the phenotype and function of cardiac side population cells. Circ Res 2008;103:825–35.

[47] Liang SX, Tan TYL, Gaudry L, Chong B. Differentiation and migration of Sca1+/CD31- cardiac side population cells in a murine myocardial ischemic model. Int J Cardiol 2010;138:40–9.

[48] Sandstedt J, Jonsson M, Kajic K, Sandstedt M. Left atrium of the human adult heart contains a population of the side population cells. Basic Res Cardiol 2012;107:255.

[49] Cai CL, Liang X, Shi Y, Chu PH, Pfaff SL, Chen J, et al. Isl1 identifies a cardiac progenitor population that proliferates prior to differentiation and contributes a majority of cells to the heart. Dev Cell 2003;5:877–89.

[50] Bu L, Jiang X, Martin-Puig S, Caron L, Zhu S, Shao Y, et al. Human ISL1 heart progenitors generate diverse multipotent cardiovascular cell lineages. Nature 2009;460:113–7.

[51] Domian IJ, Chiravuri M, van der Meer P, Feinberg AW, Shi X, Shao Y, et al. Generation of functional ventricular heart muscle from mouse ventricular progenitor cells. Science 2009;326:426–9.

[52] Genead R, Danielsson C, Andersson AB, Corbascio M, Franco-Cereceda A, Sylvén C, et al. Islet-1 cells are cardiac progenitors present during the entire lifespan: from the embryonic stage to adulthood. Stem Cells Dev 2010;19:1601–15.

[53] Khattar P, Friedrich FW, Bonne G, Carrier L, Eschenhagen T, Evans SM, et al. Distinction between two populations of islet-1-positive cells in hearts of different murine strains. Stem Cells Dev 2011;20:1043–52.

[54] Weinberger F, Mehrkens D, Friedrich FW, Stubbendorff M, Hua X, Müller JC, et al. Localization of islet-1-positive cells in the healthy and infarcted adult murine heart. Circ Res 2012;110:1303–10.

[55] Takahashi K, Yamanaka S. Induction of pluripotent stem cells from mouse embryonic and adult fibroblast cultures by defined factors. Cell 2006;126:663–76.

[56] Ieda M, Fu JD, Delgado-Olguin P, Vedantham V, Hayashi Y, Bruneau BG, et al. Direct reprogramming of fibroblasts into functional cardiomyocytes by defined factors. Cell 2010;142:375–86.

[57] Song K, Nam Y, Luo X, Qi X, Tan W, Huang G, et al. Heart repair by reprogramming non-myocytes with cardiac transcription factors. Nature doi:10.1038/nature11139.

[58] Jayawardena T, Egemnazarov B, Finch E, Zhang L, Payne J, Pandya K, et al. MicroRNA-mediated in vitro and in vivo direct reprogramming of cardiac fibroblasts to cardiomyocytes. Circ Res doi:10.1161/circresaha.112.269035.

[59] Qian L, Huang Y, Spencer C, Foley A, Vedantham V, Liu L, et al. In vivo reprogramming of murine cardiac fibroblasts into induced cardiomyocytes. Nature 2012;485:593–8.

[60] Chen J, Krane M, Deustch M, Wang L, Rav-Acha M, Gregorie S, et al. Inefficient reprogramming of fibroblasts into cardiomyocytes using Gata4, Mef2c, and Tbx5. Circ Res doi:10.1161/circresaha.112.270264.

Chapter 9

The Neural Crest and the Stem Cells of Neural Crest

Andréa Gonçalves Trentin, Giordano Wosgrau Calloni
Departamento de Biologia Celular, Embriologia e Genética, Centro de Ciências Biológicas, Universidade Federal de Santa Catarina, Campus Universitário-Trindade, 88040-900, Florianópolis, Santa Catarina, Brazil

Introduction

The neural crest (NC) has attracted developmental biologists and also evolutionary and evolutionary-developmental biologists for a long time. The NC is a unique structure of the vertebrate embryo. It disappears soon after being formed and, later, its derivatives become widespread all over the body. In addition, the NC plays a major role in head development [1-3]. Although it is derived from ectoderm, some authors suggest that the NC corresponds to the fourth germ layer, illustrating the significance of this structure to the development of the animal body [2-3].

In 1868, the Swiss embryologist Wilhelm His identified a band of cells named *Zwischenstrang* (the intermediate cord), localized between the developing neural tube and the future epidermal ectoderm, as the source of spinal and cranial ganglia. Some years later, in 1879, Arthur Milnes Marshall used for the first time the term *neural crest* for these cells. During the first half of the 20th century, most of the work on the NC was carried out in amphibian embryos [1-2].

Resident Stem Cells and Regenerative Therapy. http://dx.doi.org/10.1016/B978-0-12-416012-5.00009-8

During the 1960s, James Weston and Mac Johnston evaluated the mechanisms of NC cell migration [4-5]. Detailed fate maps of NC cells were obtained during the 1970s and were promoted by the discovery and exploitation of the quail nuclear marker by Nicole Le Douarin, introducing the quail-chick chimeras [6]. The micro-environment encountered by NC cells in their migratory pathways and at the final destination was revealed as a major determinant of their migration, differentiation, and morphogenesis. Syndromes involving one or more cell types derived from NC cells were recognized during this period and classified as *neurocristopathies* [1,7].

In the 1980s and 1990s, significant advances were made concerning the mechanisms of cell diversification into the various derivatives of the NC. Owing to the development of elegant in vivo lineage tracing techniques of single progenitors by Marianne Bronner-Fraser and of in vitro cloning procedures by Nicole Le Douarin, it is now clear that NC stem cells exist and that diversification of NC derivatives are similar to the hematopoietic system. The existence of highly multipotent NC progenitors endowed with both neural and mesenchymal potentials was demonstrated in the early 2000s, although their self-replicating capacity had to be proved. On the other hand, the self-renewal capacity has been demonstrated in some restricted NC progenitors in both mammals and birds [8-12].

Recently, surprising results have been obtained concerning the discovery of neural crest-like stem cells (NCSC) in postmigratory targets of embryonic NC, such as sciatic nerve [12], dorsal root ganglia (DRG) [13-14], the gut [15-16], the skin [17-18], and the tooth [19-20]. These NCSC resemble their embryonic counterparts in their ability to differentiate into a variety of cell types. Considering the broad potential of NCSC and the possibility of their isolation from easily accessible tissues, NC cells have become an ideal model system to study stem cell biology in development and disease [21-22].

In recent years, we have witnessed what could be considered a boom in the stem cell field. It is, of course, the result of the hopes deposited in these cells as a source of healing for a large spectrum of diseases. Nowadays, we have enough knowledge not only to isolate and purify stem cells from different adult sources but also to induce differentiated cells to return to an embryonic state. The knowledge about in vivo and in vitro behavior of hematopoietic, neural, and mesenchymal stem cells advances at the speed of light. Despite enormous insights generated during the early 2000s and constant methodological improvements, the actual nature of the cells that give rise to the so-called mesenchymal stem cells, for instance, sometimes is far from being completely understood. In an attempt to clarify some of these aspects, we discuss in this chapter some insights about what we have been learning in the last few years regarding NC cells. In the first part, we discuss the recent data that we and others have obtained on embryonic NC; and in the second part we consider how NCSC cells contribute to the stem cells population in various NC-derived tissues in fetal and adult organisms. It is possible that at least some of the so-called mesenchymal stem cells, are in fact, originated from NC cells. The knowledge of the real nature of these cells could be crucial to improve several future medical applications.

The Neural Crest

Up to the middle of the 20th century, most of the studies of the NC were performed in fish and amphibians as nicely reviewed by Sven Hörstadius in 1950 [23]. Meanwhile, the comprehension of the paramount features of the NC during ontogeny of vertebrates started just around the 1960s. When the investigations moved away from amphibian toward avian embryos as the organisms of choice, the major problem in NC studies was the absence of permanent cell labeling. This question was partially solved by the use of ^{3}H-thymidine-labeled grafts introduced by James Weston (1963) and Pierre Chibon (1964), which allowed following NC cells during a short period of time in chick embryos [4,24]. In 1969, a landmark in the NC field was achieved when Nicole Le Douarin discovered the quail nuclear marker and developed the quail-chick chimera system to pursue NC cells during avian development. Le Douarin observed differences between quail and chick chromatin organization that allowed distinguishing the cells of one organism from another. When fragments of quail neural tube were grafted in chick embryos, it was possible to analyze several days later not only the routes of NC cell migration but also the contribution of quail NC cells to the different chick organs [6]. These studies were confirmed more recently with the use of mice genetically engineered and cell labeling techniques that identified the cell derivatives as well as the organs populated by NC cells [25-26].

The Derivatives and Plasticity of NC Cells

In fact, the number of cell types that arise from the NC is as remarkable as is the number of tissues and organs to which the NC contributes. The NC is the source of melanocytes, the neurons and glial cells of the peripheral nervous system, and some endocrine cells [1].

The fate map of the NC revealed that the developmental potentials of the cephalic NC cells are greater than those of the trunk NC, as they provide the head with mesenchymal cells. In 1893, Julia Platt designated these cells *mesectoderm* or *ectomesenchyme*, in order to distinguish them from the mesenchyme that is derived from the mesodermal germ layer. The mesectoderm, which turned out to play a crucial role in vertebrate head development, has many unique developmental characteristics. Mesectodermal cells are able to differentiate into tendon, connective, cartilage, bone, and adipose tissues [1,27].

Equally remarkable is the plasticity exhibited by NC cells during in vivo and in vitro experiments, making NC a wonderful model for studying the mechanisms of cell fate decisions and differentiation. The fate of NC cells is not fixed at the time that they arise from the neural folds and rather depends on the environment encountered during migration and at sites of arrest. For example, experiments performed in chick embryo revealed that trunk NC cells are unable to give rise to mesenchymal phenotypes when transplanted to the cephalic level [28]. Despite this lack of mesenchymal fate in vivo, upon special culture conditions trunk NC cells are able to give rise to chondrocytes and osteocytes in vitro [29-31]. Therefore, NC cells are endowed with a plentiful and versatile potential that could be modulated through the clues found by these cells during their process of differentiation [1].

The Multipotentiality of NC Cells

The wide variety of cell types derived from the NC raises an important question about the multipotentiality of NC cells: Are NC cells in fact multipotent stem cells or do the various phenotypes arise from precursors already committed? This issue was exhaustively addressed in the past decades by analyzing the NC clonal progeny generated in vivo or in vitro. In 1975, Cohen and Konigsberg developed a pioneering work culturing single NC cells that emigrated from explanted neural primordium of quails [32]. Sieber-Blum and Cohen in 1980, using clonal cultures, provided the first evidence of NC cell multipotency. They identified colonies formed by pigmented, unpigmented, or both types of cells, showing that NC cells could give rise to different cell phenotypes originating from a common progenitor [33]. When cultured quail NC cell colonies were reintroduced back into embryos, these cells were able to contribute to NC derivatives such as the sympathetic ganglia, adrenal gland, and aortic plexus as well as various populations of neurons [34].

Beyond the 1980s, a series of elegant in vivo studies performed by Bronner-Fraser and Fraser consisted of injecting single cells of the trunk NC cells of avian embryos with lysinated rhodamine dextran to monitor migration and cell fate [35-36]. These studies together with the in vitro clonal analysis (Fig. 9.1) performed by the laboratory of Le Douarin revealed that the NC consists of a collection of multipotent and intermediate progenitors endowed with different proliferation and differentiation potentials, suggesting a model of progressive restrictions in NC development as observed in the hematopoietic system [8]. Experiments that randomly selected single quail cells from the mesencephalic NC yielded clones of different sizes and phenotypic compositions. Rare highly multipotent progenitors, able to generate neurons (of different types), glial cells, melanocytes, and mesenchymal derivatives (e.g., cartilage nodules and fibroblasts) were recorded and designated *mesenchymal-neural* progenitors [9].

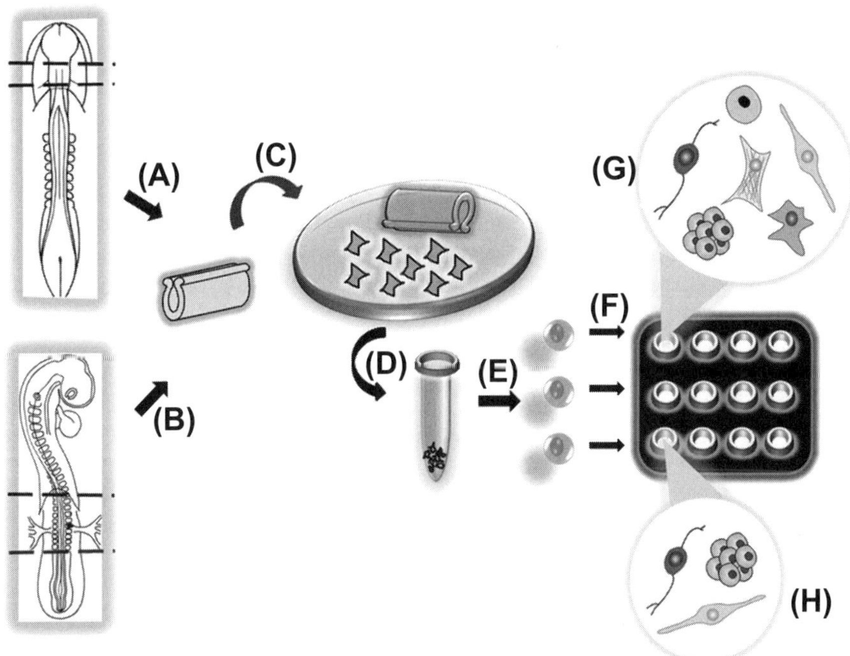

Figure 9.1 Methodological approach to study neural crest (NC) potentiality in an avian model [10,41]. Neural tubes at mesencephalic **(A)** or trunk **(B)** levels are carefully excised from quail embryos at 6 somite or 20 to 25 somite stages, respectively. Neural tubes are submitted to enzymatic digestion and microsurgical procedures and plated into plastic dishes **(C)**. After 15 or 24 hours, the NC cells that migrate out of the neural tubes to the culture dish are collected by trypsinization, centrifuged, and individually seeded on mitomycin arrested 3T3 fibroblast feeder-layers **(C-F)**. Clonal cultures are maintained for an additional 10 days in complex medium (enriched with growth factors and hormones), fixed, and stained by immunocytochemistry to specific phenotypic markers. The potentiality of the founder cell of the colony is identified *a posteriori* by the presence of cell markers. The highest multipotent progenitors are able to give rise to glial cells, neurons, melanocytes, smooth muscle cells, cartilage nodules, and osteoblasts **(G)**. More restricted progenitors are also found **(H)**.

Extensive studies performed in avian NC during the 1990s up to the middle 2000s [9-10,37-40] confirmed these findings, nevertheless a question persisted: Why were just a few mesenchymal-neural progenitors observed? How could the rarity of the chondrocytic phenotype in NC cell cultures in spite of the enormous in vivo contribution of cranial NC to vertebrate skull be explained? To address these questions, the classical methodology adopted to isolate and culture NC cells was optimized to obtain a high number of cartilage nodules and also osteoblastic derivatives. Exhaustive single cell culture analysis performed by one of the authors of this chapter revealed that in fact 92% of the cranial NC progenitors are endowed with mesenchymal-neural potentials, 7% of which are able to give rise to six different phenotypes (i.e., glial cells, neurons, melanocytes, smooth muscle cells, chondrocytes, and osteoblasts, named GNMFCO). To date, the GNMFCO progenitor is considered the highest multipotent NC cell described so far (Fig. 9.2) [41-42]. Therefore, the majority of NC cells at the cephalic level correspond in fact to highly multipotent precursors, which are comparable to the multipotent hematopoietic stem cells (see Fig. 9.2) [31,41-42]. One striking result obtained by the analysis of thousands of clones along all these years was that most of the precursors were able to

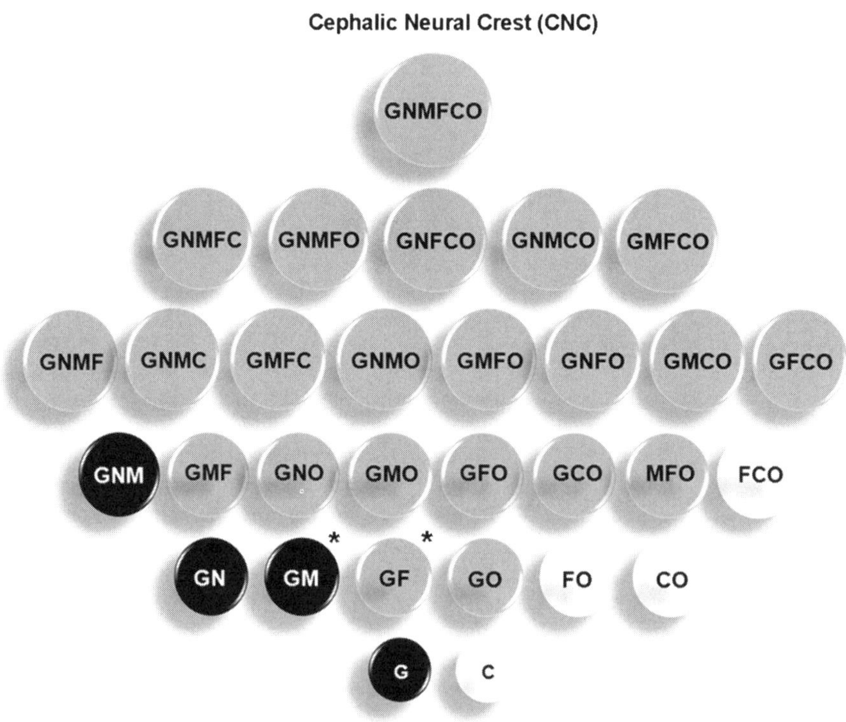

Figure 9.2 Schematic representation of NC progenitors obtained from the analysis of avian mesencephalic NC single cell cultures. Data from in vitro clonal analysis [41] indicate that the majority of mesencephalic NC are composed of neural-mesenchymal progenitors (gray), while only-neural (black) or only-mesenchymal (white) progenitors are rare. The highly multipotent NC progenitor (GNMFCO) is able to give rise to glial cells (G), neurons (N), melanocytes (M), smooth muscle cells (F), chondrocytes (C) and osteocytes (O). Glial cells and osteocytes are the most common phenotypes that represent neural and mesenchymal lineages, respectively. The bipotent GM and GF progenitors (marked by asterisks) are able to self-renew and behave as stem cells [10].

yield glial cells, suggesting that the gliogenic differentiation potential could constitute a general marker of uncommitted NC cells [10,41,43].

In mammals the study of the developmental capabilities of individual NC cells started in rats with the identification of a surface marker for rat NC cells, the low-affinity nerve growth factor receptor (p75). The p75-positive NC cells also express nestin (an intermediate filament marker that is expressed by all neuroepithelial precursor cells) and do not express some lineage-specific markers for glial cells, neurons, or other NC derivatives. When grown in feeder-layer free cultures, these p75-positive and nestin-positive progenitors eventually differentiate into neurons, glial cells, and smooth muscle cells [11]. In the mouse, the development of a clonal system for NC cells was proposed a year later, in 1993, by the Sieber-Blum group [44].

Because of ethical and methodological difficulties, human embryonic NC cells were not isolated until 2008, when Etchevers and collaborators obtained NC primary cultures from explanted neural tubes. These cells are morphologically and phenotypically similar to murine or avian NC cells and can be propagated and frozen for many cycles. In addition, the molecular profile of human embryonic NC cells is similar to that of embryonic pluripotent stem cells, expressing markers such as Sox2, Nanog, and POU5F1 [45].

Self-Renewal Capacity of NC Cells

In addition to the potential to differentiate into different cell types, a stem cell is defined by its capacity to proliferate and self-renew. By this concept, are NC cells true stem cells? Their multipotentiality has been extensively demonstrated, but what about the self-renewal?

The first evidence of NC self-renewal came from experiments in rat NC obtained by Stemple and Anderson (1992). They showed that trunk NC cells produce in vitro autonomic neurons, glial cells, and smooth muscle cells. These oligopotent progenitors can be propagated in culture during some generations, maintaining their potential, and thus are able to self-renew [11].

Further evidence of NC self-renewal was obtained several years later by one of the authors of this chapter using serial subcloning experiments in quail NC. At that time, it was demonstrated that two subpopulations of bipotent NC progenitors (glial-melanocyte and glial-smooth muscle progenitors) are also endowed with self-renewal capacity [10].

Although the self-renewal ability of the most multipotent NC cells has not been demonstrated yet, these findings indicate that at least some intermediate bi- and tripotent progenitors of the NC display the stem cell property of self-renewal. Despite the methodological difficulty to properly address this issue, massive efforts should be made to finally establish the stemness of NC cells.

The Microenvironment and NC Cells

NC cells are highly responsive to microenvironment factors that they encounter during their migratory pathways and at their final destination. In vivo and in vitro studies revealed that environmental signals regulate fate decisions in NC cells. These factors could act selectively, favoring the development of a given lineage or, alternatively, instructing an NC cell to adopt a certain fate at the expense of other possibilities [46-47].

Some growth factors have been suggested as instructive to NC cells: depending on the context, transforming growth factor β (TGFβ) promotes smooth muscle differentiation or autonomic neurogenesis. Bone morphogenetic protein type 2 (BMP2) promotes autonomic neurogenesis by up-regulating the expression of the helix-loop-helix (bHLH) transcription factor Mash-1 at the expense of gliogenesis [47-48]. The acquisition of a glial phenotype is controlled by Neuregulin-1 [46] and Notch activation [49]. Signals that drive multipotent NC cells to form sensory neurons involve the activation of Wnt/β-catenin signaling, which is not only required but also sufficient to trigger sensory neurogenesis. Together with BMP, Wnt factors play an important role in the maintenance of multipotent NC stem cells, as shown in the mouse [50-51].

Among the factors that have instructive influence on NC cells is sonic hedgehog (Shh), which strongly promotes the chondrogenic cell fate and chondrocyte differentiation [31]. Moreover, Shh also enhances the development of common progenitors for bone and cartilage (designated as osteochondrogenic progenitors) at the expense of progenitors yielding dermal-like osteoblasts [41]. These data support the important roles exerted by Shh during in vivo skeletogenic development [52-53]. However, the most striking finding was that Shh favors the development of the multipotent mesenchymal-neural progenitor, GNMFCO, arguing that Shh enhances NC cell multipotentiality [31,41].

On the other hand, stem cell factor (SCF) promotes the survival of NC and, in combination with nerve growth factor (NGF), brain-derived neurotrophic factor (BDNF), and neurotrophin 3 (NT-3), is trophic for the melanocytes [54]. Furthermore, endothelin 3 (ET3) was reported to promote the self-renewal of the bipotent glial-melanogenic progenitor and their differentiation to melanocyte phenotype [10]. ET3 was also shown to promote phenotypic reversion of both differentiated skin melanocytes and differentiated sciatic nerve Schwann cells to the bipotent state [55-56]. On the other hand, fibroblast growth factor-type 2 (FGF2) was shown to promote glial differentiation, whereas the epidermal growth factor (EGF) drives the NC to the neuronal and melanogenic differentiation pathways [57].

Extracellular matrix molecules have also been shown to influence NC cell migration, proliferation, differentiation, and fate. A number of extracellular molecules with permissive influence on NC cell adhesion and migration, such as fibronectin, laminin, type I and type IV collagen, and tenascin, have been identified in NC cell migratory pathways [58-61]. Recently in our laboratory we demonstrated that fibronectin influences the differentiation of NC cells to smooth muscle phenotype by stimulating the survival of oligopotent progenitors and thus selectively influencing NC cell fate [62].

Neural Crest-Derived Stem Cells (NCSCs)

Although the multipotency of NC cells has been demonstrated, some important questions remained unsolved until recently: How long do these cells retain stem cell properties during development? Are NC cells restricted to the embryos, or can they also be found in adult tissues? It is now apparent that some cells capable of forming a broad range of NC derivatives seem to persist into adulthood. Cells with similar properties to the premigratory NC cells have been isolated from the post-migratory phase of NC development, postnatal tissues, and even from the adult [21-22]. These tissues contain precursor cells that either retain or recapitulate features of NC cells, and we shall refer to them as *neural crest-derived stem cells* (NCSCs) (Fig. 9.3).

Sciatic Nerve and Gut

In 1999, Morrison and coworkers obtained the first evidence of persistence of NC cells in latter phases of development. Cells positives to the NC cell marker p75 were isolated from the rat embryonic sciatic nerve (using fluorescence-activated cell sorting [FACS]). In clonogenic assays, these cells were able to self-renew and to

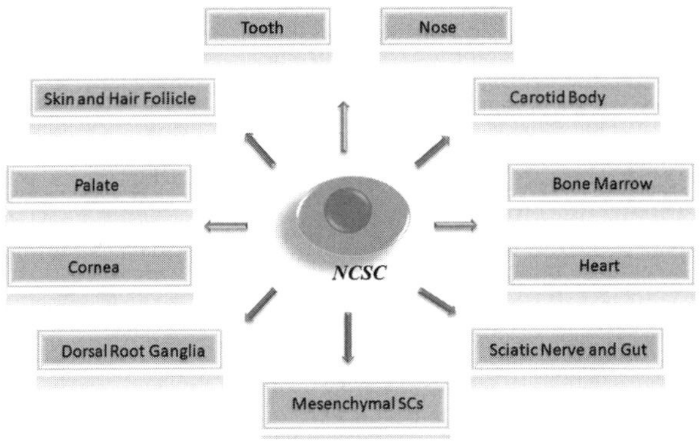

Figure 9.3 Neural crest-derived stem cells (NCSCs) have been identified and isolated from the postmigratory phase of NC development, postnatal tissues, and even in the adult. These cells represent important alternative sources of viable and autologous cells for regenerative medicine.

generate neurons, glial cells, and smooth muscle cells, the main derivatives of trunk NC [12].

In the following decade, Morrison's group also isolated NCSCs from the gut of fetal and adult animals, again by selection of p75-positive cells. They demonstrated that adult gut NCSCs self-renew less efficiently and differentiate into a narrower range of neuronal subtypes than fetal gut NCSCs. Interestingly, NCSCs from gut were sensitive to neurogenic factors, whereas NCSCs from sciatic nerve were sensitive to gliogenic factors in vitro. When transplanted into developing peripheral nerve in vivo, gut NCSCs gave rise mainly to neurons, whereas sciatic NCSCs gave rise to glial cells suggesting that postmigratory microenvironment factors control cell phenotype [16,63-64]. In addition, NCSCs from the gut transplanted in rat stomach exhibited enhanced migration and expression of enteric markers when compared with NCSCs from sciatic nerve [65].

Dorsal Root Ganglion (DRG)

In 1999, Sommer's group isolated cells similar to the premigratory NC cells from DRG of rat embryos, again by FACS and based in the expression of p75. As for sciatic nerve and gut, these cells display clonogenic ability and are able to generate neurons, glial cells, and smooth muscle cells [66]. Subsequent studies revealed that the boundary cap, a cluster of cells derived from the NC that will colonize the DRG during embryogenesis, has a population of multipotent NCSCs that are able to differentiate into sensory neurons in vitro [67-68] and in vivo after transplantation [69]. Importantly, these cells can differentiate into myelin-forming cells and promote remyelination of central nervous system (CNS) lesions, suggesting a potential therapeutic application [70]. In addition, recent studies also isolated NCSCs from mammalian adult DRG [13-14,71].

Skin

Among the various tissues reported to contain NCSCs, the skin is perhaps the most intriguing and therapeutically significant. The skin is undoubtedly the most accessible tissue of the body for autologous cell replacement therapies and regenerative medicine. The skin is a large, complex, and highly regenerative organ that houses a number of different precursors and stem cell populations.

In 2001, a seminal paper by Toma and coworkers described a novel stem cell population in the skin of both mouse and human, which was named *skin-derived precursor cells* (SKPs) [17]. These cells were isolated and expanded by the methodology normally used to culture stem cells of the CNS. SKPs are cultured as spheres, express a range of NC cell markers such as Slug, Snail, Twist, Pax3, and Sox9, and are able to self-renew and to generate neurons and glial cells.

Mesenchymal derivatives such as smooth muscle cells, adipocytes, osteocytes, and chondrocytes can also be obtained by specific differentiation of SKPs. In addition, it was demonstrated that SKPs derive from a dermal precursor that resides within the dermal papilla and dermal sheath of hair and whisker follicles [72-74]. A report from the same group revealed that facial and trunk SKPs have different developmental origins. Whereas facial SKPs are derived from the NC, trunk SKPs have a mesodermal origin. Although trunk SKPs do not express the NC marker p75 [17], they form spheroids, are able to self-renew, and exhibit neural and non-neural potentials. Thus, in spite of the different developmental origin, both facial and trunk SKPs generate precursors with highly similar phenotype [75].

Similar results were obtained from Wong and coworkers (2006) that isolated NCSCs from adult trunk skin of both mouse and human. These skin-derived cells are also multipotent and able to generate neural and non-neural lineages in clonal cultures. However, in contrast to the findings obtained by Toma and colleagues [17], they express the NC cell marker p75 in addition to Sox10. Furthermore, in vitro and in vivo fate map experiments revealed the origin of these cells in the face and trunk skin of the mouse, whereas in whisker follicles the NC derived cells are associated with mesenchymal structures; in the trunk, they are related to the melanocytes and glial lineages of the hair follicle [76].

Using a different experimental approach (adherent cultures of bulge explants), Sieber-Blum's group described in 2004 the presence of multipotent NC cells in the adult hair follicle of mouse whiskers, which were named *epidermal neural crest cells* (EPI-NCSCs). These cells are localized in the bulge region of hair follicle and extensively migrate out of the tissue explants to the culture dish as in vitro embryonic NC cells do. Their NC characteristics were demonstrated by lineage tracing experiments using transgenic mice and by the expression of NC markers such as Sox 10. In addition, they have clonogenic and self-renewing capacities and can originate neurons, melanocytes, glial cells, and smooth muscle cells in vitro [77]. Cells with similar characteristics were also isolated from the bulge area of adult human hair follicles [78]. Moreover, the molecular signature of both mouse and human EPI-NCSCs include NC markers and stem cell genes such as those used to create induced pluripotent stem (iPS) cells [79].

Melanocyte Stem Cell

Melanocytes are originated during vertebrate development from a subpopulation of NC cells that migrate along the dorsolateral pathway (between epidermis and dermis) and will colonize the skin [1,3]. Glial cells and melanocytes share a common NC progenitor [55,80] that can self-renew in vitro stimulated by ET3 and thus display stem cell properties [10]. In addition, glial-melanocyte progenitors were also obtained by dedifferentiation of sciatic nerve Schwann cells or skin pigmented melanocytes after exposure to ET3, demonstrating the high plasticity of NC cells and NC derivatives [55-56].

Upon mammalian development, melanoblasts distribute along the entire epidermis and colonize the developing hair follicle. A population of melanocyte stem cells resides in the lower permanent portion of the hair follicle (the bulge region) and is responsible for the maintenance of the hair pigmentary system during hair cycles [81]. These melanocytic stem cells are immature, slow cycling, and self-renewing; in fact, impairment in melanocyte stem cell maintenance leads to hair graying [82]. In spite of the growing interest in this stem cell population, its precise characteristics remain to be elucidated.

Tooth

Many tissues of the cranial region are derived from the NC. In the developing head, the facial NC migrates from the posterior midbrain and hindbrain regions to the branchial arches where they populate almost the entire mesenchyme adjacent to the oral epithelium. Interactions of these ectomesenchymal NC cells with the surrounding tissues induce the differentiation of NC cells to form mandibular bones, cartilages, dentin, cementum, periodontal ligament, and other connective tissues. Elegant experiments of chick and mouse chimera performed by Mitsiadis and coworkers (2003) illustrate the importance of cranial NC in odontogenesis. Transplantation of mouse NC into chick embryos allowed the growth of tooth germs in chick [83].

In addition, multipotent NCSCs have been identified in a number of mammalian and human odontogenic lineages, including dental follicle, dental pulp, periodontal ligament, and the mandibular process of embryos and adults. These cells express NC cell markers such as nestin and p75 and are able to differentiate into neural and mesenchymal cells [19-20,84]. Moreover, NCSCs isolated from human tooth respond to osteogenic stimulation to form a simply ordered bonelike matrix [85]. Further studies are necessary to better characterize and isolate these cells.

Oral Mucosa

The connective tissue of the face originates from NC. Cells of the oral mucosa lamina propria (OMLP) constitute the connective tissue that lines the oral cavity of mammals. In 2010, two independent groups demonstrated the existence of NCSCs in humans OMLP. Marynka-Kalmani and coworkers observed the expression of p75, Sox2, and Oct4 in cells constituting cordlike structures in the human OMLP. Interestingly, some of these cells showed superposed expression of Oct4 and p75, suggesting that these cords might constitute a niche of NCSCs in the adult oral mucosa [86]. Moreover, Davies and colleagues also found OMLP cells expressing NC markers like Slug, Snail, Twist, and Sox10 within developing colonies. OMLP cells also display endogenous expression of the embryonic stem cells markers Sox2, Klf-4, and Oct4 [87]. In both studies, NCSCs were able to differentiate into neural and mesenchymal derivatives.

Palate

In the head of vertebrates, the secondary palate is a highly regenerative and innervated tissue that receives direct contribution of NC cells. In 2009, Widera and coworkers described the presence of NCSCs in the hard palate of adult rats. These cells are able to form neurospheres, are highly plastic, and display NC markers such as nestin and p75. In addition, they also express Klf4, Sox2, Oct4, and c-Myc [88].

Nose

The olfactory ensheathing cells (OECs) are a unique class of vertebrate glial cells that envelop bundles of olfactory axons, both peripherally in the olfactory nerve and within the olfactory nerve layer of the olfactory bulb. The embryologic origin of OECs was initially attributed to the placodes [89]. However, a series of elegant fate-mapping experiments using transgenic chick and mouse models have recently demonstrated that OECs are of NC origin [90]. In addition, using different molecular techniques, Delorme and colleagues demonstrated that the human nose harbors a niche of olfactory stem cells that express neural and NC-related genes, which were named as olfactory ectomesenchymal stem cells (OE-MSC) [91].

Cornea

The adult mouse cornea is another source of NCSCs that displays sphere-forming ability, expresses the NC markers nestin, Twist, Snail, Slug, Sox9, and Musashi-1, and is able to self-renew and differentiate into neural and mesenchymal derivatives [92].

Heart

NCSCs were also identified in the heart of adult mice. These cells express nestin and Musashi-1, display sphere-forming ability, and differentiate into neurons, glial cells, and smooth muscle cells. Upon transplantation into chick embryos, heart NCSCs migrate to the cardiac outflow tract, DRG, sympathetic ganglia, and spinal nerves [93].

Carotid Body

The carotid body (CB) corresponds to an oxygen-sensing organ located at the carotid bifurcation. The CB is derived from the NC, as demonstrated by Nicole Le Douarin in 1986 [94]. In 2007, Pardal and coworkers, using cell fate mapping experiments in transgenic mice, confirmed the neuroectodermal origin of this structure [95]. They also showed that adult CB contains a population of multipotent and self-renewing NCSCs able to form clonogenic colonies. These cells express nestin and the glial cell marker, glial fibrillary acidic protein.

Bone Marrow

One of the most surprising pieces of data is the evidence of NCSC in the aorta-gonad-mesonephros region, circulating blood, and liver of embryonic rats. These findings suggest that NCSCs migrate through the bloodstream to the bone marrow. In vitro, bone marrow–derived NCSCs also originate clonal spheres, self-renew, and differentiate into neurons, glial cells, and smooth muscle cells in vitro [14].

Are NC Cells a Source of Mesenchymal Stem Cells (MSCs)?

In 2007, Takashima and coworkers demonstrated that the earliest wave of MSCs in the mice embryonic trunk originates from the neuroepithelium instead of the mesoderm. The Sox1-expressing neuroepithelium gives rise to MSCs in part through a transient NC intermediate stage that is later replaced by MSCs from unknown sources. Using genetically engineered mice, the authors concluded that the MSCs are a definite in vivo entity of multiple developmental origins, including the NC [96].

Concluding Remarks

Extensive in vitro and in vivo experiments performed in avian and mammalian embryos have demonstrated the high multipotency, plasticity, and self-renewal capacities of NC cells, attesting to their truly stem cell nature. Importantly, the majority of these embryonic NC cells are able to differentiate into several phenotypes of both neural and mesenchymal lineages. In recent years, cells with characteristics of embryonic NC cells have been demonstrated in several adult organs.

Therefore, the identification of NC-derived progenitor cells in diverse accessible mammalian tissues, such as skin, hair follicle, tooth, and bone marrow, opens new strategies for developing autologous cell replacement therapies in regenerative medicine. NCSCs are now considered promising sources for cellular therapy of several degenerative diseases. Hirschsprung's disease, for instance, one of the most common diseases affecting the enteric nervous system, is characterized by the absence of enteric ganglia due to defects in gut colonization by NC progenitor cells and is a potential candidate for future cell replacement therapies [97-98].

In addition, the discovery that NCSCs can be isolated from skin and expanded in vitro raises a number of exciting biotechnologic and therapeutic prospects. In fact, NCSCs from this accessible location have been utilized with some success in animal models in attempts to repair lesions of sciatic nerve [99] and spinal cord without evidence of tumorigenesis [100-101]. Furthermore, a number of studies have demonstrated the possibility of using myelinating glial cells derived from NCSCs in remyelinating and reparative strategies for congenital and acquired disorders of myelin [70,102-103]. In addition, both human and rodent NCSCs isolated from skin were demonstrated to contribute to wound healing [73] and repair of injured bone [74].

Acknowledgments

The authors would like to thank Nicole Le Douarin and Elisabeth Dupin for their constant support and encouragement.

References

[1] Le Douarin NM, Kalcheim C. The neural crest. 2nd ed. Cambridge: Cambridge University Press; 1999.

[2] Hall BK. The neural crest and neural crest in vertebrate development and evolution. New York: Springer 400; 2009.

[3] Gilbert SF, editor. Developmental biology. 9th ed. Sinauer Associates, Inc; 2010. p. 711.

[4] Weston JA. A radioautographic analysis of the migration and localization of trunk neural crest cells in the chick. Dev Biol 1963;6:279–310.

[5] Johnston MC. A radioautographic study of the migration and fate of cranial neural crest cells in the chick embryo. Anat Rec 1966;156(2):143–55.

[6] Le Douarin N. Details of the interphase nucleus in Japanese quail (Coturnix coturnix japonica). Bull Biol Fr Belg 1969;103(3):435–52.

[7] Johnston MC. The neural crest in abnormalities of the face and brain. Birth Defects Orig Artic Ser 1975;11(7):1–18.

[8] Baroffio A, Dupin E, Le Douarin NM. Clone-forming ability and differentiation potential of migratory neural crest cells. Proc Natl Acad Sci USA 1988;85(14):5325–9.

[9] Baroffio A, Dupin E, Le Douarin NM. Common precursors for neural and mesectodermal derivatives in the cephalic neural crest. Development 1991;112(1):301–5.

[10] Trentin A, et al. Self-renewal capacity is a widespread property of various types of neural crest precursor cells. Proc Natl Acad Sci USA 2004;101(13):4495–500.

[11] Stemple DL, Anderson DJ. Isolation of a stem cell for neurons and glia from the mammalian neural crest. Cell 1992;71(6):973–85.
[12] Morrison SJ, et al. Prospective identification, isolation by flow cytometry, and in vivo self-renewal of multipotent mammalian neural crest stem cells. Cell 1999;96(5):737–49.
[13] Li HY, Say EH, Zhou XF. Isolation and characterization of neural crest progenitors from adult dorsal root ganglia. Stem Cells 2007;25(8):2053–65.
[14] Nagoshi N, et al. Ontogeny and multipotency of neural crest-derived stem cells in mouse bone marrow, dorsal root ganglia, and whisker pad. Cell Stem Cell 2008;2(4):392–403.
[15] Young HM, et al. A single rostrocaudal colonization of the rodent intestine by enteric neuron precursors is revealed by the expression of Phox2b, Ret, and p75 and by explants grown under the kidney capsule or in organ culture. Dev Biol 1998;202(1):67–84.
[16] Kruger GM, et al. Neural crest stem cells persist in the adult gut but undergo changes in self-renewal, neuronal subtype potential, and factor responsiveness. Neuron 2002;35(4): 657–69.
[17] Toma JG, et al. Isolation of multipotent adult stem cells from the dermis of mammalian skin. Nat Cell Biol 2001;3(9):778–84.
[18] Sieber-Blum M, Grim M. The adult hair follicle: cradle for pluripotent neural crest stem cells. Birth Defects Res C Embryo Today 2004;72(2):162–72.
[19] Coura GS, et al. Human periodontal ligament: a niche of neural crest stem cells. J Periodontal Res 2008;43(5):531–6.
[20] Mitsiadis TA, Graf D. Cell fate determination during tooth development and regeneration. Birth Defects Res C Embryo Today 2009;87(3):199–211.
[21] Delfino-Machin M, et al. The proliferating field of neural crest stem cells. Dev Dyn 2007;236(12):3242–54.
[22] Shakhova OSL. Neural crest-derived stem cells. In: Gage, F FHW, editor. StemBook. The Stem Cell Community; 2010. p. 1–13.
[23] Hörstadius S. The neural crest: its properties and derivatives in the light of experimental research. London: Oxford University Press; 1950.
[24] Chibon P. Analysis by the method of nuclear labelling with tritiated thymidine of derivatives of the cephalic neural crest in the Urodele Pleurodeles waltlii Michah. C R Hebd Seances Acad Sci 1964;259:3624–7.
[25] Pietri T, et al. The human tissue plasminogen activator-Cre mouse: a new tool for targeting specifically neural crest cells and their derivatives in vivo. Dev Biol 2003;259(1):176–87.
[26] Yamauchi Y, et al. A novel transgenic technique that allows specific marking of the neural crest cell lineage in mice. Dev Biol 1999;212(1):191–203.
[27] Billon N, et al. The generation of adipocytes by the neural crest. Development 2007; 134(12):2283–92.
[28] Le Lievre CS, et al. Restrictions of developmental capabilities in neural crest cell derivatives as tested by in vivo transplantation experiments. Dev Biol 1980;77(2):362–78.
[29] McGonnell IM, Graham A. Trunk neural crest has skeletogenic potential. Curr Biol 2002;12(9):767–71.
[30] Abzhanov A, et al. Dissimilar regulation of cell differentiation in mesencephalic (cranial) and sacral (trunk) neural crest cells in vitro. Development 2003;130(19):4567–79.
[31] Calloni GW, et al. Sonic hedgehog promotes the development of multipotent neural crest progenitors endowed with both mesenchymal and neural potentials. Proc Natl Acad Sci USA 2007;104(50):19879–84.
[32] Cohen AM, Konigsberg IR. A clonal approach to the problem of neural crest determination. Dev Biol 1975;46(2):262–80.

[33] Sieber-Blum M, Cohen AM. Clonal analysis of quail neural crest cells: they are pluripotent and differentiate in vitro in the absence of noncrest cells. Dev Biol 1980;80(1):96–106.
[34] Bronner-Fraser M, Sieber-Blum M, Cohen AM. Clonal analysis of the avian neural crest: migration and maturation of mixed neural crest clones injected into host chicken embryos. J Comp Neurol 1980;193(2):423–34.
[35] Bronner-Fraser M, Fraser S. Developmental potential of avian trunk neural crest cells in situ. Neuron 1989;3(6):755–66.
[36] Bronner-Fraser M, Fraser SE. Cell lineage analysis reveals multipotency of some avian neural crest cells. Nature 1988;335(6186):161–4.
[37] Dupin E, et al. Schwann-cell differentiation in clonal cultures of the neural crest, as evidenced by the anti-Schwann cell myelin protein monoclonal antibody. Proc Natl Acad Sci USA 1990;87(3):1119–23.
[38] Dupin E, Le Douarin NM. Development of melanocyte precursors from the vertebrate neural crest. Oncogene 2003;22(20):3016–23.
[39] Lahav R, et al. Endothelin 3 selectively promotes survival and proliferation of neural crest-derived glial and melanocytic precursors in vitro. Proc Natl Acad Sci USA 1998;95(24):14214–9.
[40] Baroffio A, Dupin E, Le Douarin NM. Cell lineage studies in avian neural crest ontogeny. Adv Exp Med Biol 1990;265:53–62.
[41] Calloni GW, Le Douarin NM, Dupin E. High frequency of cephalic neural crest cells shows coexistence of neurogenic, melanogenic, and osteogenic differentiation capacities. Proc Natl Acad Sci USA 2009;106(22):8947–52.
[42] Dupin E, Calloni GW, Le Douarin NM. The cephalic neural crest of amniote vertebrates is composed of a large majority of precursors endowed with neural, melanocytic, chondrogenic and osteogenic potentialities. Cell Cycle 2010;9(2):238–49.
[43] Dupin E, et al. Neural crest progenitors and stem cells. CR Biol 2007;330(6-7):521–9.
[44] Ito K, Morita T, Sieber-Blum M. In vitro clonal analysis of mouse neural crest development. Dev Biol 1993;157(2):517–25.
[45] Thomas S, et al. Human neural crest cells display molecular and phenotypic hallmarks of stem cells. Hum Mol Genet 2008;17(21):3411–25.
[46] Shah NM, et al. Glial growth factor restricts mammalian neural crest stem cells to a glial fate. Cell 1994;77(3):349–60.
[47] Shah NM, Groves AK, Anderson DJ. Alternative neural crest cell fates are instructively promoted by TGFbeta superfamily members. Cell 1996;85(3):331–43.
[48] Le Douarin NM, Dupin E. Multipotentiality of the neural crest. Curr Opin Genet Dev 2003;13(5):529–36.
[49] Morrison SJ, et al. Transient Notch activation initiates an irreversible switch from neurogenesis to gliogenesis by neural crest stem cells. Cell 2000;101(5):499–510.
[50] Lee HY, et al. Instructive role of Wnt/beta-catenin in sensory fate specification in neural crest stem cells. Science 2004;303(5660):1020–3.
[51] Kleber M, et al. Neural crest stem cell maintenance by combinatorial Wnt and BMP signaling. J Cell Biol 2005;169(2):309–20.
[52] Chiang C, et al. Cyclopia and defective axial patterning in mice lacking sonic hedgehog gene function. Nature 1996;383(6599):407–13.
[53] Brito JM, Teillet MA, Le Douarin NM. An early role for sonic hedgehog from foregut endoderm in jaw development: ensuring neural crest cell survival. Proc Natl Acad Sci USA 2006;103(31):11607–12.
[54] Langtimm-Sedlak CJ, et al. Multiple actions of stem cell factor in neural crest cell differentiation in vitro. Dev Biol 1996;174(2):345–59.

[55] Dupin E, et al. Endothelin 3 induces the reversion of melanocytes to glia through a neural crest-derived glial-melanocytic progenitor. Proc Natl Acad Sci USA 2000;97(14):7882–7.

[56] Dupin E, et al. Reversal of developmental restrictions in neural crest lineages: transition from Schwann cells to glial-melanocytic precursors in vitro. Proc Natl Acad Sci USA 2003;100(9): 5229–33.

[57] Garcez RC, et al. Epidermal growth factor (EGF) promotes the in vitro differentiation of neural crest cells to neurons and melanocytes. Cell Mol Neurobiol 2009;29(8):1087–91.

[58] Perris R, Perissinotto D. Role of the extracellular matrix during neural crest cell migration. Mech Dev 2000;95(1-2):3–21.

[59] Newgreen D, Thiery JP. Fibronectin in early avian embryos: synthesis and distribution along the migration pathways of neural crest cells. Cell Tissue Res 1980;211(2):269–91.

[60] Perris R, et al. Molecular mechanisms of neural crest cell attachment and migration on types I and IV collagen. J Cell Sci 1993;106(Pt 4):1357–68.

[61] Bronner-Fraser M. Distribution and function of tenascin during cranial neural crest development in the chick. J Neurosci Res 1988;21(2-4):135–47.

[62] Costa-Silva B, et al. Fibronectin promotes differentiation of neural crest progenitors endowed with smooth muscle cell potential. Exp Cell Res 2009;315(6):955–67.

[63] Bixby S, et al. Cell-intrinsic differences between stem cells from different regions of the peripheral nervous system regulate the generation of neural diversity. Neuron 2002;35(4): 643–56.

[64] Kruger GM, et al. Temporally distinct requirements for endothelin receptor B in the generation and migration of gut neural crest stem cells. Neuron 2003;40(5):917–29.

[65] Geisbauer CL, et al. Transplantation of enteric cells expressing p75 in the rodent stomach. J Surg Res 2012;174(2):257–65.

[66] Hagedorn L, Suter U, Sommer L. P0 and PMP22 mark a multipotent neural crest-derived cell type that displays community effects in response to TGF-beta family factors. Development 1999; 126(17):3781–94.

[67] Maro GS, et al. Neural crest boundary cap cells constitute a source of neuronal and glial cells of the PNS. Nat Neurosci 2004;7(9):930–8.

[68] Hjerling-Leffler J, et al. The boundary cap: a source of neural crest stem cells that generate multiple sensory neuron subtypes. Development 2005;132(11):2623–32.

[69] Aldskogius H, et al. Regulation of boundary cap neural crest stem cell differentiation after transplantation. Stem Cells 2009;27(7):1592–603.

[70] Zujovic V, et al. Boundary cap cells are highly competitive for CNS remyelination: fast migration and efficient differentiation in PNS and CNS myelin-forming cells. Stem Cells 2010; 28(3):470–9.

[71] Singh RP, et al. Retentive multipotency of adult dorsal root ganglia stem cells. Cell Transplant 2009;18(1):55–68.

[72] Fernandes KJ, et al. A dermal niche for multipotent adult skin-derived precursor cells. Nat Cell Biol 2004;6(11):1082–93.

[73] Biernaskie J, et al. SKPs derive from hair follicle precursors and exhibit properties of adult dermal stem cells. Cell Stem Cell 2009;5(6):610–23.

[74] Lavoie JF, et al. Skin-derived precursors differentiate into skeletogenic cell types and contribute to bone repair. Stem Cells Dev 2009;18(6):893–906.

[75] Jinno H, et al. Convergent genesis of an adult neural crest-like dermal stem cell from distinct developmental origins. Stem Cells 2010;28(11):2027–40.

[76] Wong CE, et al. Neural crest-derived cells with stem cell features can be traced back to multiple lineages in the adult skin. J Cell Biol 2006;175(6):1005–15.

[77] Sieber-Blum M, et al. Pluripotent neural crest stem cells in the adult hair follicle. Dev Dyn 2004;231(2):258–69.

[78] Yu H, et al. Stem cells with neural crest characteristics derived from the bulge region of cultured human hair follicles. J Invest Dermatol 2010;130(5):1227–36.

[79] Hu YF, Zhang ZJ, Sieber-Blum M. An epidermal neural crest stem cell (EPI-NCSC) molecular signature. Stem Cells 2006;24(12):2692–702.

[80] Mollaaghababa R, Pavan WJ. The importance of having your SOX on: role of SOX10 in the development of neural crest-derived melanocytes and glia. Oncogene 2003;22(20):3024–34.

[81] Nishimura EK, et al. Dominant role of the niche in melanocyte stem-cell fate determination. Nature 2002;416(6883):854–60.

[82] Steingrimsson E, Copeland NG, Jenkins NA. Melanocyte stem cell maintenance and hair graying. Cell 2005;121(1):9–12.

[83] Mitsiadis TA, et al. Development of teeth in chick embryos after mouse neural crest transplantations. Proc Natl Acad Sci USA 2003;100(11):6541–5.

[84] Luan X, et al. Neural crest lineage segregation: a blueprint for periodontal regeneration. J Dent Res 2009;88(9):781–91.

[85] Degistirici O, et al. Using human neural crest-derived progenitor cells to investigate osteogenesis: an in vitro study. Matrix Biol 2010;29(3):219–27.

[86] Marynka-Kalmani K, et al. The lamina propria of adult human oral mucosa harbors a novel stem cell population. Stem Cells 2010;28(5):984–95.

[87] Davies LC, et al. A multipotent neural crest-derived progenitor cell population is resident within the oral mucosa lamina propria. Stem Cells Dev 2010;19(6):819–30.

[88] Widera D, et al. Adult palatum as a novel source of neural crest-related stem cells. Stem Cells 2009;27(8):1899–910.

[89] Couly GF, Le Douarin NM. Mapping of the early neural primordium in quail-chick chimeras. I. Developmental relationships between placodes, facial ectoderm, and prosencephalon. Dev Biol 1985;110(2):422–39.

[90] Barraud P, et al, Neural crest origin of olfactory ensheathing glia. Proc Natl Acad Sci USA. 107(49):21040–5.

[91] Delorme B, et al, The human nose harbors a niche of olfactory ectomesenchymal stem cells displaying neurogenic and osteogenic properties. Stem Cells Dev 19(6):853–66.

[92] Yoshida S, et al. Isolation of multipotent neural crest-derived stem cells from the adult mouse cornea. Stem Cells 2006;24(12):2714–22.

[93] Tomita Y, et al. Cardiac neural crest cells contribute to the dormant multipotent stem cell in the mammalian heart. J Cell Biol 2005;170(7):1135–46.

[94] Le Douarin NM. Cell line segregation during peripheral nervous system ontogeny. Science 1986;231(4745):1515–22.

[95] Pardal R, et al. Glia-like stem cells sustain physiologic neurogenesis in the adult mammalian carotid body. Cell 2007;131(2):364–77.

[96] Takashima Y, et al. Neuroepithelial cells supply an initial transient wave of MSC differentiation. Cell 2007;129(7):1377–88.

[97] Heanue TA, Pachnis V. Prospective identification and isolation of enteric nervous system progenitors using Sox2. Stem Cells 2011;29(1):128–40.

[98] Tsai YH, Murakami N, Gariepy CE. Postnatal intestinal engraftment of prospectively selected enteric neural crest stem cells in a rat model of Hirschsprung disease. Neurogastroenterol Motil 2010.

[99] Lin H, et al. Pluripotent hair follicle neural crest stem-cell-derived neurons and Schwann cells functionally repair sciatic nerves in rats. Mol Neurobiol 2009;40(3):216–23.

[100] Sieber-Blum M, et al. Characterization of epidermal neural crest stem cell (EPI-NCSC) grafts in the lesioned spinal cord. Mol Cell Neurosci 2006;32(1-2):67–81.

[101] Vitacolonna M, et al. Improved T and B cell recovery by the transfer of slowly dividing human hematopoietic stem cells. Leuk Res 2010;34(5):622–30.

[102] Biernaskie J, et al. Skin-derived precursors generate myelinating Schwann cells that promote remyelination and functional recovery after contusion spinal cord injury. J Neurosci 2007;27(36):9545–59.

[103] Biernaskie JA, et al. Isolation of skin-derived precursors (SKPs) and differentiation and enrichment of their Schwann cell progeny. Nat Protoc 2006;1(6):2803–12.

Chapter 10

Liver Resident Stem Cell

Christina M. Takiya[*,†,‡,§,**], Bruno Diaz Paredes[*,†,§,**,††],
Luiz Fernando Quintanilha de Mesquita[*,††,§,**,‡‡,§§],
Grazielle Suhett Dias[*,§,**,††],
Lanuza Alaby Pinheiro Faccioli[§,††,**], Taro Takami[‡‡,§§],
Shuji Terai[‡‡,§§], Isao Sakaida[‡‡,§§],
Regina Coeli dos Santos Goldenberg[*,‡,§,**,††]

[*] *Instituto de Biofísica Carlos Chagas Filho, Rio de Janeiro, RJ, Brazil*
[†] *Laboratório de Patologia Celular, Rio de Janeiro, RJ, Brazil*
[‡] *Programa de Terapia Celular e Bioengenharia, Rio de Janeiro, RJ, Brazil*
[§] *Instituto Nacional de Ciência e Tecnologia de Biologia Estrutural e Bioimagem-INBEB, Rio de Janeiro, RJ, Brazil*
[**] *Universidade Federal do Rio de Janeiro, Rio de Janeiro, RJ, Brazil*
[††] *Laboratório de Cardiologia Celular e Molecular, Rio de Janeiro, RJ, Brazil*
[‡‡] *Division of Laboratory, Yamaguchi University Hospital, Yamaguchi University School of Medicine, Japan*
[§§] *Department of Gastroenterology and Hepatology, Yamaguchi University, Graduate School of Medicine Yamaguchi University, Japan*

Introduction

The liver is an important multifunctional organ within the body. It has a central role in several metabolic processes such as regulation of blood glucose levels, synthesis, storage, and redistribution of nutrients, carbohydrates, fats, and vitamins. The liver also produces large numbers of serum proteins including albumin and acute-phase proteins, enzymes, and cofactors, and it is responsible for the detoxification of urea, removal of wastes and xenobiotics by metabolic conversion, and biliary excretion.

The liver is constituted of plates of hepatocytes lined by sinusoidal capillaries that radiate from the portal space toward a central efferent vein. The portal space contains branches of the portal vein, hepatic artery, bile ducts, lymphatic vessels, and fibroblasts, which are joined together by connective tissue [1]. The hepatic-specific

Resident Stem Cells and Regenerative Therapy. http://dx.doi.org/10.1016/B978-0-12-416012-5.00010-4

capillaries or sinusoids are lined by fenestrated endothelial cells and contain hepatic stellate cells (Ito cells), resident macrophages (Kupffer cells), and large granular lymphocytes (pit cells) [2].

The hepatic circulation is unique in that it has a dual blood supply, namely, the portal vein and the hepatic artery. Blood flow in liver parenchyma runs from the portal triad through the hepatic sinusoids to the central efferent vein. While the portal vein distributes the venous blood coming from the intestines, pancreas and spleen, the hepatic artery delivers oxygen to the liver [3].

Hepatocytes are highly polarized epithelial cells with a hexagonal shape. They are connected to each other through tight junctions. Joining hepatocytes create in the central area the biliary canaliculus in which bile salts produced by the hepatocytes are excreted. Bile canaliculi are linked to bile ducts at the portal triad. Bile ducts are formed by a specialized type of epithelial cell called a biliary epithelial cell or cholangiocyte. The basolateral surfaces of hepatocytes face the sinusoidal fenestrated endothelial cells, which do not have a lining basement membrane in order to facilitate the transfer of materials between hepatocytes and blood flow.

Hepatocytes are the major parenchymal cells carrying out most of the metabolic functions of the organ and constitute about 80% of the liver population. The other 20% comprise the nonparenchymal cells, which include endothelial cells, Kupffer cells, stellate cells, and pit cells. Hepatic stellate cells have various functions, including the storage of vitamin A and the production of the extracellular matrix. Kupffer cells, the resident macrophages, are essential for the phagocytosis of foreign particles and microorganisms, as well as the production of cytokines. Pit cells are also part of the liver innate immune system [2].

However, if the ability of the mature differentiated hepatocytes to divide is impaired, a situation that occurs with certain hepatotoxic chemicals or diets as well as during chronic liver damage, progenitor cell–dependent liver regeneration is set in motion [4].

Liver Development

Liver development occurs through a stepwise series of events, beginning in the gastrula stage of embryogenesis when the endoderm is first segregated from the ectoderm and mesoderm [5]. During foregut closure, the medial and lateral domains fuse together and the endoderm cells are specified to a hepatic fate that depends on the response of competent endoderm cells to extracellular signals. Fibroblast growth factors (FGFs) secreted by the cardiogenic mesoderm induce hepatic gene expression [6], are mediated by a mitogen-activated protein kinase signaling cascade [7], and are dependent on low concentrations of FGF (Fig. 10.1A and B). At this point, the developing heart is adjacent to the prehepatic endoderm and produces low amounts of FGF. Soon after, the cardiogenic mesoderm begins to produce higher

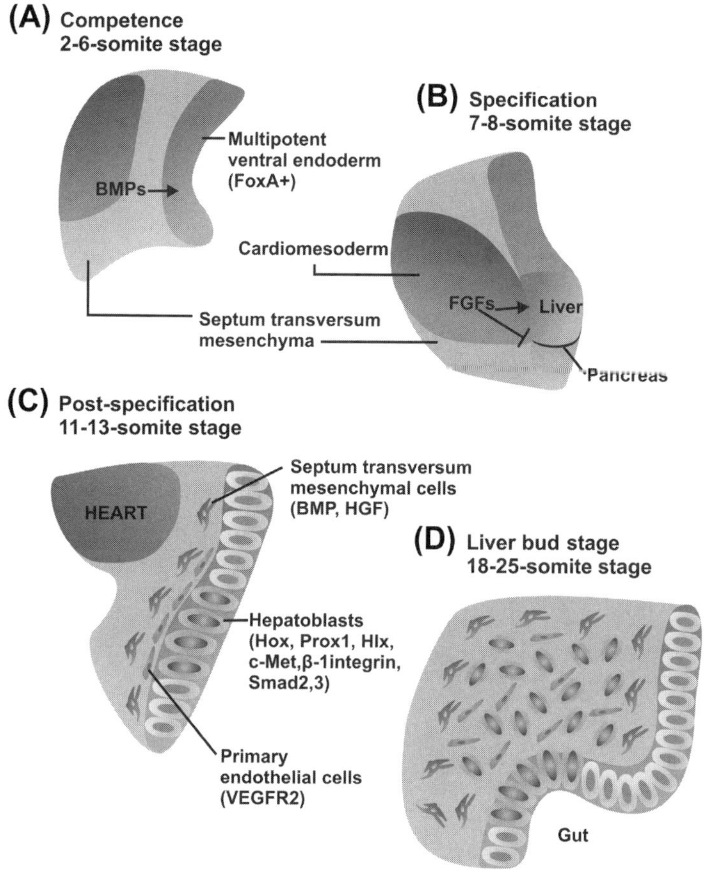

Figure 10.1 Schematic representation of liver bud formation. **(A, B)** In the 2-6 somite embryo, cells of the foregut endoderm (FoxA+) become competent: they react to BMPs secreted by septum transverse mesenchyme cells and FGFs coming from cardiac mesoderm activating liver-specific genes. (B) With gut closure, the endoderm is pulled to the midgut and the early hepatocytes begin to proliferate within the endoderm layer. **(C)** The hepatoblasts form a columnar sheet delineated by a basement membrane that contains laminin, collagen IV, nidogen, fibronectin, and heparan sulfate proteoglycan. **(D)** Hepatoblasts become a pseudostratified epithelia and migrate into the septum transversum. The early hepatocytes will then coalesce around sinusoids in the mesenchyme, forming the liver organ.

amounts of FGF and becomes separated from the prehepatic endoderm by the septum transversum [7] (Fig. 10.1C). However, an FGF signaling from the cardiogenic mesoderm is not sufficient for hepatic specification. In mouse, the septum transversum produces bone morphogenetic protein (BMP)-2 and BMP-4, which cooperate with FGF to induce hepatic gene expression. BMP signaling is necessary for the maintenance of the expression of the GATA-4 transcription factor, which, in turn contributes to competence and specification of the prehepatic endoderm [8] and early liver gene expression [9]. At the time that the primitive gut becomes patterned along its anteroposterior axis, Wnt signaling is repressed in order to allow the maintenance of the foregut identity and sustain liver development.

This suppression depends on the appropriate secretion of Wnt inhibitors by the endoderm [10,11]. At E8.5 to 9.0, the newly specified hepatic cells proliferate within the endodermal epithelium, forming the hepatic diverticulum, which is surrounded by angioblasts, the precursors of vascular structures (Fig. 10.1C). These angioblasts remain isolated from one another, but closed vascular structures or hematopoietic cells are not yet present [12]. At E9.0 in mice and at day 22 of embryonic life in humans, the liver diverticulum is lined by the specified endoderm cells, which at this stage are called hepatoblasts (Fig. 10.1D), which will give rise to definitive hepatocytes and bile duct cells (cholangiocytes) [13,14]. At this time point hepatoblasts already express some genes specific to fully differentiated hepatocytes, such as serum albumin, transthyretin, and α-fetoprotein [6,15], and form a simple epithelial layer, which is anchored to a basement membrane that contains laminin, collagen IV, nidogen, fibronectin, and heparan sulfate proteoglycan [16]. Hepatoblasts continue to proliferate, resulting in a pseudostratified epithelia [17] that is also closely associated with adjacent endothelial cells, which have a role in controlling hepatoblast proliferation [12,18,19].

At E9.0, hepatic specified endodermal cells emerge from the foregut endoderm and migrate as cords into the surrounding septum transversum mesenchyme (STM) [20,21]. At this time and up, condensed foci of angioblasts surround the hepatic endoderm as it invades the mesenchyme toward the STM. These angioblasts remain isolated from one another and interact with nascent hepatic cells prior to blood vessel formation and function, persisting from the hepatic endoderm stage (E8.5) through the formation of the liver proper (E10.5) [12]. However, closed vascular structures or hematopoietic cells are not yet present. By E9.5 to E10.5, a mechanism reminiscent of the epithelial-mesenchymal transition allows hepatoblasts to leave the endoderm epithelium, migrate through the basement membrane, and invade the septum transversum [17,22] (Fig. 10.1D). The high levels of FGF produced by the cardiogenic mesoderm [23] drives the migratory movement of the hepatoblasts in the direction of the septum transversum [24], which is dependent on down-regulation of E-cadherin expression and on the action of metalloproteinases (MMPs) secreted by the septum transversum mesenchymal cells (MMP-2) and by hepatoblasts (MMP-14) [25]. These steps are controlled by a gene network that involves the transcription factors prospero-related homeobox 1 (Prox1), hepatocyte nuclear factor-6/Onecut-1 (HNF-6/OC-1), and OC-2 [20,26].

Once the hepatoblasts have invaded the septum transversum, they continue proliferating and the liver bud expands. Several growth factors are involved not only in liver bud expansion but also in liver specification [27]. Hepatocyte growth factor (HGF) and its receptor c-met initiate a signaling cascade mediated by SEK1/MKK4 and possibly by c-jun [28,29]. The transforming growth factor (TGF) β Smad2/Smad3 pathway also stimulates proliferation and interacts functionally with HGF. Both pathways converge on β1-integrin expression, which is necessary for proliferation [30]. The hepatoblasts soon face a decision to differentiate into either hepatocytes or biliary epithelial cells. Those that follow the hepatocyte cell fate gradually express genes and acquire hepatocyte-specific functions [31], which remain during the rest of both embryonic and postnatal development [32].

After the liver bud stage, the hepatic cords surround spaces lined by mesenchymal cells, which will become the hepatic sinusoids of the liver [33], which are the first blood vessels to form during hepatogenesis. They are formed from existing vessels where they develop by angiogenesis from existing vessels in the septum transversum mesenchyme through angiogenesis [34-36]. As the liver progresses, the sinusoidal endothelial cells gradually adopt the functional and structural characteristics of mature sinusoids. This endothelial remodeling correlates with changes in the expression of extracellular matrix components, which in turn act on the maturation process [35,37]. Although angiogenesis appears to be the primary mechanism through which the sinusoids are formed, there is evidence that, at least in avians, the growth of sinusoids during embryogenesis may be partially facilitated by the incorporation of mesenchymal cells derived from septum transversum mesothelium [38]. In rodents the angioblast and in humans the hemangioblast progenitors present in fetal liver give origin to differentiated endothelial cells and to new vessels [39]. The STM-derived mesothelium is likely to be the major source for hepatic stellate cells and other fibrogenic cells such as portal fibroblasts, pericentral venous fibroblasts, and portal venous smooth muscle cells [40].

Around E10.5, the nearby vitellinic veins anastamose and provide vascularization to the tissues including the liver. The liver mass increases dramatically, and hematopoietic cells become resident until birth. At E11.5, SOX9 becomes reexpressed in specific liver epithelial cells located nearby the branches of the portal vein in the portal mesenchyme, and SOX9 expression remains restricted to the biliary lineage throughout development. SOX9 is seen in endodermal cells of the hepatic diverticulum, but its expression is no longer present when hepatoblasts begin to leave the hepatic diverticulum and invade the septum transversum [41]. The newly formed cholangiocytes become arranged as a monolayered structure called the *ductal plate* [42] and express several markers of the biliary lineage including SOX9, osteopontin, and CK19 [43].

Around E15.5 in the mouse and in humans from approximately the 12th week of gestation and beyond, focal portions of the ductal plates become lined by a second layer of cells, which then dilate, creating a lumen [44]. These structures are the so-called primitive ductal structures (PDS) and have a unique asymmetrical organization. These nascent tubules are composed of ductal plate cells resembling cholangiocytes, positive for SOX9 and cytokeratin-19 (K19) on the side facing the portal tract, and of ductal plate cells resembling hepatoblasts, positive for hepatocyte nuclear factor 4 (HNF4) and transforming growth factor receptor type II (TGF-βRII) on the side facing the parenchyma. The portal side layer shows higher levels of E-cadherin and is in contact with laminin, whereas the parenchymal side layer expresses low levels of E-cadherin, HNF-4α, and shows osteopontin expression at the apical pole of cells [41]. When the PDS mature into bile ducts, their lumens become entirely lined by biliary cells, which express SOX9, osteopontin, high levels of E-cadherin, and SgIGSF [41,45]. Each ductal plate gives on average rise to two bile ducts per portal tract [46], and only a minority of ductal plate cells participates in bile duct morphogenesis. Ductal plate cells not involved in bile duct formation suffer apoptosis [47]. There are still some controversies about the fate of segments of ductal plate not involved in ductal formation. It has been shown that during mouse liver development from the beginning of ductal plate formation to the stage at its remodeling (E15.5 to E18.5), ductal plate cells identified by staining for SOX9 or E-cadherin do not suffer apoptosis. The same was seen in human fetal liver at 11 weeks of gestation [48]. Genetic lineage tracing approaches using a mouse line that expresses inducible CreERT2 in SOX9-expressing cells (SOX9-CreER T2 mice) [49] and its intercrosses with ROSA26RYFP mice originate the permanent expression of yellow fluorescent protein (YFP) in the progeny of the cells. SOX9+ ductal plate cells give rise to cholangiocytes lining the bile ducts, including the most proximal biliary structures, as well as to periportal hepatocytes. The induction of atypical ductular structures by a CDE diet given to transgenic mice also allowed researchers to determine that the ductal plate cells are the embryonic precursors of oval cells [48].

Commitment of hepatoblasts to biliary lineage depends not only on SOX9 [41] but also on the notch signaling pathway [50]. Data from patients of Alagille syndrome—a disease caused by mutations in jagged1 [51,52] or notch2 [53] in which portal spaces exhibit a ductal paucity—indicate that notch signaling is involved in biliary commitment. In fact, notch signaling is a critical regulator of PDS formation and maturation; jagged1 and Hes1, a notch effector, are differentially expressed in biliary cells of the PDS [50,54-56]. Moreover, TGF-β is also involved in biliary differentiation. Before the ductal plate formation, a gradient of Activin/TGFβ signaling is present in nascent liver. High levels of Activin/TGFβ occur in the periportal region, where biliary cells differentiate, whereas low levels are seen in the parenchymal region, where hepatocytes differentiate [57,58]. TGFβ1 is expressed widely in the liver; both TGFβ2 and TGFβ3 are expressed predominantly in the periportal region [41,58]. In cultured hepatoblasts, TGFβ-1, -2, and -3 are all able to induce the expression of biliary markers and repress the expression of hepatocyte markers (HNF4α, albumin, apolipoprotein A, and transthyretin) [41]. Furthermore, deletion of one copy of Smad2 and Smad3—mediators of transcription induced by

TGFβ signaling—disrupts hepatic architecture [30]. TGF-β sequentially induces differentiation of the biliary cells of the PDS [41] through the expression of TGF-β receptor II (TGF-βRII). Multiple TGF-β ligands are expressed in the developing mouse liver bud mesenchyme [59].

Liver Regeneration: The Hierarchy for Cell Response

Hepatocyte and Cholangiocyte: The Forefront in Liver Regeneration

Postnatal hepatocytes are highly differentiated cells with a life expectance of about 200 days in normal mice liver [60] and 400 days in rats [61]. They are mitotically inactive under physiologic conditions (G0 phase) but show tremendous proliferative properties after liver damage [62]. In response to injury, hepatocytes proliferate and can divide up to 100 times [63-65]. Liver regeneration after partial hepatectomy and after most chemical acute injuries occurs by the division of fully differentiated hepatocytes and cholangiocytes [65,66-69]. In order to clarify controversies about the origin of liver cells during liver homeostasis and injuries, a hepatocyte fate-tracing model was generated. This in situ fate-tracing model confirmed that in normal liver homeostasis, the newly formed hepatocytes derive from preexisting hepatocytes [70].

Stem/Progenitor Cells: The Second Front in Liver Regeneration

Hepatocytes and cholangiocytes are not exclusively responsible for liver regeneration. During severe or chronic liver injury, mature liver cells are frequently incapable of entering the cell cycle because of cell arrest or senescence and cannot sufficiently restore lost parenchyma. Therefore, facultative hepatic resident stem/progenitor cells can be activated to participate in the regeneration process [4,71-73]. Oval cells are considered as hepatic progenitor cells (HPCs), the progeny of hepatic stem cells. They express both albumin and cytokeratin 19 (CK19), which are, respectively, hepatocytic and cholangiocytic markers and differentiate to hepatocytic and biliary lineages, recapitulating hepatoblast differentiation in fetal liver. In this context, this important population has received great attention as potential candidates for liver-directed therapy and as a tool for regenerative medicine prior to liver transplantation.

Upon activation, the oval cells grow as extensions of terminal biliary ductules from where they originate, form ductular structures that communicate with the biliary system at one end, and terminate at a hepatocyte-forming blind end [74]. Proliferating oval cells constitute a heterogeneous population that results in varied denominations such as ductular progenitor cells [74,75], atypical ductular cells [76], periductular liver progenitor cells, or individual progenies [77,78]. The oval cell population is characterized by the presence of oval cells (undifferentiated cells) associated with cells of intermediate characteristics between the hepatocytic and biliary lineages. They are considered as the amplifying proliferative transit cells of a facultative stem cell lineage [79,80].

The name *oval cells* came from the cytologic characteristic of these cells, that is, their small oval shaped-nuclei [81]. They are about 10 μm in size and express markers of immature liver cells such as α-fetoprotein (AFP) and of both the biliary epithelium (CK19) and hepatocyte lineages (albumin). The differentiation of HPC toward mature hepatocytes or cholangiocytes is characterized by the appearance of cells with intermediate phenotypes: polygonal cells with a size intermediate between that of HPCs and mature hepatocytes, as well as cells similar to immature cholangiocytes. The HPCs initially form strands of cells that expand either into liver parenchyma along the liver plates toward the central vein and differentiate into hepatocytes or as branching ducts moving from the canals of Hering to the center of the portal space to form the bile ductules and larger bile ducts. In this way HPCs are capable of restoring liver function and cell mass [82,83].

The oval cell-mediated liver regeneration occurs not only in liver regeneration but also during carcinogenesis [84-86]. In carcinogen-treated rat models, oval cells differentiate into mature hepatocytes via intermediate hepatocyte-like cells, named *ductular hepatocytes* and *small hepatocytes,* respectively [87-89]. In contrast, in a hamster cholangiocarcinogenesis model, the differentiation of oval cells into biliary cells occurs initially through the formation of atypical bile ductules and culminates in the development of different types of cholangiocarcinomas [89,90].

Oval cells have been most extensively studied in rodents. Oval cell proliferation is prominent in many models of liver injury including carcinogenesis induced by azo dyes and choline deficient/ethionine-containing diets (CDE diet), d-galactosamine, acetyl aminofluorene (AAF), dipin, or CCl_4 treatment in combination with PH [91,92]. In the AAF/PH and galactosamine models as well as in rodents fed the CDE diet, oval cells may constitute more than 50% of cells in the liver. Oval cells are detected in mice after providing a diet deficient in 3,5-diethoxycarbonyl-1, 4-dihydrocollidine (DDC) [93], a choline-deficient, ethionine-supplemented (CDE) diet [94], and acetaminophen (APAP) [95].

Several lines of evidence suggest the presence of a similar phenomenon in various human liver diseases, as well as in tumors [96]. Upon activation of the stem cell compartment, expansion of HPCs culminates in a proliferative reaction called the *atypical ductular reaction* [97,98]. The term *atypical reaction* is intended to differentiate it from the typical ductular reaction, which occurs in diseased liver. For an extensive HPC activation, it seems that a threshold of a 50% loss of hepatocytes, together with a significant decrease in proliferation of the remaining mature hepatocytes, is required. In this case, HPC proliferation occurs within 1 week, and their differentiation into hepatocytes starts after 2 weeks [99].

Although the oval cell is widely considered as the HPC, there is no consensus between investigators on the phenotype and molecular signature of these cells [100]. Much of these controversies could be attributed to the fact that they are a heterogeneous population [101,102], that the expression of the molecules indicating "stemness" varies in humans and animals, and even that it could also depend on the type of insult inducing the appearance of the HPC population.

In the normal liver, HPCs and their immediate biliary and hepatocytic progeny have a distinct morphology and immunohistochemical phenotype [103-105] (Tables 10.1 and 10.2). HPCs are small cells with an oval nucleus and scant cytoplasm, and they are barely recognizable in routine histologic staining. Immunohistochemistry using a panel of antibodies against markers of HPCs and their progeny are necessary to detect these cell types. The most reliable and acceptable markers for HPCs are cytokeratin (CK)-7 and CK-19, typically expressed by mature cholangiocytes of interlobular bile ducts. HPCs can also express proteins in common with mature hepatocytes (albumin, CK-8), immature fetal hepatoblasts (α-fetoprotein), hematopoietic stem cells (CD133, c-kit), and neuroepithelial cells (neural cell adhesion molecule and chromogranin-A) [4,97,103] (see Tables 10.1 and 10.2; Fig. 10.2). The monoclonal antibody, OV-6, has been found to be a useful marker for rat oval cells and human fetal and adult HPC [103,106-108]. It was prepared against rat oval cells,

TABLE 10.1 Murine Cell Markers

Cell Type	Markers	References
Hepatoblast	Albumin; α-fetoprotein (AFP)	[185]
HPC (oval cells) mouse	Albumin; α-fetoprotein (AFP); EpCam; Cytokeratin 19; OV-6; c-kit; Thy-1; Sca-1; OV-1; CD 133 (Prominin 1)	[86,114,125,182,185, 187,188]
HPC (oval cells) rat	Albumin; EpCam; Cytokeratin 19; A6; TROP2; CD 133 (Prominin 1)	[114,124,125,189,190]

TABLE 10.2 Human Cell Markers

Cell Type	Markers	References
hHpSC	Cytokeratin 19; EpCAm	[98,121,123,191]
hHB	Albumin; α-fetoprotein (AFP); Cytokeratin 19; EpCAm	[98,121,191,192]

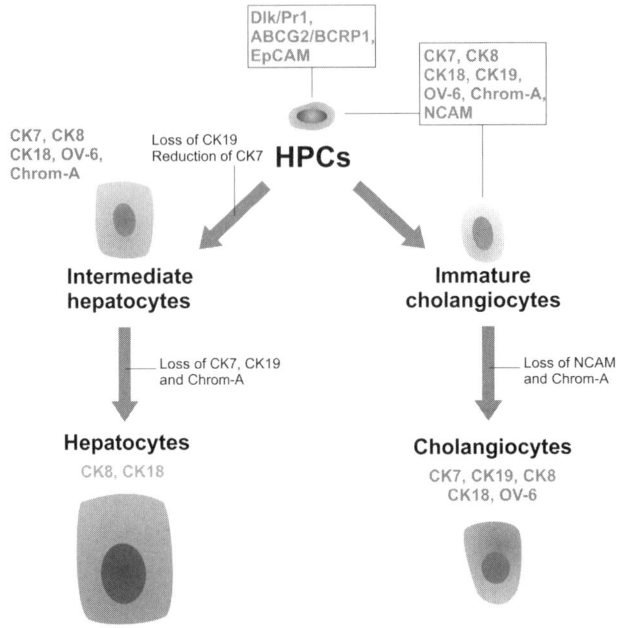

Figure 10.2 Schematic representation of the liver cell lineages differentiated from hepatic progenitor cells (HPCs): hepatocytes or cholangiocytes, and their markers.

and it has been shown to react with CK-14 and -19 [109]. OV6 stains some cells in the ductal plate and the bile ducts of human fetal liver [108,110] and in adult biliary liver diseases such as undifferentiated oval-like cells in focal nodular hyperplasia [111], proliferating ductules in primary biliary cirrhosis, and lobular hepatocytes in primary sclerosing cholangitis [108]. Another accepted marker of HPC is A6, an uncharacterized epitope recognized by mouse hepatic oval cells [112-114]. A6 is not liver specific, does not bind a cell surface marker [102], and is not commercially available and thus cannot be used for viable cell isolation [115]. It is expressed in various normal tissues and organs, especially in epithelia. In studies of mouse liver, lineages of A6 antigen are used as common markers of biliary epithelial and oval cells, hepatocytes at certain stages of differentiation, and hepatoma cells [116]. Other HPC markers have been characterized and considered as a utile tool to isolate and purify this cell population [76]. The Delta-like protein/preadipocyte factor 1 (DLK/Pref 1), a notch ligand, is expressed in hepatoblasts, and on birth, hepatocytes lack DLK/Pref 1. After liver injury, DLK/Pref 1 is reexpressed initially in a few scattered cells in the existing duct and ductular structures and later in all cells of the ductular reaction as well as their immediate hepatocyte-like progeny. However, cells in well-defined bile ducts are consistently negative [76]. Furthermore, oval cells express ABCG2/BCRP1 mRNA and exhibit the SP phenotype [77]

(Fig. 10.2). The antigen epithelial cell adhesion molecule (EpCAM) is expressed by many carcinomas and in certain normal epithelia including all of those derived from endoderm (liver, lung, pancreas, and intestine) [117,118]. EpCAM is a cell-surface glycoprotein expressed on some normal as well as neoplastic epithelial cells [119]. In human fetal liver EpCAM highlights multipotent progenitor cells located in ductal plates [120,121]. Furthermore, EpCAM is the earliest and a transient marker for hepatoblasts during hepatogenesis. EpCAM is present in hepatic cords located at the septum transversum mesenchyme at E10.5 and disappears at E14.0 being reexpressed at 14.5 in ductal plate [122]. In normal adult livers, some cholangiocytes are EpCAM + as well as in HPC/oval cells and some ductular cells in cases of nodular hyperplasia or biliary cirrhosis [118,123-125]. However, TROP2, a member of the EpCAM family, is not expressed at all in normal liver but only in the oval cells of injured liver [124]. Therefore, TROP2 is a marker of oval cells and allows distinguishing oval cells from normal cholangiocytes [124].

In the early stages of the differentiation toward mature hepatocytes, the intermediate hepatocytes are positive for CK19 and CK7, becoming negative for biliary markers at the late stages. CK19 is lost much earlier than CK7, but they maintain positivity for OV6, CK8, CK18, and Chromogranin-A (see Fig. 10.2). Finally, with full maturation hepatocytes maintain CK-8 and -18 but are completely negative for biliary markers, neuroepithelial markers, and OV6 [73] (see Fig. 10.2). On the other hand, the differentiation of HPCs toward mature cholangiocytes originates cells with the appearance of immature cholangiocytes, which ultimately will give rise to small and then large cholangiocytes, moving from the canals of Hering to the bile ductules and the larger ducts [71,105,126]. The immature cholangiocytes maintain OV6, CK-7, CK-8, CK-18, CK-19, NCAM, and Chromogranin-A immunoreactivities until fully differentiated. The immunostaining profile at this stage is characterized by the disappearance of neuroendocrine markers such as NCAM or Chromogranin-A. However, the markers OV6, CK-7, and -19 are maintained [73,127] (see Tables 10.1 and 10.2; Fig. 10.2).

Fetal Hepatic Stem Cell Niche

Hepatoblasts (fetal hepatic progenitor cell) are the progeny of hepatic stem cells during development. In humans, the hepatic stem cells are present in developing portal spaces in a limited number. They steadily give rise to hepatoblasts, which therefore could be considered as transit-amplifying cells, precursors to committed hepatocytic and biliary progenitors [121]. They are present in the ductal plates in fetal and neonatal livers expressing AFP, albumin, and CK19 [128] and in the proximal branches of the intrahepatic biliary tree [121]. Human hepatoblasts present in ductal plates are small cells (7 to 10 μm), with scant cytoplasm, and are intensely positive for CK19 and EpCAM both cytoplasmically and at the surface, weakly reactive for albumin, but negative for AFP [121]. The co-expression of CK19 and albumin is consistent with the bipotentiality of this population and corroborates the hypothesis that the canals of Hering make up a niche for HpSCs [129] (see Table 10.2).

Suzuki et al. (2000) reported that cells in embryonic day (ED) 13.5 mouse liver, which co-express CD49f and CD29 ($\alpha6\beta1$ integrin subunits) but do not express c-kit (stem cell factor receptor), CD45 (leukocyte common antigen), or TER119, were the best candidate for the hepatic stem/progenitor population [130]. Later on they showed that a single cell in the c-Met+ CD49f+/low c-kit−CD45−Ter119− fraction obtained from midgestational mouse fetal liver has the capacity to self-renew in vitro and for bipotential differentiation, which indicates that this defined fraction contains hepatic stem cells [131].

The human hepatic stem cell can be isolated efficiently by selective culture conditions and by immunoselection for EpCAM (CD326) or NCAM (CD56), having features typical of stem cells including sonic and Indian hedgehog signaling [132]. They are capable of self-renewal, as shown by clonogenic expansion for >150 population doublings, and are pluripotent, with the ability to give rise directly to committed biliary progenitors and hepatoblasts and hence to hepatocytic and biliary lineages [121].

Because ductal plates are directly antecedent to the canals of Hering, which have been identified as the reservoir of hepatic stem cells in postnatal livers [98,133], the ductal plates are considered to be the niche of fetal hepatic stem/progenitor population.

Adult Liver Stem/Progenitor Cell Niche

The basic properties of stem cells are (1) capacity for self-renewal or self-maintenance (generally slowly cycling), (2) multipotency (capable of producing progeny in at least two lineages), (3) undifferentiation, (4) functional, long-term tissue reconstitution, and (5) serial transplantability. In adult tissues the maintenance of stem cells depends on its asymmetric division, such that one of the progeny remains undifferentiated, while the other proliferates and differentiates to generate new tissue mass [134,135]. They are the source of progenitor cells committed to one or several lineages. The committed progenitor cells exhibit a capacity for active proliferation and supply abundant daughter cells, which in turn give rise to terminally differentiated cells. They can divide rapidly but in contrast to authentic stem cells do not possess the ability to self-renew, are capable of only short-term tissue reconstitution, and have been termed *transit amplifying cells* [136]. They have the potential to generate more than one differentiated cell type but cannot be serially transplanted [137,138]. Every progenitor cell could give rise to a progeny composed of transit-amplifying cells fated for differentiation or initially not committed and retaining self-renewal capabilities [135]. They constitute distinct subpopulations, some with multilineage potential (early progenitor or stem/progenitor cells) and others (late progenitor cells) that have differentiated further and give rise to progeny in only a single lineage.

Adult tissue-specific stem cells are thought to reside within a specialized microenvironment, known as the niche. The stem cell niche is a complex structure capable of integrating signals for the regulation of stem cell activities in a spatially and temporally defined manner [139]. The niche concept specifies a microenvironment comprising stem/progenitor cells, stromal cells, and extracellular matrix. The intrahepatic stem cell niche contains type III collagen, $\alpha6\beta4$ integrin-binding form of laminin, hyaluronans, and a minimally sulfated chondroitin sulfate proteoglycan (CS-PG) [140].

Interactions between these various compartments, usually mediated by direct cell-cell contact, accomplish the homeostatic regulation of stem/progenitor cell functioning [141].

It is currently considered that there are three putative locations for resident liver stem/progenitor cells: within the canals of Hering (CoH) [98], cells located within the interlobular bile ducts [142], and periductular cells. Studies dealing with trace lineage assays reinforce the presence of these three stem cell niches. A transgenic GFP animal in which expression of the reporter gene is driven by the promoter and the second intron of the nestin gene [143] was used to localize HPCs in a model of liver regeneration mediated by oval cells [144]. In quiescent liver rare GFP+ cells were seen within canals of Hering and within typical bile ducts located in periportal areas, and the majority of them co-express EpCAM and A6 [144]. These GFP+ cells correspond to the oval cells present in the CoH [98] and in bile ductules or ducts [142,145]. In the regenerating liver after single or multiple treatments with CCl_4, cells located in the ductular structures and in small clusters in the periportal area co-expressed GFP and EpCAM or Thy-1, and c-kit [144], markers of adult liver stem/progenitor cell [146].

In 2008, Kuwahara and coworkers, using label-retaining cell assay, confirmed the previous findings that the (1) canals of Hering are a niche of intrahepatic stem cells. They also could demonstrated that (2) peribiliary hepatocytes present in periportal zones represent a true LRC population. These data seem to support the "streaming liver" hypothesis, but slightly modified to reflect current understanding that the hepatocyte acinus begins not at the limiting plate but at the interface between the hepatocyte canaliculi and the biliary tree, at the CoH [147]. The (3) intraductal cholangiocytes, which could represent either labeled cells in the stream of repopulating a normal biliary system or a response to physical injury secondary to the APAP-induced parenchymal [147], and the (4) peribiliary "null" cells, which represent a subpopulation of oval cells negative for typical oval cell markers such as alpha-fetoprotein, biliary-type cytokeratins albumin, and also negative for leukocyte common antigen (CD45) and desmin. Because this pattern of cellular reaction was more prominent in models of periportal injury, rather than the more typical centrilobular APAP injury, it suggests that they may be related to hepatobiliary regeneration when the canal of Hering stem cell niche is disrupted or obliterated along with the destruction of the periportal hepatocytes [147].

Liver Stem/Progenitors Cells and Cell Therapy

In spite of the extensive regenerative capacity of the liver against diverse types of injuries, alternative methodologies to treat end-stage liver diseases are still urgently needed. Liver transplantation is the standard of care for end-stage liver disease and many liver-based metabolic conditions. Techniques involve whole organ replacement, split or reduced donor liver, and auxiliary liver transplantation. However, transplantation has serious limitations, such as donor scarcity, immunologic incompatibilities, high cost, significant morbidity and mortality associated with the procedure, and death while waiting for the transplant [148]. Furthermore, considerable long-term side effects have been reported [149-152]. Hepatocyte transplantation (HT) was thought to be a promising alternative to orthotopic liver transplantation (OLT) for treating liver-based inborn errors of metabolism where the aim is to replace a single deficient enzyme or its product [153-156]. The aim of this kind of procedure is to maintain liver function while the patient awaits OLT or until regeneration of the native liver occurs. The procedure is less invasive than OLT and can be performed repeatedly. The number of cells transplanted usually represents approximately 5% of theoretic liver mass, and either fresh or cryopreserved cells have been used. The safety of the procedure has been well established, and the clinical results are encouraging with clear improvement in disease phenotype. However, cell function often declines after about 9 months with the result that patients then undergo OLT. Problems with immunosuppression and rejection may be an important factor. Intraportal injection is the main cell delivery route for clinical HT with the portal venous system accessed by percutaneous transhepatic puncture or inferior mesenteric vein catheterization [157]. However, mature hepatocyte transplantation has been performed for more than 15 years in humans and there is still lack of evidence of success and reproducibility in large scale [158,159]. The main problems related to this approach are the fact that these harvested cells normally do not show optimal condition as well as the lack of standardized protocols to assess the cell's quality, the low proliferation/engraftment rate, the poor cell viability after cryopreservation methods, and the lack of hepatic metabolic functions after routine culture [160-162]. Moreover, it is known that mature hepatocytes exhibit increased rates of polyploidy that contribute to proliferation decrease and cell senescence. Furthermore the latter events might presumably impair the regenerative capacity of these cells [163,164]. On the other hand, extrahepatic stem cells have been exhaustively tested. Stem cells obtained from different tissues (i.e., fetal annex, adipose tissue, bone marrow) have been successfully utilized in diverse settings of experimental chronic liver diseases [165-169].

A number of animal studies show that adult bone marrow cells could be applied to therapeutic purposes in certain liver diseases. Transplantation of adult bone marrow stem cells (BMSCs), either the mononuclear/hematopoietic cell fraction or mesenchymal stem cells, has therapeutic effects of restoration of liver function and mass, alleviation of fibrosis, and correction of inherited liver diseases. Although some controversial issues exist in relation to the results obtained by the different groups, mainly in relation with the beneficial effect on fibrosis, the restoration of liver function is evident in almost all animal studies [165,168,170,171-174]. Some of the discrepancies are thought to lie either in the differences between the experimental protocols or in the techniques employed to validate the effects [170].

Other sources of extrahepatic stem cells, such as embryonic stem cells and umbilical cord blood cells, have been tested and have demonstrated a potential for hepatic repopulation [166,170]. However, because of the ethics controversy and source shortages, their availability is limited. Therefore, BMSCs have unique advantages over other stem cell sources, particularly those BMSCs from the autologous source.

It must be noticed that the high prevalence of chronic liver disease and the increased number of patients reaching end-stage disease and requiring OLT may lead to a shortage of donor livers. This clinical scenario has driven forward a number of trials of autologous stem cell therapy. Cell therapy has several potential advantages when compared to OLT, because transplantable cells can be expanded in vitro and cryopreserved, genetically manipulated to correct inborn errors of metabolism, cryopreserved for future use and infused without major surgery, or obtained from the same patient, thereby avoiding risk of rejection and the need for lifelong immune-suppression.

Many of the clinical trials for liver diseases are still pilot studies and are therefore unrandomized and uncontrolled, but they show some interesting results. Studies from Terai and colleagues (2012) in Japan and Lyra and collaborators (2007) have confirmed the safety and efficacy of autologous bone marrow cell infusion (ABMi) therapy applied to patients with liver cirrhosis [175,176].

Terai and Sakaida, (2003) et al. have developed an in vivo murine model (the green fluorescent protein (GFP)/carbon tetrachloride (CCl_4) model) and reported that GFP-positive bone marrow cells infused via a tail vein (peripheral vein) efficiently repopulated cirrhotic liver. Repopulated bone marrow cells ameliorated liver fibrosis through higher expression of matrix metalloproteinase-9, consistent with improved liver functions and survival rate [165,177]. They also confirmed that the number of A6-positive cells in GFP-positive bone marrow cell infused livers increased, suggesting the activation of the HPC compartment by the bone marrow cell infusion [170]. Based on these findings, they started a clinical trial using autologous bone marrow cell infusion (ABM*i*) therapy for decompensated liver cirrhotic patients. As a result, at 6 months after ABM*i*, the average levels of serum albumin and Child-Pugh score significantly improved in nine patients (hepatitis B virus–related: three cases, hepatitis C virus–related: five cases, unknown: one case). The average proliferating cell nuclear antigen (PCNA)-labeling index also increased in biopsied livers after ABM*i*, suggesting induced proliferation of resident hepatocytes by ABM*i* [178]. In addition, Kim et al. confirmed that ABM*i* improved serum albumin levels, Child-Pugh score, liver volume measured by abdominal magnetic resonance imaging (MRI), and accumulation of ascites in 10 patients with hepatitis B virus–related decompensated liver cirrhosis, and histologic observations of liver biopsies taken over time showed increased CK-7 positive cells after ABM*i*, suggesting the possibility of HPC activation as the underlying mechanism [171].

In this scenario, studies focusing on intrahepatic stem/progenitors cells have shown promising results to overcome the present limitations. Because they are able to proliferate and give rise to hepatocytes and cholangiocytes [164,184], liver stem/progenitor cells could make a better choice for long-term repopulation and sustained metabolic activity as well as an efficient alternative for treating liver disorders.

Tanimizu et al. showed that Dlk-1 (delta-like1, a cell surface transmembrane protein highly expressed in human and rodent fetal liver, but not in the adult) is useful for enriching a progenitor population harvested from fetal liver. They described culture condition standardization for these cells (which they called hepatoblasts by their characteristic AFP expression), evidencing the important role of extracellular matrix proteins for cell behavior. Furthermore, they proved successful engraftment of Dlk+ cells harvested from GFP+ mice in recipient damaged livers [179,180]. In accordance with Tanamizu's reports, Oertel et al. also isolated Dlk-1+ cells from fetal liver and injected them in hepatectomized rats. These cells (but not Dlk-1− cells) were able to repopulate damaged liver [181].

Likewise, aiming at future clinical applications, Weiss et al. isolated Thy-1/CD90+ cells from human adult liver and transplanted them into immunodeficient mice. The group was able to verify engraftment and human hepatic marker production [182]. It is important to note that the Thy-1 (CD90) marker is absent in Dlk-1+ cells, as previously reported by Oertel M et al. [183].

Another recent and interesting study discussed other advantages of the hepatic stem/progenitor cells for future use in therapy. Steatotic livers, discarded for orthotopic liver transplantation, could be a good source of large number of these cells [184]. In this study, Tolosa et al. used EpCAM, Thy-1, and OV-6 markers to select cells from both human and rat livers and verified significant proliferation of these cells. The group suggests that steatotic liver could be used to isolate stem/progenitor liver cells and transplant them in large scale. EpCAM has proved to be an important marker of stem/progenitor liver cells. It was demonstrated that purified EpCAM+/AFP− cells from fetal and postnatal livers are able to engraft the livers of immunodeficient adult mice and give rise to mature human liver parenchymal cells. Interestingly, these cells showed multipotencity and self-renewal [121].

In 2008, McClelland et al. demonstrated that the use of differential culture conditions can successfully isolate HpSCs, but not hepatoblasts, their immediate descendants, which died after few days. The tools used to differentiate them were size (HpSCs ~ 7 to 9 μm; HBs ~10 to 12 μm), morphology (HpSCs have high nucleus/cytoplasm ratio; HBs produce colonies with cordlike morphology), and markers (HBs express AFP and ICAM-1, but not NCAM or claudin 3). Furthermore, they identified high telomerase activity in their HpSC cultures, suggesting self-replication and proliferation [185].

More recently, new advances were achieved in understanding the relationship between HpSCs and their niche [184,185]. Wang et al. elicited the relationship among HpSCs [140] (EpCAM+/NCAM+) and their neighbors (angioblasts, endothelial and stellate cells) and focused on the paracrine signals, in particular those elicited by the ECM components able to regulate the parenchymal lineage stages. Co-culture of the hHpSCs with the different subpopulations of mesenchymal cells elicited distinct biologic responses. The hHpSCs co-cultured with angioblasts resulted in the maintenance of stem cell phenotype, whereas the co-culture of hHpSCs with endothelia and precursors of stellate cells led to hepatoblasts. Moreover, the most extensive effect on differentiation was found in the culture conditions that produced the highest levels of heparan sulfate proteoglycans and was also correlated with tri-dimensionality, the ratio of type I collagen to other collagen types, the ratio of fibronectin to laminin isoforms, the presence of proteoglycans with moderate to high levels of sulfation such as HS-PGs isoforms, and the rigidity of the hydrogels.

Yet another relevant clinical approach was proposed: the use of tissue scaffolds to seed stem cells. The main goal of this approach is to load cells onto a synthetic or natural three-dimensional scaffold in order to induce hepatic differentiation with enhanced cell viability, proliferation, and function before transplantation [186]. This methodology, however, needs more long-term studies to verify feasibility and efficacy.

We conclude that the use of liver stem cells in clinical practice still faces obstacles. It is necessary to identify good markers to isolate the appropriate cell fractions. Moreover, methodologies to maintain and expand these cells in culture have to be developed. But these hurdles do not diminish the excitement about the future use of HpSCs to reduce the suffering of patients waiting for liver transplantation.

References

[1] Trutmann M, Sasse D. The lymphatics of the liver. Anat Embryol (Berl) 1994;190:201–9.

[2] Wisse E, Braet F, Luo D, et al. Structure and function of sinusoidal lining cells in the liver. Toxicol Pathol 1996;24:100–11.

[3] Rappaport AM. Hepatic blood flow: morphologic aspects and physiologic regulation. Int Rev Physiol 1980;21:1–63.

[4] Shafritz DA, Oertel M, Menthena A, et al. Liver stem cells and prospects for liver reconstitution by transplanted cells. Hepatology 2006;43:89–99.

[5] Le Douarin NM. An experimental analysis of liver development. Med Biol 1975;53:427–55.

[6] Zaret KS. Genetic programming of liver and pancreas progenitors: lessons for stem-cell differentiation. Nat Rev Genet 2008;9:329–40.

[7] Calmont A, Wandzioch E, Tremblay KD, et al. An FGF response pathway that mediates hepatic gene induction in embryonic endoderm cells. Dev Cell 2006;11:339–48.

[8] Rossi JM, Dunn NR, Hogan BLM, Zaret KS. Distinct mesodermal signals, including BMPs from the septum transversum mesenchyme, are required in combination for hepatogenesis from the endoderm. Genes Dev 2001;15:1998–2009.

[9] Bossard P, Zaret KS. GATA transcription factors as potentiators of gut endoderm differentiation. Development 1998;125:4909–17.

[10] McLin VA, Rankin SA, Zorn AM. Repression of Wnt/beta-catenin signaling in the anterior endoderm is essential for liver and pancreas development. Development 2007;134:2207–17.

[11] Finley KR, Tennessen J, Shawlot W. The mouse secreted frizzled-related protein 5 gene is expressed in the anterior visceral endoderm and foregut endoderm during early post-implantation development. Gene Expr Patterns 2003;3:681–4.

[12] Matsumoto K, Yoshitomi H, Rossant J, Zaret KS. Liver organogenesis promoted by endothelial cells prior to vascular function. Science 2001;294:559–63.

[13] Shiojiri N. Analysis of differentiation of hepatocytes and bile duct cells in developing mouse liver by albumin immunofluorescence. Dev Growth Differ 1984;26:555–61.

[14] Gualdi R, Bossard P, Zheng M, Hamada Y, Coleman JR, Zaret KS. Hepatic specification of the gut endoderm in vitro: cell signaling and transcriptional control. Genes Dev 1996;10:1670–82.

[15] Lemaigre F, Zaret KS. Liver development update: new embryo models, cell lineage control, and morphogenesis. Cur Opinion Genet Dev 2004;14:582–90.

[16] Shiojiri N, Sugiyama Y. Immunolocalization of extracellular matrix components and integrins during mouse liver development. Hepatology 2004;40:346–55.

[17] Bort R, Signore M, Tremblay K, Martinez Barbera JP, Zaret KS. Hex homeobox gene controls the transition of the endoderm to a pseudostratified, cell emergent epithelium for liver bud development. Dev Biol 2006;290:44–56.

[18] Zaret KS. Regulatory phases of early liver development: paradigms of organogenesis. Nat Rev Genet 2002;3:499–512.

[19] Zhao R, Duncan SA. Embryonic development of the liver. Hepatology 2005;41:956–67.

[20] Sosa-Pineda B, Wigle JT, Oliver G. Hepatocyte migration during liver development requires Prox1. Nat Genet 2000;25:254–5.

[21] Zaret K. Early liver differentiation: genetic potentiation and multilevel growth control. Curr Opin Genet Dev 1998;8:526–31.

[22] Medlock ES, Haar JL. The liver hemopoietic environment: I. Developing hepatocytes and their role in fetal hemopoiesis. Anat Rec 1983;207:31–41.

[23] Jung J, Zheng M, Goldfarb M, Zaret KS. Initiation of mammalian liver development from endoderm by fibroblasts growth factors. Science 1999;284:1998–2003.

[24] Serls AE, Doherty S, Parvatiyar P, et al. Different thresholds of fibroblast growth factors pattern the ventral foregut into liver and lung. Development 2005;132:35–47.

[25] Margagliotti S, Clotman F, Pierreux CE, et al. Role of metalloproteinases at the onset of liver development. Dev Growth Differ 2008;50:331–8.

[26] Margagliotti S, Clotman F, Pierreux CE, et al. The Onecut transcription factors HNF-6/OC-1 and OC-2 regulate early liver expansion by controlling hepatoblast migration. Dev Biol 2007;311:579–89.

[27] Tanimizu N, Miyajima A. Molecular mechanism of liver development and regeneration. Int Rev Cytol 2007;259:1–48.

[28] Hilberg F, Aguzzi A, Howells N, et al. C-jun is essential for normal mouse development and hepatogenesis. Nature 1993;365:179–81.

[29] Schmidt C, Bladt F, Goedecke S, et al. Scatter factor/hepatocyte growth factor is essential for liver development. Nature 1995;373:699–702.

[30] Weinstein M, Monga SP, Liu Y, et al. Smad proteins and hepatocyte growth factor control parallel regulatory pathways that converge on beta1-integrin to promote normal liver development. Mol Cell Biol 2001;21:5122–31.

[31] Petkov PM, Zavadil J, Goetz D, Chu T, Carver R, Rogler CE, et al. Gene expression pattern in hepatic stem/progenitor cells during rat fetal development using complementary DNA microarrays. Hepatology 2004;39:617–27.

[32] Jochheim A, Cieslak A, Hillemann T, et al. Multi-stage analysis of differential gene expression in BALB/C mouse liver development by high-density microarrays. Differentiation 2003;71:62–72.

[33] Du Bois AM. The embryonic liver. In: Rouiller C, editor. The Liver, Morphology, Biochemistry, Physiology, vol. I. New York: Academic Press; 1963. p. 1–39.

[34] Collardeau-Frachon, Scoazec J-Y. Vascular development and differentiation during human liver organogenesis. Anat Rec 2008;291:614–27.

[35] Couvelard A, Scoazec J-Y, Dauge M-C, et al. Structural and functional differentiation of sinusoidal endothelial cells during liver organogenesis in humans. Blood 1996;87:4568–80.

[36] Enzan H, Hara H, Yamashita Y, et al. Fine structure of hepatic sinusoids and their development in human embryos and fetuses. Acta Pathol Jpn 1983;33:447–66.

[37] Nonaka H, Tanaka M, Suzuki K, Miyajima A. Development of murine hepatic sinusoidal endothelial cells characterized by the expression of hyaluronan receptors. Dev Dyn 2007;236:2258–67.

[38] Perez-Pomares JM, Carmona R, Gonzalez-Iriarte M, Macias D, Guadix JA, Munoz-Chapuli R. Contribution of mesothelium-derived cells to liver sinusoids in avian embryos. Dev Dyn 2004;229:465–74.

[39] Cherqui S, Kurian SM, Schussler O, Hewel JA, Yates JR, Salomon DR. Isolation and angiogenesis by endothelial progenitors in the fetal liver. Stem Cells 2006;24:44–54.

[40] Asahina K. Hepatic stellate cell progenitor cells. J Gastroenterol Hepatol 2012;27:80–4.

[41] Antoniou A, Raynaud P, Cordi S, et al. Intrahepatic bile ducts develop according to a new mode of tubulogenesis regulated by the transcription factor SOX9. Gastroenterology 2009;136:2325–33.

[42] Van Eyken P, Sciot R, Callea F, Van der Steen K, Moerman P, Desmet VJ. The development of the intrahepatic bile ducts in man: a keratin-immunohistochemical study. Hepatology 1988;1988(8):1586–95.

[43] Lemaigre FP. Molecular mechanisms of biliary development. Prog Mol Biol Transl Sci 2010;97:103–26.

[44] Tanimizu N, Miyajima A, Mostov KE. Liver progenitor cells fold up a cell monolayer into a double-layered structure during tubular morphogenesis. Mol Biol Cell 2009;20:2486–94.

[45] Ito A, Nishikawa Y, Ohnuma K, Ohnuma I, Koma Y, Sato A, et al. SgIGSF is a novel biliary-epithelial cell adhesion molecule mediating duct/ductule development. Hepatology 2007;45:684–94.

[46] Crawford AR, Lin XZ, Crawford JM. The normal adult human liver biopsy: a quantitative reference standard. Hepatology 1998;28:323–31.

[47] Terada T, Nakanuma Y. Detection of apoptosis and expression of apoptosis-related proteins during human intrahepatic bile duct development. Am J Pathol 1995;146:67–74.

[48] Carpentier R, Suñer RE, van Hul N, Kopp JL, Beaudry JB, Cordi S, et al. Embryonic ductal plate cells give rise to cholangiocytes, periportal hepatocytes, and adult liver progenitor cells. Gastroenterology 2011;141:1432–8.

[49] Kopp JL, Dubois CL, Schaffer AE, et al. Sox9+ ductal cells are multipotent progenitors throughout development but do not produce new endocrine cells in the normal or injured adult pancreas. Development 2011;138:653–65.

[50] Zong Y, Panikkar A, Xu J, Antoniou A, Raynaud P, Lemaigre F, et al. Notch signaling controls liver development by regulating biliary differentiation. Development 2009;136:1727–39.

[51] Li L, Krantz ID, Deng Y, Genin A, Banta AB, Collins CC, et al. Alagille syndrome is caused by mutations in human Jagged1, which encodes a ligand for Notch1. Nat Genet 1997;16:243–51.

[52] Oda T, Elkahloun AG, Pike BL, Okajima K, Krantz ID, Genin A, et al. Mutations in the human Jagged1 gene are responsible for Alagille syndrome. Nat Genet 1997;16:235–42.

[53] McDaniell R, Warthen DM, Sanchez-Lara PA, Pai A, Krantz ID, Piccoli DA, et al. NOTCH2 mutations cause Alagille syndrome, a heterogeneous disorder of the notch signaling pathway. Am J Hum Genet 2006;79:169–73.

[54] Kodama Y, Hijikata M, Kageyama R, Shimotohno K, Chiba T. The role of notch signaling in the development of intrahepatic bile ducts. Gastroenterology 2004;127:1775–86.

[55] Lozier J, McCright B, Gridley T. Notch signaling regulates bile duct morphogenesis in mice. PLoS One 2008;3:e1851.

[56] McCright B, Lozier J, Gridley T. A mouse model of Alagille syndrome: Notch2 as a genetic modifier of Jag1 haploinsufficiency. Development 2002;129:1075–82.

[57] Clotman F, Jacquemin P, Plumb-Rudewiez N, Pierreux CE, Van der Smissen P, Dietz HC, et al. Control of liver cell fate decision by a gradient of TGF beta signaling modulated by Onecut transcription factors. Genes Dev 2005;19:1849–54.

[58] Clotman F, Lemaigre FP. Control of hepatic differentiation by activin/TGFbeta signaling. Cell Cycle 2006;5:168–71.

[59] Pelton RW, Saxena B, Jones M, et al. Immunohistochemical localization of TGF beta 1, TGF beta 2, and TGF beta 3 in the mouse embryo: expression patterns suggest multiple roles during embryonic development. J Cell Biol 1991;115:1091–105.

[60] Magami Y, Azuma T, Inokuchi H, Kokuno S, Moriyasu F, Kawai K, et al. Cell proliferation and renewal of normal hepatocytes and bile duct cells in adult mouse liver. Liver 2002;22:419–25.

[61] MacDonald RA, Rogers AE. Control of regeneration of the liver. Gastroenterology 1961;41:33–8.

[62] Fausto N. Liver regeneration. J Hepatol 2000;(Suppl. 32):19–31.

[63] Cressman DE, Greenbaum LE, DeAngelis RA, et al. Liver failure and defective hepatocyte regeneration in interleukin-6-deficient mice. Science 1996;274:1379–83.

[64] Hata S, Namae M, Nishina H. Liver development and regeneration: from laboratory study to clinical therapy. Dev Growth Differ 2007;49:163–70.

[65] Overturf K, al-Dhalimy M, Ou CN, et al. Serial transplantation reveals the stem-cell-like regenerative potential of adult mouse hepatocytes. Am J Pathol 1997;151:1273–80.

[66] Fausto N, Campbell JS, Riehle KJ. Liver regeneration. Hepatology 2006;43:45–53.

[67] Michalopoulos GK, DeFrances MC. Liver regeneration. Science 1997;276:60–6.

[68] Michalopoulos GK. Liver regeneration: alternative epithelial pathways. Int J Biochem Cell Biol 2011;221:173–9.

[69] Riehle KJ, Dan YY, Campbell JS, Fausto N. New concepts in liver regeneration. Journal Gastroenterol Hepatol 2011;1(Suppl. 26):203–12.

[70] Malato Y, Naqvi S, Schürmann N, Ng R, Wang B, Zape J, et al. Fate tracing of mature hepatocytes in mouse liver homeostasis and regeneration. J Clin Invest 2011;121(12):4850–60.

[71] Roskams TA, Libbrecht L, Desmet VJ. Progenitor cells in diseased human liver. Semin Liver Dis 2003;23:385–96.

[72] Zhou H, Rogler LE, Teperman L, et al. Identification of hepatocytic and bile ductular cell lineages and candidate stem cells in bipolar ductular reactions in cirrhotic human liver. Hepatology 2007;45:716–24.

[73] Gaudio E, Carpino G, Cardinale V, Franchitto A, Onori P, Alvaro D. New insights into liver stem cells. Dig Liver Dis 2009;41:455–62.

[74] Paku S, Schnur J, Nagy P, Thorgeirsson SS. Origin and structural evolution of the early proliferating oval cells in rat liver. Am J Pathol 2001;158:1313–23.

[75] Paku S, Dezso K, Kopper L, et al. Immunohistochemical analysis of cytokeratin 7 expression in resting and proliferating biliary structures of rat liver. Hepatology 2005;42:863–70.

[76] Jensen CH, Jauho EI, Santoni-Rugiu E, Holmskov U, Teisner B, Tygstrup N, et al. Transit-amplifying ductular (oval) cells and their hepatocytic progeny are characterized by a novel and distinctive expression of delta-like protein/preadipocyte factor 1/fetal antigen 1. Am J Pathol 2004;164:1347–59.

[77] Shimano K, Satake M, Okaya A, Kitanaka J, Kitanaka N, Takemura M, et al. Hepatic oval cells have the side population phenotype defined by expression of ATP-binding cassette transporter ABCG2/BCRP1. Am J Pathol 2003;163:3–9.

[78] Preisegger KH, Factor VM, Fuchsbichler A, Stumptner C, Denk H, Thorgeirsson SS. Atypical ductular proliferation and its inhibition by transforming growth factor beta1 in the 3,5-diethoxycarbonyl-1,4-dihydrocollidine mouse model for chronic alcoholic liver disease. Lab Invest 1999;79:103–9.

[79] Dabeva MD, Shafritz DA. Activation, proliferation and differentiation of progenitor cells into hepatocytes in the D-galactosamine model of liver regeneration. Am J Pathol 1993;143:1606–20.

[80] Sell S, Osborn K, Leffert HL. Autoradiography of "oval cells" appearing rapidly in the livers of rats fed N-2-fluorenylacetamide in a choline devoid diet. Carcinogenesis 1981;2:7–14.

[81] Farber E. Similarities in the sequence of early histological changes induced in the liver of the rat by ethionine, 2-acetylamino-fluorene, and 3'-methyl-4-dimethylaminoazobenzene. Cancer Res 1956;16:142–8.

[82] Alison MR, Vig P, Russo F, Bigger BW, Amofah E, Themis M, et al. Hepatic stem cells: from inside and outside the liver? Cell Prolif 2004;37:1–21.

[83] Fausto N, Campbell JS. The role of hepatocytes and oval cells in liver regeneration and repopulation. Mech Dev 2003;120(1):17–130.

[84] Evarts RP, Nagy P, Nakatsukasa H, Marsden E, Thorgeirsson S. In vivo differentiation of rat liver oval cells into hepatocytes. Cancer Res 1989;49:1541–7.

[85] Lazaro CA, Rhim JA, Yamada Y, Fausto N. Generation of hepatocytes from oval cell precursors in culture. Cancer Res 1998;58:5514–22.

[86] Yoon BI, Choi YK, Kim DY. Differentiation processes of oval cells into hepatocytes: proposals based on morphological and phenotypical traits in carcinogen-treated hamster liver. J Comp Pathol 2004;131:1–9.

[87] Lemire JM, Shiojiri N, Fausto N. Oval cell proliferation and the origin of small hepatocytes in liver injury induced by D-galactosamine. Am J Pathol 1991;139:535–52.

[88] Sirica AE, Williams TW. Appearance of ductular hepatocytes in rat liver after bile duct ligation and subsequent zone 3 necrosis by carbon tetrachloride. Am J Pathol 1992;140:129–36.
[89] Yoon BI, Jung SY, Hur K, Lee JH, Joo KH, Lee YS, et al. Differentiation of hamster liver oval cells following Clonorchis sinensis infection. J Veter Med Science 2000;62:1303–10.
[90] Lee JH, Rim HJ, Sell S. Heterogeneity of the "oval-cell" response in the hamster liver during cholangiocarcinogenesis following Clonorchis sinensis infection and dimethylnitrosamine treatment. J Hepatol 1997;26:1313–23.
[91] Shafritz DA, Dabeva MD. Liver stem cells and model systems for liver repopulation. J Hepatol 2002;36:552–64.
[92] Thorgeirsson SS. Stem cells and hepatocarcinogenesis. In: Jirtle RL, editor. Liver Regeneration and Carcinogenesis: Molecular and Cellular Mechanisms. San Diego, CA: Academic Press; 1995. p. 99–112.
[93] Wang X, Foster M, Al-Dhalimy M, Lagasse E, Finegold M, Grompe M. The origin and liver repopulating capacity of murine oval cells. Proc Natl Acad Sci USA 2003;100:11881–8.
[94] Akhurst B, Croager EJ, Farley-Roche CA, Ong JK, Dumble ML, Knight B, et al. A modified choline-deficient, ethionine-supplemented diet protocol effectively induces oval cells in mouse liver. Hepatology 2001;34:519–22.
[95] Kofman AV, Morgan G, Kirschenbaum A, Osbeck J, Hussain M, Swenson S, et al. Dose- and time-dependent oval cell reaction in acetaminophen-induced murine liver injury. Hepatology 2005;41:1252–61.
[96] Lee JS, Heo J, Libbrecht L, Chu IS, Kaposi-Novak P, Calvisi DF, et al. A novel prognostic subtype of human hepatocellular carcinoma derived from hepatic progenitor cells. Nat Med 2006;12:410–6.
[97] Roskams T, De Vos R, Van Eyken P, Myazaki H, Van Damme B, Desmet V, et al. Hepatic OV-6 expression in human liver disease and rat experiments: evidence for hepatic progenitor cells in man. J Hepatol 1998;29:455–63.
[98] Theise ND, Saxena R, Portmann BC, Thung SN, Yee H, Chiriboga L, et al. The canals of Hering and hepatic stem cells in humans. Hepatology 1999;30:1425–33.
[99] Katoonizadeh A, Nevens F, Verslype C, Pirenne J, Roskams T. Liver regeneration in acute severe liver impairment: a clinicopathological correlation study. Liver Int 2006;26:1225–33.
[100] Dollé L, Best J, Mei J, Al Battah F, Reynaert H, Leo A, et al. The quest for liver progenitor cells: A practical point of view. J Hepatol 2010;52:117–29.
[101] Zheng YW, Taniguchi H. Diversity of hepatic stem cells in the fetal and adult liver. Semin Liver Dis 2003;23:337–48.
[102] Dorrell C, Erker L, Lanxon-Cookson KM, Abraham SL, Victoroff T, Ro S, et al. Surface markers for the murine oval cell response. Hepatology 2008;48:1282–91.
[103] Roskams T, Desmet V. Ductular reaction and its diagnostic significance. Sem Diag Pathol 1998;15:259–69.
[104] Desmet V, Roskams T, Van Eyken P. Ductular reaction in the liver. Pathol Res Pract 1995;191:513–24.
[105] Roskams T. Different types of liver progenitor cells and their niches. J Hepatol 2006;45:1–4.
[106] Dunsford HA, Sell S. Production of monoclonal antibodies to preneoplastic liver cell populations induced by chemical carcinogens in rat and to transplantable Morris hepatomas. Cancer Res 1989;49:4887–93.
[107] Sell S, Dunsford HA. Evidence for the stem cell origin of hepatocellular carcinoma and cholangiocarcinoma. Am J Pathol 1989;134:1347–63.

[108] Crosby HA, Hubscher S, Fabris L, Joplin R, Sell S, Kelly D, et al. Immunolocalization of putative human liver progenitor cells in livers from patients with end-stage primary biliary cirrhosis and sclerosing cholangitis using the monoclonal antibody OV-6. Am J Pathol 1998;152:771–9.

[109] Bisgaard HC, Parmelee DC, Dunsford HA, Sechi S, Thorgeirsson SS. Keratin 14 protein in cultured nonparenchymal rat hepatic epithelial cells: characterization of keratin 14 and keratin 19 as antigens for the commonly used mouse monoclonal antibody OV-6. Mol Carcinog 1993;7:60–6.

[110] Blakolmer K, Jaskiewicz K, Dunsford HA, Robson SC. Hemopoietic stem cell markers are expressed by ductal plate and bile duct cells in developing human liver. Hepatology 1995;21:1510–6.

[111] Roskams T, De Vos R, Desmet V. 'Undifferentiated progenitor cells' in focal nodular hyperplasia of the liver. Histopathology 1996;28:291–6.

[112] Engelhardt NV, Factor VM, Medvinsky AL, Baranov VN, Lazareva MN, Poltoranina VS. Common antigen of oval and biliary epithelial cells (A6) is a differentiation marker of epithelial and erythroid cell lineages in early development of the mouse. Differentiation 1993;55:19–26.

[113] Factor VM, Radaeva SA. Oval cells—hepatocytes relationships in Dipin-induced hepatocarcinogenesis in mice. Exp Toxicol Pathol 1993;45:239–44.

[114] Petersen BE, Grossbard B, Hatch H, Pi L, Deng J, Scott EW. Mouse A6-positive hepatic oval cells also express several hematopoietic stem cell markers. Hepatology 2003;37:632–40.

[115] Conigliaro A, Colletti M, Cicchini C, Guerra MT, Manfredini R, Zini R, et al. Isolation and characterization of a murine resident liver stem cell. Cell Death Differ 2008;15:123–33.

[116] Engelhardt NV, Factor VM, Yasova AK, Poltoranina VS, Baranov VN, Lasareva MN. Common antigens of mouse oval and biliary epithelial cells. Expression on newly formed hepatocytes. Differentiation 1990;45:29–37.

[117] Balzar M, Bakker HA, Briaire-de-Bruijn IH, Fleuren GJ, Warnaar SO, Litvinov SV. Cytoplasmic tail regulates the inter-cellular adhesion function of the epithelial cell adhesion molecule. Mol Cell Biol 1998;18:4833–43.

[118] Balzar M, Winter MJ, de Boer CJ, Litvinov SV. The biology of the 17-1A antigen (Ep-CAM). J Mol Med 1999;77:699–712.

[119] Armstrong A, Eck SL. EpCAM: A new therapeutic target for an old cancer antigen. Cancer Biol Ther 2003;2:320–6.

[120] Dan YY, Riehle KJ, Lazaro C, Teoh N, Haque J, Campbell JS, et al. Isolation of multipotent progenitor cells from human fetal liver capable of differentiating into liver and mesenchymal lineages. Proc Natl Acad Sci USA 2006;103:9912–7.

[121] Schmelzer E, Zhang L, Bruce A, Wauthier E, Ludlow J, Yao HL, et al. Human hepatic stem cells from fetal and postnatal donors. J Exp Med 2007;204:1973–87.

[122] Tanaka M, Okabe M, Suzuki K, Kamiya Y, Tsukahara Y, Saito S, et al. Mouse hepatoblasts at distinct developmental stages are characterized by expression of EpCAM and DLK1: drastic change of EpCAM expression during liver development. Mech Dev 2009;126:665–76.

[123] de Boer CJ, van Krieken JH. Janssen-van Rhijn CM, Litvinov SV. Expression of Ep-CAM in normal, regenerating, metaplastic, and neoplastic liver. J Pathol 1999;188:201–6.

[124] Okabe M, Tsukahara Y, Tanaka M, Suzuki K, Saito S, Kamiya Y, et al. Potential hepatic stem cells reside in EpCAM+ cells of normal and injured mouse liver. Development 2009;136:1951–60.

[125] Yovchev MI, Grozdanov PN, Zhou H, Racherla H, Guha C, Dabeva MD. Identification of adult hepatic progenitor cells capable of repopulating injured rat liver. Hepatology 2008;47:636–47.

[126] Paku S, Nagy P, Kopper L, Thorgeirsson SS. 2-Acetylaminofluorene dose-dependent differentiation of rat oval cells into hepatocytes: confocal and electron microscopic studies. Hepatology 2004;39:1353–61.

[127] Cassiman D, Libbrecht L, Sinelli N, et al. The vagal nerve stimulates activation of the hepatic progenitor cell compartment via muscarinic acetylcholine receptor type 3. Am J Pathol 2002;161:521–30.

[128] Fausto N, Lemire JM, Shiojiri N. Cell lineages in hepatic development and the identification of progenitor cells in normal and injured liver. Proc Soc Exp Biol Med 1993;204:237–41.

[129] Roskams TA, Theise ND, Balabaud C, et al. Nomenclature of the finer branches of the biliary tree: canals, ductules, and ductular reactions in human livers. Hepatology 2004;39:1739–45.

[130] Suzuki A, Zheng Y, Kondo R, Kusakabe M, Takada Y, Fukao K, et al. Flow-cytometric separation and enrichment of hepatic progenitor cells in the developing mouse liver. Hepatology 2000;32:1230–9.

[131] Suzuki A, Zheng Y-M, Shin Kaneko S, et al. Clonal identification and characterization of self-renewing pluripotent stem cells in the developing liver. J Cell Biol 2002;156:173–84.

[132] Sicklick JK, Li YX, Melhem A, et al. Hedgehog signaling maintains resident hepatic progenitors throughout life. Am J Physiol Gastrointest Liver Physiol 2006;290:G859–70.

[133] Crawford JM. Development of the intrahepatic biliary tree. Semin Liver Dis 2002;22:213–26.

[134] Reya T, Morrison SJ, Clarke MF, Weissman IL. Stem cells, cancer, and cancer stem cells. Nature 2001;414:105–11.

[135] Smith A. A glossary for stem-cell biology. Nature 2006;441:1060.

[136] Potten CS, Loeffler M. Stem cells: attributes, cycles, spirals, pitfalls and uncertainties; lessons for and from the crypt. Development 1990;110:1101–20.

[137] Potten CS. Epithelial cell growth and differentiation. II. Intestinal apoptosis. Am J Physiol 1997;273:G253–7.

[138] Shafritz DA, Oertel M. Model systems and experimental conditions that lead to effective repopulation of the liver by transplanted cells. Int J Biochem Cell Biol 2011;43:198–213.

[139] Jones DL, Wagers AJ. No place like home: anatomy and function of the stem cell niche. Nat Rev Mol Cell Biol 2008;9:11–21.

[140] Wang Y, H-l Yao, Barbier C, Wauthier E, Cui C, Moss N, et al. Lineage-dependent epithelial-mesenchymal paracrine signals dictate growth versus differentiation of human hepatic stem cells to adult fates. Hepatology 2010;52:1443–54.

[141] Theise ND. Gastrointestinal stem cells. III. Emergent themes of liver stem cell biology: niche, quiescence, self-renewal, and plasticity. Am J Physiol Gastrointest Liver Physiol 2006;290:G189–93.

[142] Baumann U, Crosby HA, Ramani P, Kelly DA, Strain AJ. Expression of the stem cell factor receptor c-kit in normal and diseased pediatric liver: identification of a human hepatic progenitor cell. Hepatology 1999;30:112–7.

[143] Mignone JL, Kukekov V, Chiang AS, Steindler D, Enikolopov G. Neural stem and progenitor cells in nestin-GFP transgenic mice. J Comp Neurol 2004;469:311–24.

[144] Gleiberman AS, Encinas JM, Mignone JL, Michurina T, Rosenfeld MG, Enikolopov G. Expression of nestin-green fluorescent protein transgene marks oval cells in the adult liver. Dev Dyn 2005;234:413–21.

[145] Sell S. The hepatocyte: heterogeneity and plasticity of liver cells. Int J Biochem Cell Biol 2003;35:267–71.

[146] Petersen BE, Bowen WC, Patrene KD, Mars WM, Sullivan AK, Murase N, et al. Bone marrow as a potential source of hepatic oval cells. Science 1999;284:1168–70.

[147] Kuwahara R, Kofman AV, Landis CS, et al. The hepatic stem cell niche: identification by label retaining cell assay. Hepatology 2008;47:1994–2002.

[148] Lee SW, Wang X, Chowdhury NR, Roy-Chowdhury J. Hepatocyte transplantation: State of the art and strategies for overcoming existing hurdles. Ann Hepatol 2004;3:48–53.

[149] Chung H, Kim KH, Kim JG, Lee SY, Yoon YH. Retinal complications in patients with solid organ or bone marrow transplantation. Transplantation 2007;83:694–9.

[150] Francoz C, Belghiti J, Durand J. Indications of liver transplantation in patients with complications of cirrhosis. Best Pract Res Clin Gastroenterol 2007;21:175–90.

[151] Patel H, Vogl DT, Aqui N, Shaked A, Olthoff K, Markmann J, et al. Posttransplant lymphoproliferative disorder in adult liver transplant recipients: a report of seventeen cases. Leuk Lymphoma 2007;48:885–91.

[152] Tamsel S, Demirpolat G, Killi R, Aydın U, Kilic M, Zeytunlu M, et al. Abdom Imaging 2007;32:339–47.

[153] Fox IJ, Roy-Chowdhury J. Hepatocyte transplantation. J Hepatol 2004;40:878–86.

[154] Dhawan A, Mitry RR, Hughes RD. Hepatocyte transplantation for liver-based metabolic disorders. J Inherit Metab Dis 2006;29:431–5.

[155] Fisher RA, Strom SC. Human hepatocyte transplantation: Worldwide results. Transplantation 2006;82:441–9.

[156] Dhawan A, Puppi J, Hughes RD, et al. Human hepatocyte transplantation: Current experience and future challenges. Nature Rev Gastroenterol Hepatol 2010;7:288–98.

[157] Hughes RD, Mitry RR, Dhawan A. Current status of hepatocyte transplantation. Transplantation 2012;93:342–7.

[158] Wu PC, Lai VC, Fang JE, Gerber MA, Lai CL, Lau JY. Hepatocellular carcinoma expressing both hepatocellular and biliary markers also expresses cytokeratin 14, a marker of bipotential progenitor cells. J Hepatol 1999;31:965–6.

[159] Komuta M, Spee B, Vander Borght S, De Vos R, Verslype C, Aerts R, et al. Clinicopathological study on cholangiolocellular carcinoma suggesting hepatic progenitor cell origin. Hepatology 2008;47:1544–56.

[160] Grossman M, Rader DJ, Muller DW, Kolansky DM, Kozarsky K, Clark 3rd BJ, et al. A pilot study of *ex vivo* gene therapy for homozygous familial hypercholesterolaemia. Nat Med 1995;1: 1148–54.

[161] Dhawan A, Puppi J, Hughes RD, Mitry RR. Human hepatocyte transplantation: current experience and future challenges. Nat Rev Gastroenterol Hepatol 2010;7:288–98.

[162] Smets F, Najimi M, Sokal EM. Cell transplantation in the treatment of liver disease. Pediatr Transplant 2008;12:6–13.

[163] Watanabe T, Tanaka Y. Age related alterations in the size of human hepatocytes. A study of mononuclear and binucleate cells. Virchows Arch B Cell Pathol Incl Mol Pathol 1982;39:9–20.

[164] Malhi H, Irani AN, Gagandeep S, Gupta S. Isolation of human progenitor liver epithelial cells with extensive replication capacity and differentiation into mature hepatocytes. J Cell Sci 2002;115: 2679–88.

[165] Sakaida I, Terai S, Yamamoto N, Aoyama K, Ishikawa T, Nishina H, et al. Transplantation of bone marrow cells reduces CCl4-induced liver fibrosis in mice. Hepatology 2004;40:1304–11.

[166] Di Campli C, Piscaglia AC, Pierelli L, Rutella S, Bonanno G, Alison MR, et al. A human umbilical cord stem cell rescue therapy in a murine model of toxic liver injury. Dig Liver Dis 2004;36:603–13.

[167] Oyagi S, Hirose M, Kojima M, Okuyama M, Kawase M, Nakamura T, et al. Therapeutic effect of transplanting HGF-treated bone marrow mesenchymal cells into CCL4-injured rats. J Hepatol 2006;44:742–8.

[168] Banas A, Teratani T, Yamamoto Y, Tokuhara M, Takeshita F, Osaki M, et al. Takashi Kato T, Okochi H, Ochiya T. IFATS collection: in vivo therapeutic potential of human adipose tissue mesenchymal stem cells after transplantation into mice with liver injury. Stem Cells 2008;26:2705–12.

[169] Maeda M, Takami T, Terai S, Sakaida I. Autologous bone marrow cell infusions suppress tumor initiation in hepatocarcinogenic mice with liver cirrhosis. J Gastroenterol Hepatol 2012; 2(Suppl. 27):104–11.

[170] Russo FP, Parola M. Stem cells in liver failure. Best Pract Res Clin Gastroenterol 2012;26:35–45.

[171] Kim JK, Park YN, Kim JS, Park MS, Paik YH, Seok JY, et al. Autologous bone marrow infusion activates the progenitor cell compartment in patients with advanced liver cirrhosis. Cell Transplant 2010;19(10):1237–46.

[172] Kisseleva T, Uchinami H, Feirt N, Quintana-Bustamante O, Segovia JC, Schwabe RF, et al. Bone marrow-derived fibrocytes participate in pathogenesis of liver fibrosis. J Hepatol 2006;45:429–38.

[173] Russo FP, Alison MR, Bigger BW, Amofah E, Florou A, Amin F, et al. The bone marrow functionally contributes to liver fibrosis. Gastroenterology 2006;130:1807–21.

[174] Popp FC, Slowik P, Eggenhofer E. No contribution of multipotent mesenchymal stromal cells to liver regeneration in a rat model of prolonged hepatic injury. Stem Cells 2007;25:639–45.

[175] Terai S, Tanimoto H, Maeda M, Zaitsu J, Hisanaga T, Iwamoto T, et al. Timeline for development of autologous bone marrow infusion (ABMi) therapy and perspective for future stem cell therapy. J Gastroenterol 2012;47:491–7.

[176] Lyra AC, Soares MB, da Silva LF, Fortes MF, Silva AG, Mota AC, et al. Feasibility and safety of autologous bone marrow mononuclear cell transplantation in patients with advanced chronic liver disease. World J Gastroenterol 2007;13:1067–73.

[177] Terai S, Sakaida I, Yamamoto N, Omori K, Watanabe T, Ohata S, et al. An in vivo model for monitoring the transdifferentiation of bone marrow cells into functional hepatocytes. J Biochem 2003;134:551–8.

[178] Terai S, Ishikawa T, Omori K, Aoyama K, Marumoto Y, Urata Y, et al. Improved liver function in patients with liver cirrhosis after autologous bone marrow cell infusion therapy. Stem Cells 2006;24:2292–8.

[179] Tanimizu N, Saito H, Mostov K, Miyajima A. Long-term culture of hepatic progenitors derived from mouse Dlk+ hepatoblasts. J Cell Sci 2004;117:6425–34.

[180] Tanimizu N, Mitsuo Nishikawa M, Saito H, Tsujimura T, Miyajima A. Isolation of hepatoblasts based on the expression of Dlk/Pref-1. J Cell Sci 2003;116:1775–86.

[181] Oertel M, Menthena A, Chen YQ, Teisner B, Jensen CH, Shafritz DA. Purification of fetal liver stem/progenitor cells containing all the repopulation potential for normal adult rat liver. Gastroenterol 2008;134:823–32.

[182] Weiss TS, Lichtenauer M, Kirchner S, Stock P, Aurich H, Christ B, et al. Hepatic progenitor cells from adult human livers for cell transplantation. Gut 2008;57:1129–38.

[183] Oertel M, Menthena A, Chen YQ, Shafritz DA. Comparison of hepatic properties and transplantation of Thy-1+ and Thy-1− cells isolated from embryonic day 14 rat fetal liver. Hepatology 2007;46:1236–45.

[184] Tolosa L, Bonora-Centelles A, Teresa Donato M, Pareja E, Negro A, López S, et al. Steatotic liver: a suitable source for the isolation of hepatic progenitor cells. Liver Int 2011;31:1231–8.

[185] McClelland R, Wauthier E, Uronis J, Reid L. Gradients in the liver's extracellular matrix chemistry from periportal to pericentral zones: influence on human hepatic progenitors. Tissue Eng Part A 2008;14:59–70.

[186] Wang Y, Yao H-L, Cui CB, Wauthier E, Barbier C, Costello MJ, et al. Paracrine signals from mesenchymal cell populations govern the expansion and differentiation of hepatic stem cells to adult liver fates. Hepatology 2010;52:1443–54.

[187] Wang Y, Cui CB, Yamauchi M, Miguez P, Roach M, Malavarca R, et al. Lineage restriction of human hepatic stem cells to mature fates is made efficient by tissue-specific biomatrix scaffolds. Hepatology 2011;53:293–305.

[188] Orlando G, Bendala JD, Shupe T, Bergman C, Bitar KN, Booth C, et al. Cell and organ bioengineering technology as applied to gastrointestinal diseases. Gut 2012. Jan 20.

[189] Shinozuka H, Lombardi B, Sell S, Iammarino RM. Early histological and functional alterations of ethionine liver carcinogenesis in rats fed a choline-deficient diet. Cancer Res 1978;38:1092–8.

[190] Fujio K, Evarts RP, Hu Z, Marsden ER, Thorgeirsson SS. Expression of stem cell factor and its receptor, c-kit, during regeneration from putative stem cells in adult rat. Lab Invest 1994;70:511–6.

[191] Sanchez A, Factor VM, Schroeder IS, Nagy P, Thorgeirsson SS. Activation of NFkappaB and STAT3 in rat oval cells during 2-acetylaminofluorene/partial hepatectomy-induced liver regeneration. Hepatology 2004;39:376–85.

[192] Petersen BE, Goff JP, Greenberger JS, Michalopoulos GK. Hepatic oval cells express the hematopoietic stem cell marker Thy-1 in the rat. Hepatology 1998;27:433–45.

Chapter 11

Telocytes and Stem Cells

L.M. Popescu*, M.I. Nicolescu[†,‡]

* *Division of Advanced Studies, Victor Babeş, National Institute of Pathology, Bucharest, Romania*

† *Molecular Medicine Department, Victor Babeş, National Institute of Pathology, Bucharest, Romania*

‡ *Division of Cellular and Molecular Medicine, Carol Davila, University of Medicine and Pharmacy, Bucharest, Romania*

What Are Telocytes?

Telocytes (TC) are a newly described distinct type of interstitial cells [1-3]. Their presence was either overlooked or they were confused with other interstitial cell types, depending on the specific cellular repertoire of a given organ.

Their (most) striking ultrastructural feature is the presence of very long prolongations (several tens to hundreds of μm), called *telopodes* (Tp) [1] (see Figs. 11.1 and 11.2). Moreover, Tp conformation is "specific", consisting of an alternation of thin fibrillar-like segments (*podomers*) and dilated, cistern-like regions (*podoms*) [4]. Podoms accommodate mitochondria, caveolae, and elements of endoplasmic reticulum (ER) the so-called calcium accumulating/releasing units.

The term *telocyte* was coined using the Greek affix *telos*. Aristotle believed that *Telos* was an object's or individual's greatest potential [5]. The initial meaning of the word was "burden", and the most probable semantic development was from "duty"/ "task" to "execution of task", "completeness," and the most important, "power to decide"[6].

Resident Stem Cells and Regenerative Therapy. http://dx.doi.org/10.1016/B978-0-12-416012-5.00011-6

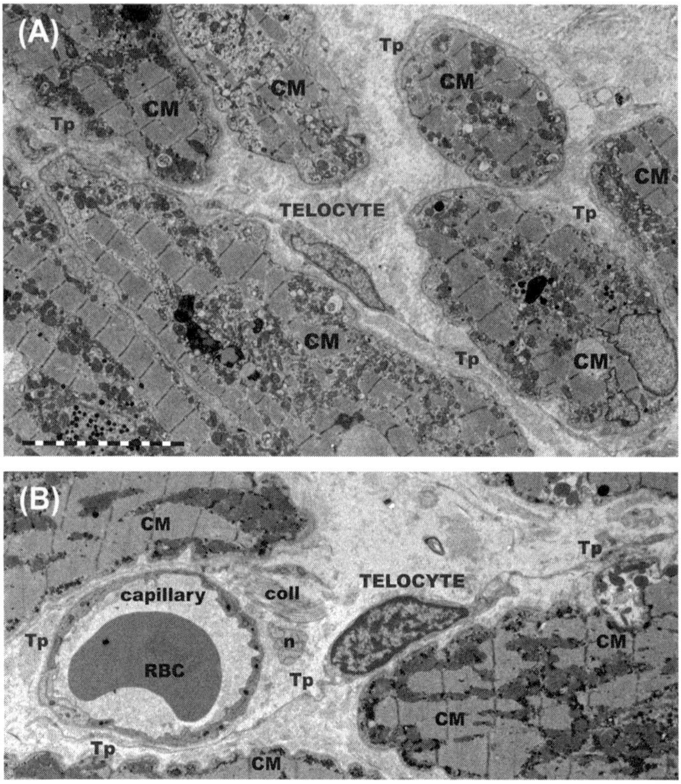

Figure 11.1 Human atrial myocardium, transmission electron microscopy. **(A)** Cardiomyocytes (CM) containing lipofuscin and atrial granules are surrounded by telopodes (Tp) of telocytes, digitally colored in blue. **(B)** Two telopodes (Tp) are bordering a myocardial capillary. RBC, red blood cell; n, nerve ending; coll, collagen. **See Plate 15**.

What Defines Telocytes as a Unique, Distinct Cell Type?

Under electron microscope (EM), the TC silhouette is distinctive, with a small, oval- or triangular-shaped body, and several (two to five) remarkably long, thin, and moniliform Tp (Figs. 11.3 to 11.5).

The identification of TC is mostly based on recognition of their Tp. Therefore, the simplest definition of telocytes is *cells with telopodes* [7]. Tp are characterized by the following:

1. **Number** (one to five, frequently two to three).
2. **Length** (several tens up to hundreds of micrometers).
3. **Moniliform aspect**: alternation of dilated segments (podoms) and thin segments (podomers–less than 200 nm thickness, below the resolving power of light

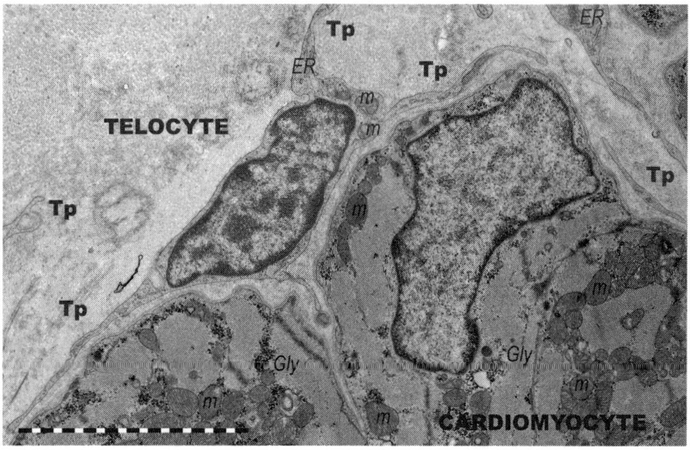

Figure 11.2 Human myocardium. Transmission electron microscopy. A telocyte (digitally colored in blue) extends several long telopodes (Tp) around and between cardiomyocytes. Gly, glycogen; m, mitochondria; ER, endoplasmic reticulum. Scale bar = 10 μm. **See Plate 16.**

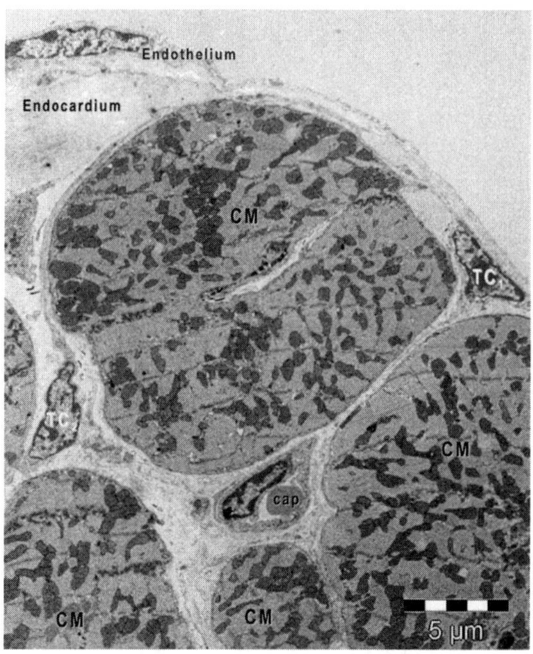

Figure 11.3 Digitally colored electron micrograph of rat ventricular endocardium (burgundy). Telocytes (blue) make an interstitial network in the heart. Subendocardial telocytes (TC_1) send telopodes among cardiomyocytes (CM) making a network with myocardial telocytes (TC_2). Tp network envelopes bundles of cardiomyocytes (CM), cross-cut here. cap, capillary. *(Reproduced with permission from [10]).* **See Plate 17.**

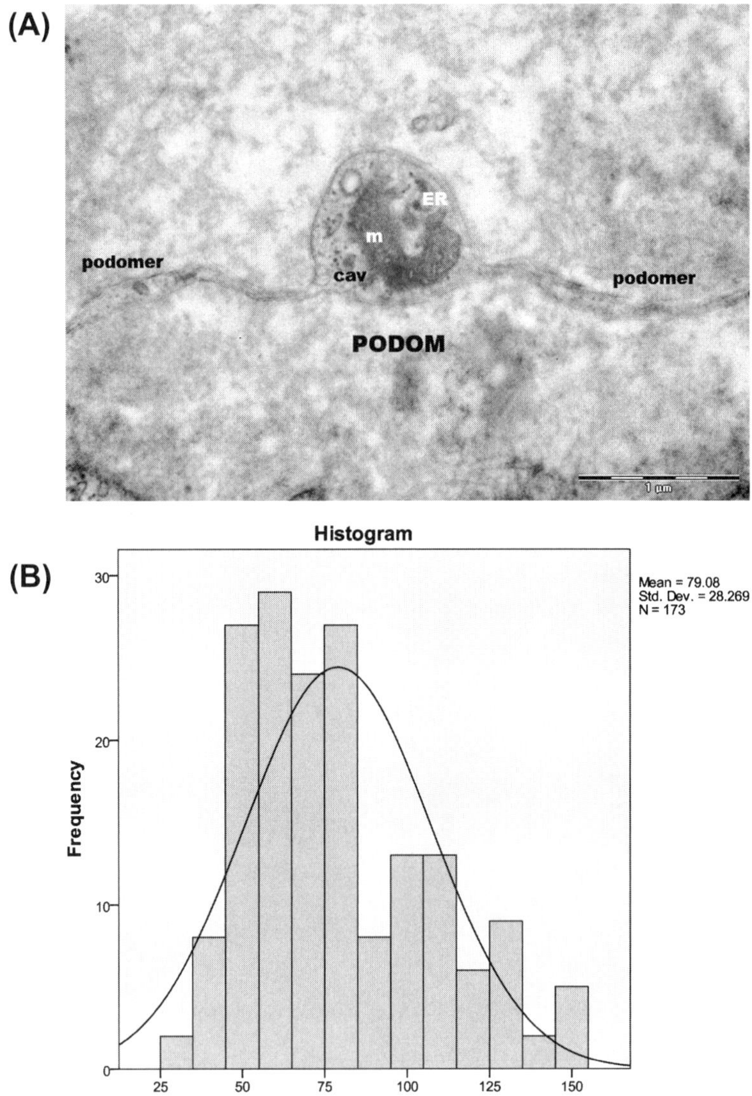

Figure 11.4 **(A)** Rat myocardium. Electron micrograph of a telopode (blue), depicting a podom between two podomers. Note that the podomer thickness is clearly below 0.2 µm, the resolving power of light microscopy. **(B)** Mean width of podomer (about 80 nm) places it under the resolving power of light microscopy. **See Plate 18.**

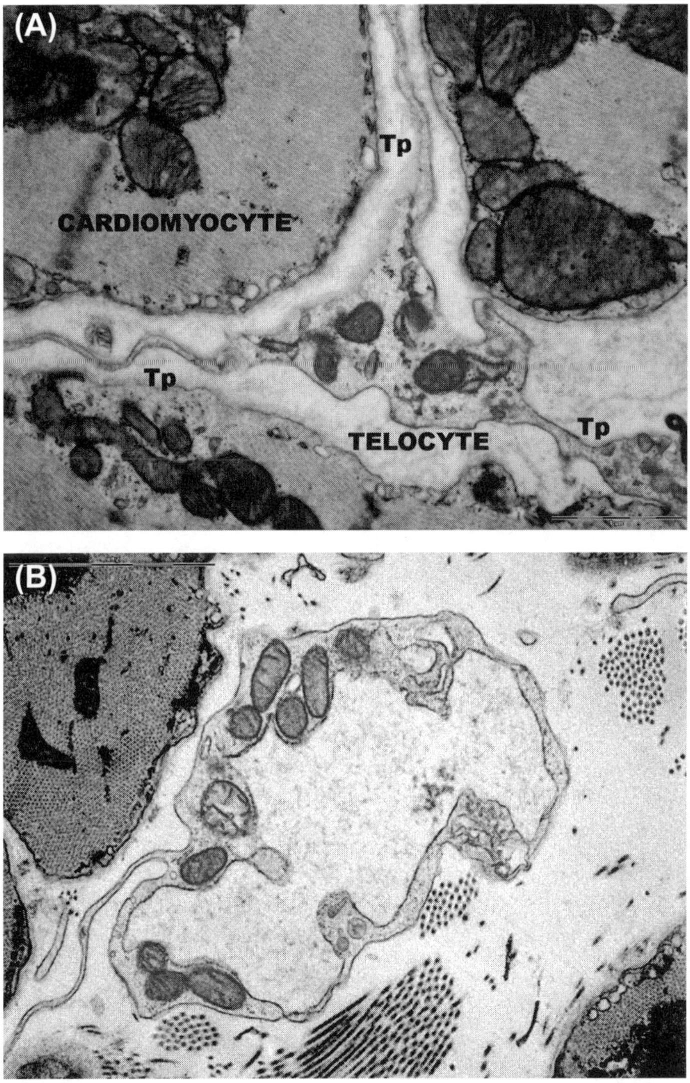

Figure 11.5 **(A)** A triangular telocyte (digitally colored in blue) with three emerging telopodes (Tp). Rat myocardium. **(B)** A particular circular appearance of a telopode, with both podoms and podomers, in rat heart interstitium. **See Plate 19**.

microscopy, explaining why TC were overlooked so far and why EM is mandatory for unequivocal TC identification).

4. **Podoms** accommodate: a) mitochondria (one or two), endoplasmic reticulum elements, either smooth or/and rough ER, and c) caveolae; the so-called Ca^{2+} uptake/release units.

5. **Dichotomous branching pattern**, making a 3D network, a labyrinthine system with particular intercellular junctions: either homocellular junctions between TC themselves or heterocellular junctions between TC and other cell types (e.g. cardiomyocytes, macrophages, stem cells etc.) [1,2,7].

To observe a Tp, one should consider its main ultrastructural features. First of all, the tortuous trajectory of a Tp requires convenient section planes, in which a larger/longer portion of a Tp is enclosed. Second, because Tp are both very long and very thin structures, both a large and detailed image is required. In other words, a higher magnification overview is needed. Several neighbor areas should be analyzed under EM and the captured images aligned and merged both horizontally and vertically. The reconstructed collage will show an area that could not have been directly captured entirely unless at lower magnification. In this way, a wider "field of view" may be analyzed at a higher magnification.

Comparison with Fibroblasts

A superficial likeness does not always reveal an essential likeness.

L. Mackenzie [8]

The interstitium (stroma) is in most cases seen as a connecting "device" for the specific structures of an organ. Usually, interstitial cells are perceived as being *mainly* (or even *only*) fibroblasts (Fig. 11.6). However, fibroblast main function is the production of collagen fibrils and some extracellular matrix proteins. Although it is obvious that TC are not fibroblasts, some are hiding behind the syntagma "fibroblast

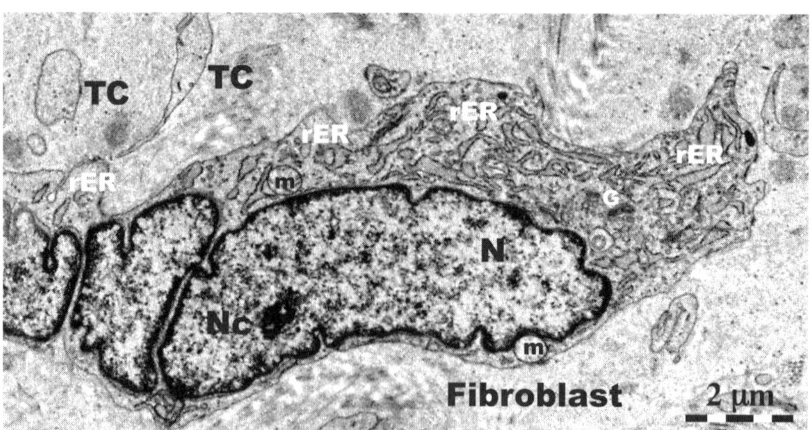

Figure 11.6 Transmission Electron Microscopy of a typical fibroblast in rat myocardium. N, nucleus; Nc, nucleolus; m, mitochondria; RER, rough endoplasmic reticulum; G, Golgi complex, TC, telocyte fragments.

like." Where should one draw the line(s) between the intrinsic variability of fibroblasts and their look-alikes? In fact, the so-called fibroblast-like cells were never clearly defined, the lack of specific information leading more or less to "a transfer of ignorance". There is a danger that a cell that is not clearly a fibroblast might be considered *a priori* fibroblast-like.

However, the distinction between TC and fibroblasts is obvious because they have different ultrastructure (Table 11.1) and phenotype. Therefore, their functions are different: fibroblasts promote collagen synthesis, while TC promote intercellular signaling either by direct contact (junctions), or remotely (via extracellular vesicles). In other words, fibroblasts are *more* structurally oriented, responsible for collagen and extracellular matrix synthesis and fibrosis, whereas TC are *more* functionally oriented, being involved in inter(trans)cellular communication via 3D network(s), and, maybe, some specific functions.

Where Are Telocytes Located?

The presence of TC has been reported in the stroma of many organs (Table 11.2). Their location is in the interstitium, where they connect with resident and nonresident cells. By branching of their long prolongations, TC are making a virtual 3D network.

Telocyte Intercellular Relationships

Extracellular Vesicles

Extracellular vesicles are: multivesicular bodies (Fig. 11.7) or shed microvesicles or exosomes. ECV might play a unique role in horizontal transfer of macromolecules between cells (e.g., proteins, various RNAs, miRs).

Intercellular Relationships

The intercellular interactions are achieved by either direct contact (intercellular junctions) or by extracelluar vesicles (ECV).

The variable Tp width might also explain the heterogeneity of the released vesicles. Thus, exosomes (with a diameter under 100 nm) could be released at the podomeric level (where average width rarely exceeds 100 nm; see Fig. 11.4B), and shed microvesicles (with a mean diameter of 180 nm) could transport macromolecular signals from podoms to neighboring cells, modifying their transcriptional activity eventually. The release of multi-vesicular bodies has been noted at both podomeric (Fig. 11.7A) and podomic (Fig 11.7B) level. The microenvironment also plays an important role in local cellular cross talk [54].

TABLE 11.1 Comparison of the Ultrastructural Characteristics of Telocytes and Fibroblasts

Feature	Telocytes	Fibroblasts
Cell body	Small; piriform/spindle/triangular/stellate shaped;	Pleomorphic (phenotypic heterogeneity)
Cytoplasm	Small amount	Large amount
Nucleus	**One, oval/rod-shaped**	**One, oval**
Chromatin	Heterochromatin	Typically euchromatic
Nucleolus	Rarely visible	1-2 nucleoli
Organelles		
Golgi complex	Small	Prominent
Mitochondria	2%-5% of cell cytoplasm Present in podoms (not in podomers)	≈ 5%
Endoplasmic reticulum (ER)	≈ 2% of cell volume; either smooth or rough; located in podoms	Smooth ER virtually absent, but rough ER prominent (8%-12% of cell volume), located mainly in cell body, but also in processes
Membrane		
Caveolae	Many; more on the cell processes versus cell body	Hardly any, *in situ**
Junctions	Homo- and heterocellular junctions	No junctions (or difficult to assess) with other cells†
Number of prolongations	**2-5 telopodes**	**Usually 2**
Branching	Dichotomic pattern, forming 3D convoluted network(s)	Randomly (?)
Conformation	Overall moniliform aspect (alternating podoms and podomers)	Usually cone shaped
Emergence from the cell body	Thin	Thick, followed by gradual thinning
Length	Very long‡ (tens, up to hundreds, of micrometers)	Usually several micrometers
Podomers	Very thin (mostly below 0.2 μm, the resolving power of light microscopy); their caliber does not allow the presence of any membrane-bound organelles inside	No
Podoms	Dilated portions ("knobs") of telopodes, with an average width of about 0.5 μm; they accommodate caveolae, mitochondria and ER	No

*Fibroblasts in situ have few caveolae. Many caveolae could be found in human/mammalian cultured fibroblasts; phenotypic modification.
†Finding connexin 43 by immunofluorescence does not necessarily imply the existence of stable "gap junctions." Electron microscope detailed studies [9] failed to find the existence of gap junctions between fibroblasts and other cell types.
‡Only some nerve cells processes (axons) appear longer than telopodes in the human/mammalian body.

TABLE 11.2 Telocyte Presence in Various Tissues/Organs

Heart	
Endocardium	[10]
Myocardium	[9,11-15]
Epicardium	[16,17]
Skeletal Muscle	**[18-20]**
Gastrointestinal Tract and Annexes	
Duodenum	[21]
Jejunum	[22]
Salivary glands	[23]
Gallbladder	[24]
Pancreas	[25,26]
Urinary System	**[27]**
Kidney	[27]
Renal pelvis	[28]
Ureter	[28]
Bladder	[29,30]
Urethra	[28]
Respiratory System	
Lungs	[31-34]
Trachea	[35,36]
Pleura	[37]
Meninges and choroid plexus	[38]
Mammary glands	[39,40]
Uterus	[41-45]
Fallopian tube	[43,44,46]
Placenta	[47]
Skin	[48,49]
Mesentery	[50]
Vasculature	[9,51-53]

Eye: Ciliary body and iris*
*Maria-Jose Luesma Bartolome (Zaragoza, Spain) and Mihaela Gherghiceanu (Bucharest, Romania), personal communication

Intercellular Junctions

Telocytes Intercellular Junctions

TC are establishing connections, via their cell body or long Tp, located very close to neighboring structures. TC are not just bystanders stromal elements, but active participants in intercellular space.

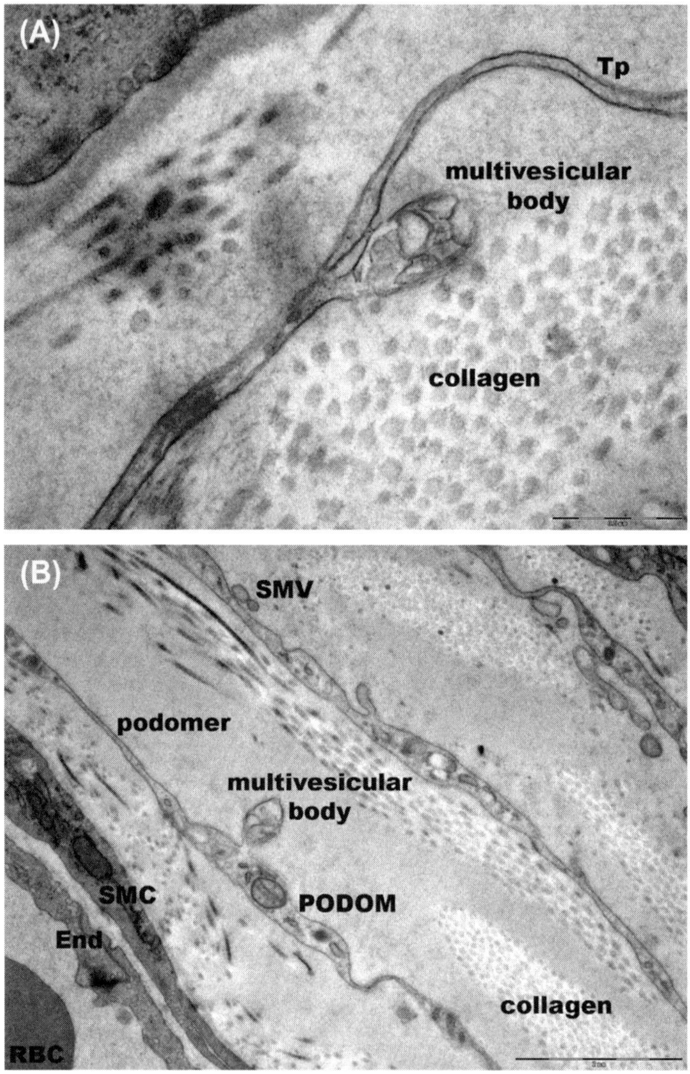

Figure 11.7 Nascent multivesicular bodies (digitally colored in violet) are budding from either podomer or podomic regions of telopodes (digitally colored in blue). Note the diameter of podomers in the range of that of collagen fibrils (cross-cut). RBC, red blood cell; End, endothelium; SMC, smooth muscle cell; SMV, shed microvesicles. *(Modified with permission from [63]).* **See Plate 20**.

Elements of the TC network are interacting with each other (**homo**-cellular connections) as well as with other cell types (**hetero**-cellular connections). The presence of homo- and hetero-tropic networks indicates the possible existence of a higher integration in a putative general interstitial system. The type of contact depends on several aspects: the distance between the TC and the target cell(s), the specific type of the target cell(s) (e.g., immune cells, neurons), and the organ where the interaction takes place.

The TC intercellular junctions have been thoroughly described in the heart [9], but examples are reported in other organs, too (e.g. pancreas [26], salivary glands [23]).

In myocardium, **homocellular** junctions might reveal different morphologies under EM: *puncta adhaerentia minima, processus adhaerentes* or *recessus adhaerentes* [9]. The homocellular junctions (e.g., Fig. 11.8) are typical and they occur at both podomeric and podomic level, either side to side (presumably for exchanging information) or end to end (probably for relaying, passing on information).

Heterocellular junctions are encountered at myocardic level between TC and cardiomyocytes, cardiomyocytes progenitors, fibroblasts, mast cells, macrophages, pericytes, endothelial or Schwann cells [9]. TC are integrating all cardiac cellular types [9] through a complex 3D framework/lattice, formed by heterocellular junctions. These networks are providing both structural and functional support for long-distance signaling, important in cardiac renewing [55].

The connections of TC with various cell types in the adult heart might be classified depending on the intermembranar distance in nanocontacts (10 nm), point or planar contacts (10 to 30 nm), and close vicinity (under 150 nm) [9].

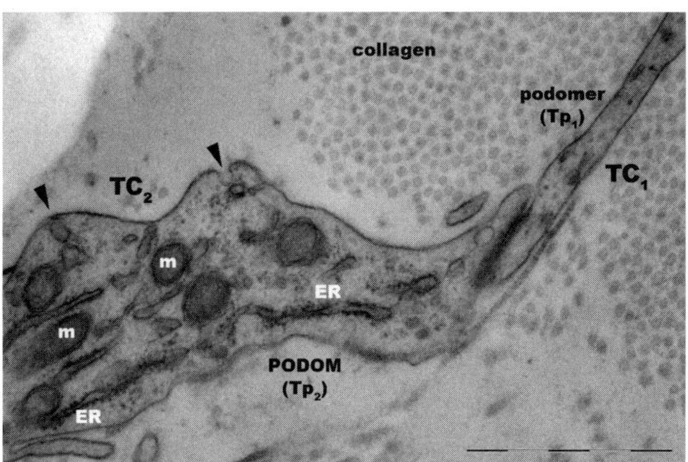

Figure 11.8 Rat myocardium. Transmission electron microscopy. Two telopodes (Tp1 and Tp2) are participating in a homocellular junction (procesus adherens, red dotted circle) between two telocytes (TC1 and TC2). The podom of Tp2 accommodates mitochondria (m), endoplasmic reticulum (ER), and caveolae *(arrowheads)*. Scale bar = 1 μm.

The heterocellular junctions are formed between TC and the following type of cells:

1. Specific cells from that tissue (e.g., cardiomyocytes [56]; Fig. 11.9).
2. Stem cells (SC) [17,19,57].
3. Immune reactive cells (mast cells – Fig. 11.10, basophils, eosinophils [58]). The connection with macrophages may indicate a certain role in local tissue remodeling. This particular connection type was labeled *stromal synapse* [58].

Frequently, Tp are observed in a close relation to nerve endings.

The junctions between TC and cardiomyocytes do not belong to any specific category but consist mostly in clusters of direct nanocontacts, free of any basal lamina interference. This would also depend, certainly, on the TC density in several heart areas (as reported in [59], TC concentration is higher in atria than ventricles—about 20 versus 9 cells/mm^2—and significantly higher in subepicardium than in endocardium—18 versus 7 cells/mm^2).

Phenotype "Portrait"

Immunocytochemical Phenotype of Telocytes

Several markers have been identified (with a variable expression) on TC, either by immunohistochemistry (IHC) or by confocal imaging. The immunophenotype of TC includes mainly CD34, CD117/c-Kit, and vimentin, but also caveolin-1, CD44,

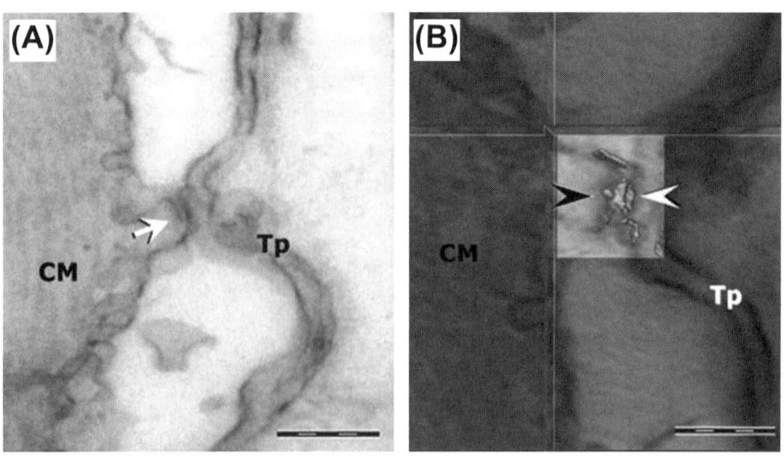

Figure 11.9 EM Tomography. **(A)** Digital section from tomographic volume shows a junction between the Tp and a discrete projection of the CM. **(B)** Top view of the 3D isosurface reconstruction of the junction shows a macromolecular complex connecting Tp and CM *(arrowheads)*. *(Reproduced with permission from [56]).*

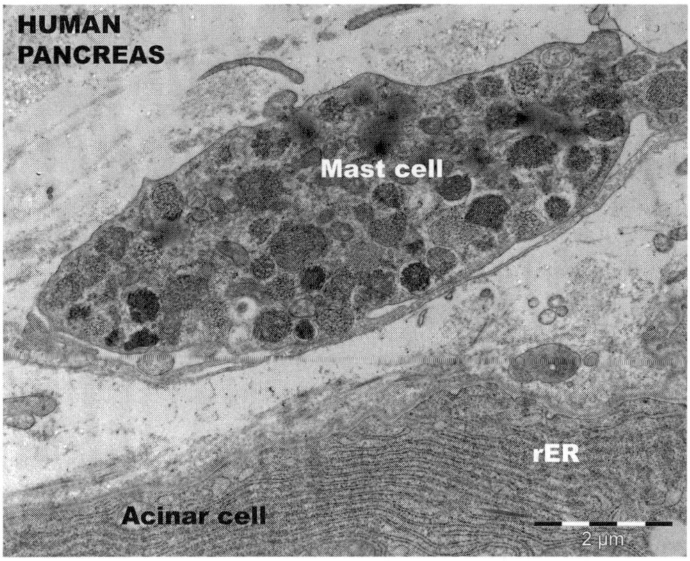

Figure 11.10 Human exocrine pancreas. A telopode (blue) establishes multiple junctions with a mast cell. RER, rough endoplasmic reticulum. **See Plate 21**.

NOS-2, desmin, cadherin-11, and PDGF-R beta [16,20,47,60]. Although for the time being, EM remains the method of choice to precisely identify TC, the double positive immunostaining with CD34/c-Kit (mainly for cell body) or CD34/vimentin (mainly for Tp) also represents a useful marker for TC. It is noteworthy that TC are immunohistochemically negative for procollagen 1 (data not published) and CD90/Thy-1, respectively [47].

Protein Secretory Profile of Telocytes

The secretory capacity of TC has been analyzed by surface-enhanced laser desorption/ionization time-of-flight mass spectrometry (SELDI-TOF-MS) and xMAP technology—Luminex 200 for cytokine/growth factor release. Several differentially expressed peptide and protein peaks were identified in telocyte-enriched myocardial cell culture supernatants (TEMC), compared to simple culture medium (no cells) or 3T3 "standard" fibroblast cultures. Using SELDI-TOF-MS (Fig. 11.11A), we were able to detect in TEMC supernatants significantly different protein peaks with estimated molecular weights of 4083, 11682, and 22596 Da, as well as several peptide peaks in the 500 to 1000 Da range (in preparation).

Using antibody-based xMAP technology (Fig. 11.11B), we detected specific secreted molecules in TEMC supernatants from both serum-free and serum-supplemented media. Detected levels of IL-6 and VEGF increased with passage number. In all experimental systems, TEMC supernatants showed a four- to five-fold

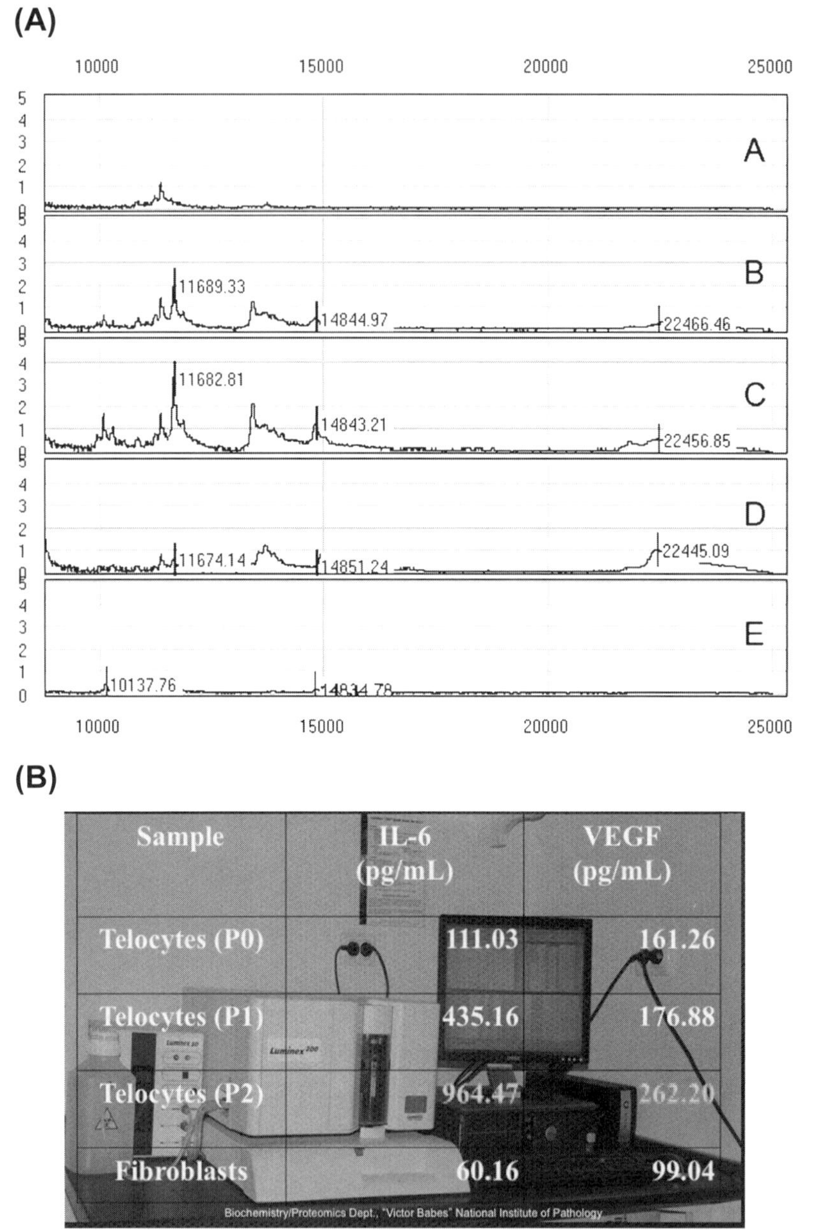

Figure 11.11 **(A)** SELDI-TOF-MS protein spectra. M/Z range: 0 to 25,000 Da, CM10 chips, SPA matrix. Characteristic proteins of TEMC at M/Z values of 4.09, 11.6 and 22.5 KDa. A: Cell culture medium, B: TEMC Passage 0, C: TEMC Passage 1, D: TEMC Passage 2, E: 3T3 Fibroblasts. **(B)** Cell culture supernatants analyzed on xMAP technology (Luminex 200) using Mouse Cytokine/Chemokine Panel.

increase of the secreted IL-6 compared to controls. The VEGF level in TEMC supernatant was also increased, but not at the same magnitude as IL-6. The secretory levels of the two cytokines suggest the potential regulatory role of TC on other cell types, including resident SC, with implications in controlling cell growth/myocyte differentiation or angiogenesis.

MicroRNA Signature of Telocytes

Seeking to identify miRs specifically expressed by TC, we have combined the laser capture microdissection method (for cell isolation) with the quantitative PCR-based microRNA assay (for miR profiling). Limited amounts of RNA (in the range of nanograms) represent a major constraint to quantify miRs by standard procedures: northern blotting, cloning, microarrays, and deep sequencing. Compared with these methods, real-time quantitative PCR (RT-qPCR) gives the highest sensitivity of miRNA quantification while producing fewer false positives and reduced bias [61]. We have found that TC express significant amounts of miR-21, 22, 29, and 199a but lack the expression of cardiomyocyte-specific miRs (miR-1 and 133a or 208a), which supports the view that TC are of mesenchymal origin. We also have shown that apart from other cardiac interstitial cells, TCs do not express miR-193 [62]. In addition, we have found recently that TC express many of the pro-angiogenic microRNAs (i.e., miR-126, miR-130, let-7e, miR-21, miR-27b, miR143, miR-503, and miR-100) [63].

The Tandem Telocytes & Stem Cells

Epicardial Stem Cell Niche

Telocytes have been spotted in the vicinity of several types of cardiac progenitors in various stages of differentiation (Fig. 11.12). Their interstitial network of telopodes builds a dynamic scaffold, a guiding framework essential for the development of new cardiomyocytes. TC, with their Tp, surround cardiac progenitors or precursors to guide them to form the coherent 3D myocardial architecture. Apparently, Tp provide "tracks" for the sliding of cardiomyocyte precursors in their development and integration as working cardiomyocytes.

TC are cells with a certain degree of mobility. This was demonstrated in cell cultures [4], but the moniliform aspect of Tp could indicate this also occurs in situ (a podom could represent either a fulcrum or an elongation reservoir).

The role of TC in nursing cardiomyocytes has also been studied in pathology. Thus, after an experimental myocardial infarction, the TC population is at first disorganized, but gradually regroups following a preestablished pattern and

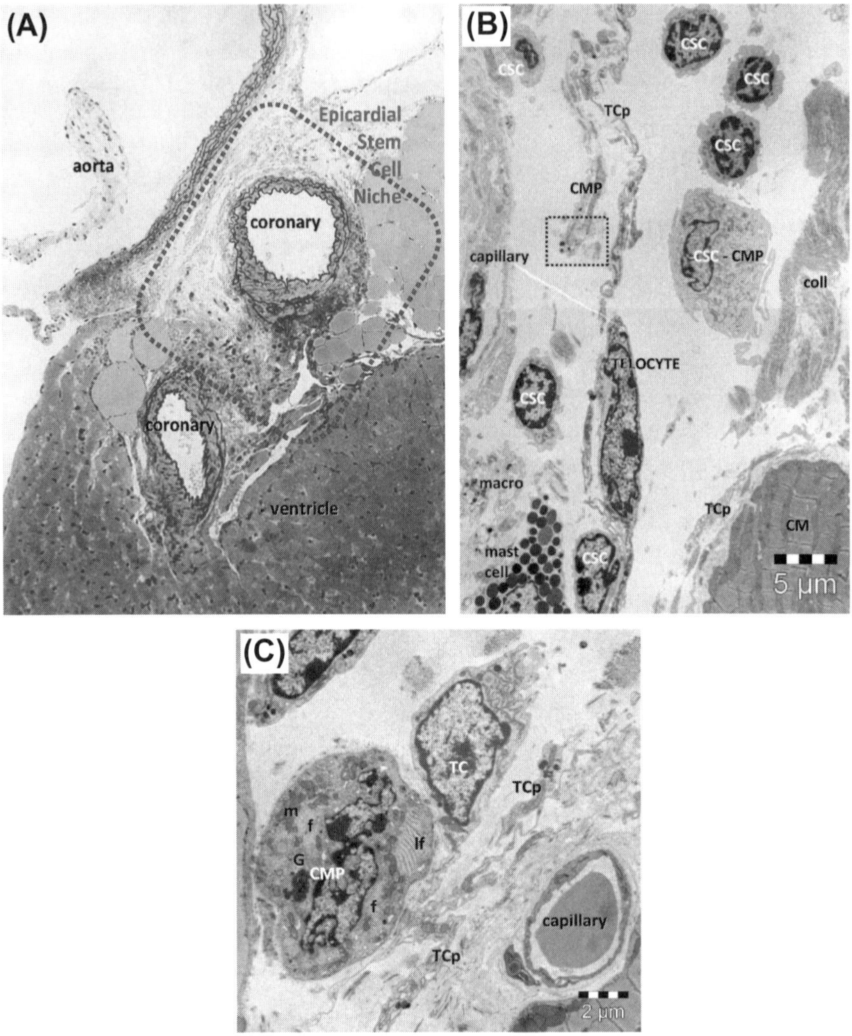

Figure 11.12 **(A)** High-resolution light microscopy on toluidine blue–stained semithin section (~1 μm thick ultramicrotome section) of Epon-embedded mouse heart (6 months old) shows the limited space where cardiomyocyte progenitors have been found by electron microscopy. The cardiac stem cell niche is located in the subepicardial area surrounding the coronary artery next to the emergence from the aorta (rectangle red mark). **(B)** Electron microscopy of the niche depicted in (A) shows the presence of putative cardiac stem cells (CSC), isolated or in small groups, cardiomyocyte progenitors (CMP), and cells with intermediate features (CSC-CMP). All of these cells are placed in a loose extracellular matrix TCp, telopodes. **(C)** A telocyte (TC) chaperone a low differentiated CMP with distinctive leptofibrils (lf), unorganized myofibrils (f), Golgi apparatus (G), and clusters of mitochondria (m). *(Reproduced with permission from [17]).* **See Plate 22**.

contributes in tandem with resident SC to an increase in the regeneration rate of the cells bordering the infarcted area and surroundings [63].

Subepithelial Lung Stem Cell Niche

The tandem arrangement of TC-SC has been identified in subepithelial niches of the bronchiolar tree (Fig. 11.13). Here, the synergy of TCs and SCs may be based on nanocontacts and shed vesicles [31].

Skeletal Muscle Stem Cell Niche

In recent studies, an interstitial nonsatellite myogenic SC niche has been identified [64]. The presence of TC has been reported in both niches (Fig. 11.14) and cell cultures [18]. The migrating capacity showed by cultured TC supports their guiding

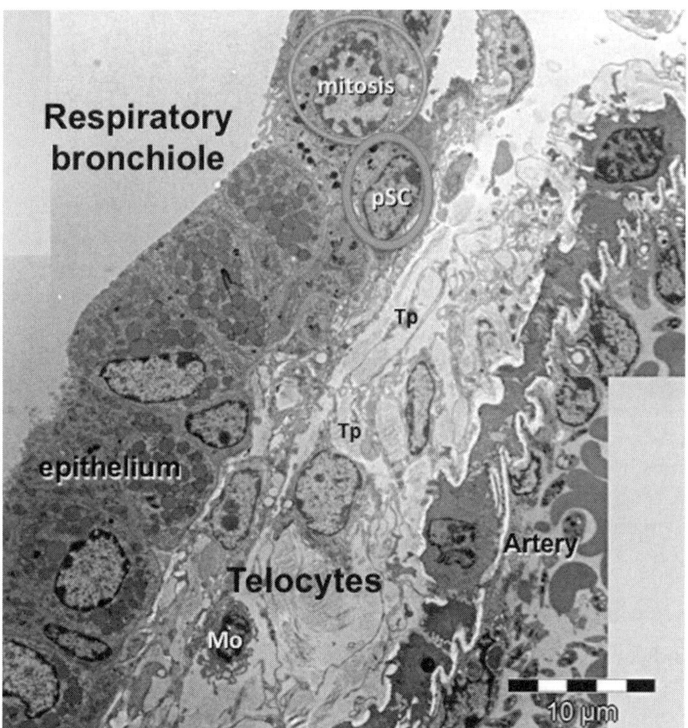

Figure 11.13 Transmission electron microscopy of mouse respiratory bronchiole. Telocytes (TC, blue) with numerous emerging telopodes (Tp). Some Tp surround a putative stem cell (pSC). Mo, macrophage. **See Plate 23.**

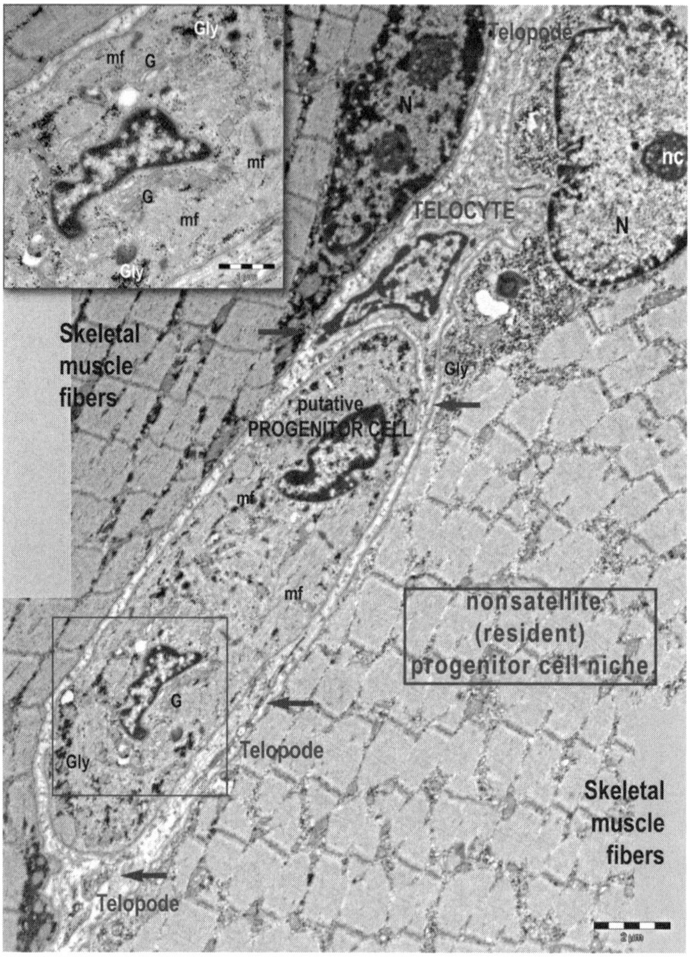

Figure 11.14 Electron micrographs of human skeletal muscle show a TC (blue colored), which extends its Tps indicated by red arrows around a striated cell, in fact, a (putative) progenitor cell. Note the tandem TC—progenitor cell making a nonsatellite (resident) progenitor stem cell niche. **(Inset)** Higher magnification of the progenitor cell shows incompletely differentiated features: unorganized myofilaments (mf), glycogen deposits (Gly), and prominent Golgi complex (G). N, nucleus; nc, nucleolus. *(Reproduced with permission from [18]).* **See Plate 24**.

features indicated by the presence of long Tp circumventing various cell types, including putative progenitor cells.

Meninges and Choroid Plexus

Telocytes establish direct intermembranar contacts with putative SC in a choroid plexus microenvironment (Fig. 11.15).

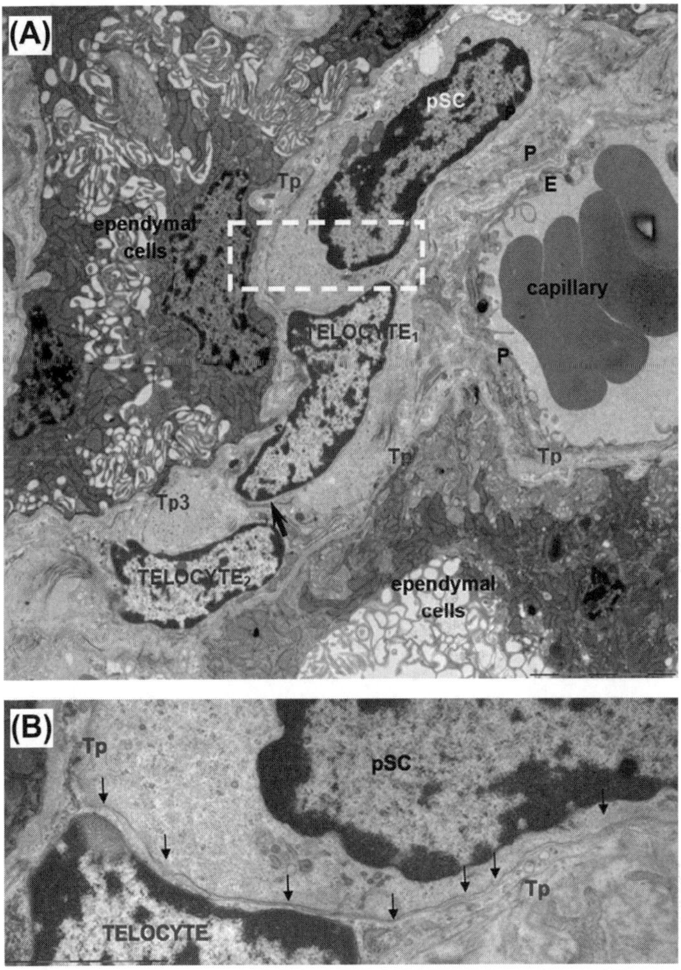

Figure 11.15 EM of adult rat choroid plexus. (A) The 3 telopodes (Tp1-Tp3) of Telocyte1 embrace a stem cell in the interstitial space between ependymal cells layer and the fenestrated capillary. Telocyte1 has a direct inter-membranar contact (arrow) with Telocyte2, whose Tp extends between capillary and ependymal cells. The telopodes width is about 100 nm. Pericytes (P) are located around the endothelial cells (E). (B) Higher magnification of a consecutive section, the rectangular area in (A). Small point contacts between TC and Tp and stem cells are indicated by arrows. The intermembranar distance is variable (mean – 28nm) on 6 μm length but direct contacts are visible (arrows). Scale bars: 5 μm (A), 2 μm (B) *Reproduced with permission from [38].* **See Plate 25**.

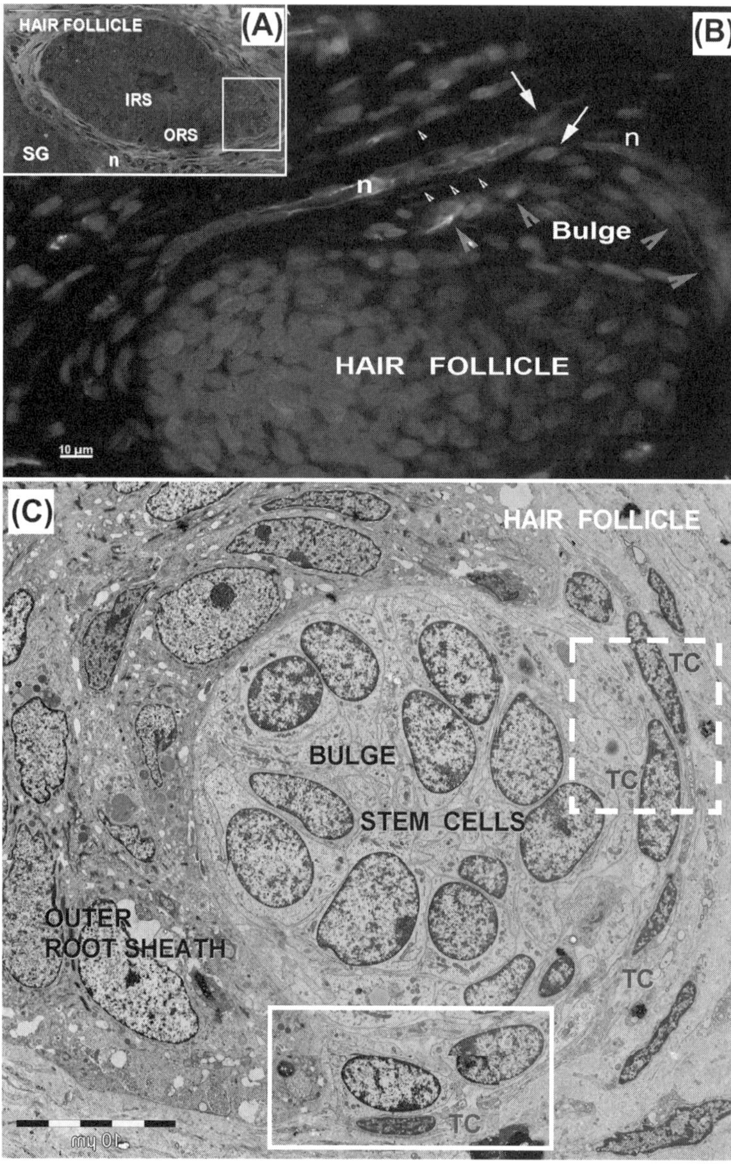

Figure 11.16 Telocytes in reticular dermis. **(A)** Light microscopy image (stained by toluidine-blue) showing a hair follicle and an adjacent sebaceous gland (SG). IRS, inner root sheath; ORS, outer root sheath; n, perifollicular nerve fibers. **(B)** Epifluorescence microscopy: double labelling revealing nestin positive cells (green), stem cells from the bulge area of hair follicle (magenta arrowheads), and perifollicular nerve fibers (N), surrounded by CD117 positive TCs (red, white arrows) with long Tps (white arrowhead). Nuclei were counterstained with DAPI (blue). Original magnification 400x. **(C)** EM of the boxed area in **(A)** shows a cluster of stem cells in the outer root sheath of a hair follicle. The stem cells are bordered by TCs with Tps, as well as other cells from the outer root sheath. *Reproduced with permission after [49].* **See Plate 26.**

Skin stem cell clusters

Telocytes presence was identified between the bulge areas of hair follicles and the adjoining nerve fibres in reticular dermis (Fig. 11.16).

A Hope for Regenerative Medicine

One debatable topic in regenerative medicine is the dedifferentiation concept. Telocytes might represent cells that are closer than others to the mesenchymal pseudo-differentiated stage and, consequently, are more likely to be the subject of dedifferentiation compared to terminally differentiated resident tissular cells.

In the adult mammalian heart, TC together with resident SC and cardiomyocyte progenitors sustain a continuous cardiac renewal process and might be key players in repairing the damaged heart. TC "nurse" the progenitor cells in SC niches. The tandem TC-SC could be a better option for therapy rather than SC alone. Finally, TC are directly (physically) and indirectly (chemically) involved in neoangiogenesis after myocardial infarction [63].

The substrate is also an important element in the regeneration process, even as an inducer or as a guiding mesh. The overall environment also plays a crucial role in the regeneration of various structures. The immunologic, mechanical, and bioelectric components are all-important in modeling, guiding, and influencing the regeneration. The plasticity degree of a tissue/organ is directly linked with its ability to adapt and to respond to putative injuries with a fast or slow repair/regeneration rate.

Our opinion is that TC presence and actions (by direct contact or indirect, via shed microvesicles) increase the efficiency and efficacy of resident local SC in the process of repair/regeneration.

The Functions of Telocytes in Physiology and Pathology

The implication of TC in both physiologic and pathologic situations may lead to a phenotype adjustment depending on extracellular conditions or the transmission/reception of information via microvesicles or direct contact. This would explain the inconsistency in the expression of some markers and why intermediary forms may coexist in the differentiation process with a high degree of involvement in organization/function of the organ of residence.

Because TC connections in exocrine glands include both neurovascular and exocrine elements (e.g., acini, ducts), it is attractive to think that TC

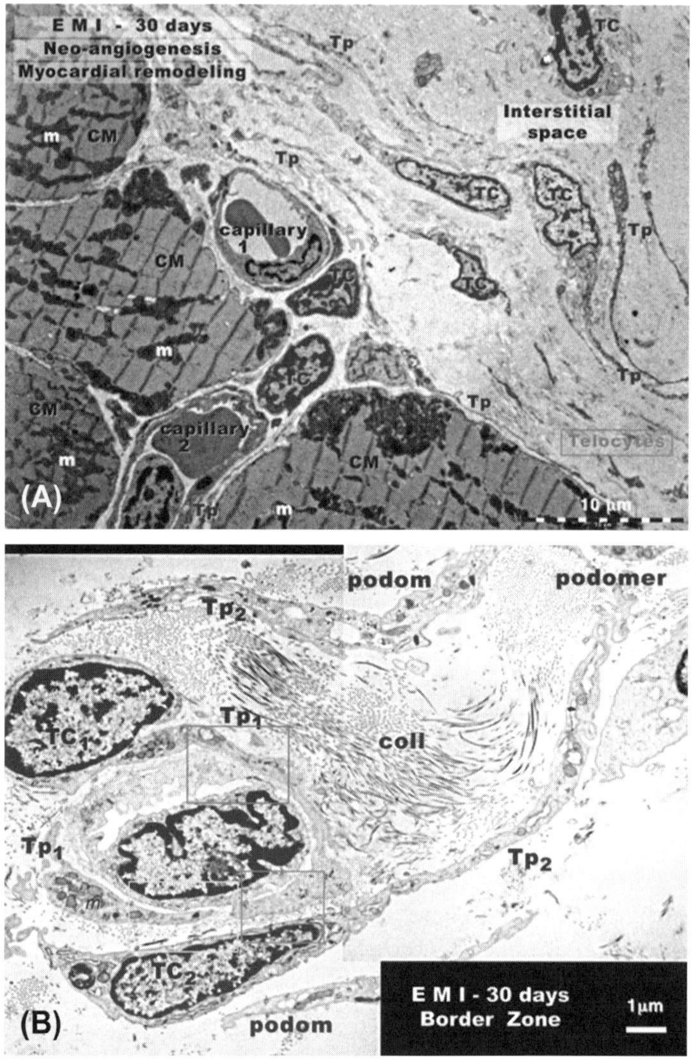

Figure 11.17 Rat experimental myocardial infarction. Border zone: 30 days old. Transmission electron microscopy. **(A)** This low-magnification view shows four cardiomyocytes (CM), two blood capillaries (1 and 2), and numerous telocytes (TC) with long and slender telopodes (Tp). Note the close spatial relationship between TC/Tp and the capillary-1 wall (endothelium). Capillary-1 is presumably a neocapillary created in the interstitial space. Capillary-2, between three cardiomyocytes (CMs), has a TC and Tps in the vicinity, but the distance between the abluminal membrane of the endothelium and the TC/Tp plasma membrane is wider. Thus, capillary-2 is probably an "old" capillary. **(B)** A new-formed blood capillary with an anfractuous and narrow lumen is shown (brown color) in the mass of collagen fibrils (coll) of the scar. This is surrounded by two telocytes (TC1 and TC2—blue color) and their

might be involved in modulating local homeostasis as well as in mediating and integrating neural or vascular input with the function of the organ of residence.

This would partially explain a functional link between different interstitial cells (e.g., fibroblasts to produce the collagen and TC to organize the extracellular matrix according to local/general changes).

Human Isolated Atrial Amyloidosis

Interstitial amyloid fibrillar deposits were revealed by EM upon analyzing atrial samples. Some amyloid fibrils appeared amid a honeycomb Tp framework, whereas others were simply wrapped by Tp, blood vessels, or cardiomyocytes [65].

Experimental Acute Myocardial Infarction

Immunocytochemistry, EM (Fig. 11.17), and microRNA analysis showed TC involvement in angiogenic processes in the late stage myocardial infarction, by direct contact with endothelial tubes, as well as by pro-angiogenic microRNAs and paracrine secretion (VEGF, NOS2) [61].

Mechanisms by which TC presumably influence the neoangiogenesis in the border zone of myocardial infarction are summarized in Fig. 11.18. Depending on the distance between TC and target cell(s), intercellular communication may follow a direct or indirect route. Direct physical contact (by either apposition or nanocontacts, without any basal membrane interference) may favor cellular bidirectional cooperation/information exchange. On the other hand, indirect chemical signaling relays para- or microcrine messages, presumably more one-way oriented.

Gastrointestinal and Extragastrointestinal Stromal Tumors (GISTs) and Perivascular Epithelioid Cell Tumors (PEComas)

Telocytes have been proposed as the common cells of origin for both GISTs and PEComas, an archetypal entity that is able to switch from a perivascular stromal cell to a differentiated cell type with contractile and signaling properties [66].

corresponding telopodes (Tp1 and Tp2). Typically, podoms (dilated portions) and the intercalary podomers (thin portions of Tp) can be observed. At the level of podoms there are many mitochondria (m), elements of endoplasmic reticulum, and caveolae. Note the close spatial relationships between telopodes and endothelial cells. The space between telopodes and the membrane of endothelial cell is occasionally less than 50 nm and there is no visible endothelial basal lamina. *(Reproduced with permission from [63]).* **See Plate 27**.

DIRECT **PHYSICAL CONTACT** **(no basal membrane)**	Aposition ("synapse-like")	planar (long distance)	length: 0.5 - 1.5 μm
		spot ("finger" - like protrusion) (short distance)	length: < 0.5 μm
	Nano-contact	nano-feet (electron dense)	height: 60 nm
INDIRECT **CHEMICAL SIGNALING** **(secretion)**	Paracrine	NO ↑↓ VEGF	Through membrane
			With membrane: Shed vesicles: ~ 100 nm & Exosomes:
	Microcrine (miricrine)	microRNAs	

Figure 11.18 Possible mechanisms of telocyte involvement in neoangiogenesis in the border zone of myocardial infarction. *(Reproduced with permission from [63].)*

Future Research Directions

There are several directions for future telocyte research. One of them is the study of miR and cytokine repertoire released by TC. Work in progress in our laboratory using red and green fluorescent proteins suggested that TC and endothelial cells (at least in heart) have a common progenitor.

Another attractive research direction is the study of the cooperation between TC and SC in postinfarcted myocardial areas. This would further detail the mechanisms of cellular cooperation between the two cellular entities. This may be achieved experimentally by injecting either TC or SC alone, both TC and SC concomitantly, or TC followed by SC or the other way around.

Lastly, the presence/absence of telocytes in various organs in development, in different fetal stages, should give us more information regarding the existence of one or several telocyte/fibroblast/mesenchymal cell precursor(s).

References

[1] Popescu LM, Faussone-Pellegrini MS. TELOCYTES - a case of serendipity: the winding way from Interstitial Cells of Cajal (ICC), via Interstitial Cajal-Like Cells (ICLC) to TELOCYTES. J Cell Mol Med 2010;14:729–40.

[2] Faussone-Pellegrini MS, Popescu LM. Telocytes. Biomol Concepts 2011;2:481–9.

[3] Popescu LM. TELOCYTES - A novel type of interstitial cells. In: Braissant O, Wakamatsu H, Kuo-Kang I, Allegaert K, Lenbury Y, Wachholtz A, editors. Recent Researches in Modern Medicine. Cambridge, United Kingdom: WSEAS Press; 2011. p. 424–32.

[4] Suciu L, Nicolescu MI, Popescu LM. Cardiac telocytes: serial dynamic images in cell culture. J Cell Mol Med 2010;14:2687–92.

[5] Misra R. Serendipity and science. Sci Rep 2008;45:9–13.

[6] Waanders FMJ. The History of [Telos] and [Teleo] in Ancient Greek. Amsterdam: John Bejamins Pub Co; 1984.

[7] Popescu LM. The tandem: telocytes - stem cells. Int J Biol Biomed Eng 2011;5:83–92.

[8] Mackenzie L. Biology: A Search for Order in Complexity. 2nd ed. Arlington Heights, IL: Christian Liberty Press; 2006.

[9] Gherghiceanu M, Popescu LM. Cardiac telocytes - their junctions and functional implications. Cell Tissue Res 2012;348:265–79.

[10] Gherghiceanu M, Manole CG, Popescu LM. Telocytes in endocardium: electron microscope evidence. J Cell Mol Med 2010;14:2330–4.

[11] Zhou J, Zhang Y, Wen X, et al. Telocytes accompanying cardiomyocyte in primary culture: two- and three-dimensional culture environment. J Cell Mol Med 2010;14:2641–5.

[12] Bani D, Formigli L, Gherghiceanu M, Faussone-Pellegrini MS. Telocytes as supporting cells for myocardial tissue organization in developing and adult heart. J Cell Mol Med 2010;14: 2531–8.

[13] Kostin S. Myocardial telocytes: a specific new cellular entity. J Cell Mol Med 2010;14:1917–21.

[14] Faussone-Pellegrini MS, Bani D. Relationships between telocytes and cardiomyocytes during pre- and post-natal life. J Cell Mol Med 2010;14:1061–3.

[15] Rusu MC, Pop F, Hostiuc S, et al. Telocytes form networks in normal cardiac tissues. Histol Histopathol 2012;27:807–16.

[16] Popescu LM, Manole CG, Gherghiceanu M, et al. Telocytes in human epicardium. J Cell Mol Med 2010;14:2085–93.

[17] Gherghiceanu M, Popescu LM. Cardiomyocyte precursors and telocytes in epicardial stem cell niche: electron microscope images. J Cell Mol Med 2010;14:871–7.

[18] Popescu LM, Manole E, Serboiu CS, et al. Identification of telocytes in skeletal muscle interstitium: implication for muscle regeneration. J Cell Mol Med 2011;15:1379–92.

[19] Bojin FM, Gavriliuc OI, Cristea MI, et al. Telocytes within human skeletal muscle stem cell niche. J Cell Mol Med 2011;15:2269–72.

[20] Suciu LC, Popescu BO, Kostin S, Popescu LM. PDGFR-beta positive telocytes in skeletal muscle interstitium. J Cell Mol Med 2012;16:701–7.

[21] Cantarero Carmona I, Luesma Bartolome MJ, Junquera Escribano C. Identification of telocytes in the lamina propria of rat duodenum: transmission electron microscopy. J Cell Mol Med 2011;15: 26–30.

[22] Cretoiu D, Cretoiu SM, Simionescu AA, Popescu LM. Telocytes, a distinct type of cell among the stromal cells present in the lamina propria of jejunum. Histol Histopathol 2012.

[23] Nicolescu MI, Bucur A, Dinca O, et al. Telocytes in parotid glands. Anat Rec 2012;295:378–85.

[24] Hinescu ME, Ardeleanu C, Gherghiceanu M, Popescu LM. Interstitial Cajal-like cells in human gallbladder. J Mol Histol 2007;38:275–84.

[25] Nicolescu MI, Manole CG, Gherghiceanu M, et al. Telocytes - players of the pancreatic stromal puzzle. Eur J Cancer Suppl 2010;8:51.

[26] Nicolescu MI, Popescu LM. Telocytes in the interstitium of human exocrine pancreas: ultrastructural evidence. Pancreas 2012;41:946–56.

[27] Guisheng Q, Miao L, Ming X, et al. Telocytes in human kindney cortex. J Cel Mol Med 2012; doi:10.1111/j.1582-4934.2012.01582.x.

[28] Gevaert T, De Vos R, Van Der Aa F, et al. Identification of telocytes in the upper lamina propria of the human urinary tract. J Cell Mol Med 2012. doi:10.1111/j.582-4934.2011.01504.x.

[29] Gevaert T, De Vos R, Everaerts W, et al. Characterization of upper lamina propria interstitial cells in bladders from patients with neurogenic detrusor overactivity and bladder pain syndrome. J Cell Mol Med 2011;15:2586–93.

[30] Koh BH, Roy R, Hollywood MA, et al. PDGFRalpha cells in mouse urinary bladder: A new class of interstitial cells. J Cell Mol Med 2012;16:691–700.
[31] Popescu LM, Gherghiceanu M, Suciu LC, et al. Telocytes and putative stem cells in the lungs: electron microscopy, electron tomography and laser scanning microscopy. Cell Tissue Res 2011;345:391–403.
[32] Zheng Y, Bai C, Wang X. Telocyte morphologies and potential roles in diseases. J Cell Physiol 2012;227:2311–7.
[33] Zheng Y, Bai C, Wang X. Potential significance of telocytes in the pathogenesis of lung diseases. Expert Rev Respir Med 2012;6:45–9.
[34] Zheng Y, Manole CG, Bai C, Wang X. Potential significance of telocytes in pulmonary diseases. Respirology 2011;16:22.
[35] Zheng Y, Li H, Manole CG, et al. Telocytes in trachea and lungs. J Cell Mol Med 2011;15:2262–8.
[36] Rusu MC, Jianu AM, Mirancea N, et al. Tracheal telocytes. J Cell Mol Med 2012;16:401–5.
[37] Hinescu ME, Gherghiceanu M, Suciu L, Popescu LM. Telocytes in pleura: two- and three-dimensional imaging by transmission electron microscopy. Cell Tissue Res 2011;343:389–97.
[38] Popescu BO, Gherghiceanu M, Ceafalan L, et al. Telocytes in meninges and choroid plexus. Neurosci Lett 2012;516:265-9.
[39] Gherghiceanu M, Popescu LM. Interstitial Cajal-like cells (ICLC) in human resting mammary gland stroma. Transmission electron microscope (TEM) identification. J Cell Mol Med 2005;9: 893–910.
[40] Popescu LM, Andrei F, Hinescu ME. Snapshots of mammary gland interstitial cells: methylene-blue vital staining and c-kit immunopositivity. J Cell Mol Med 2005;9:476–7.
[41] Popescu LM, Vidulescu C, Curici A, et al. Imatinib inhibits spontaneous rhythmic contractions of human uterus and intestine. Eur J Pharmacol 2006;546:177–81.
[42] Cretoiu SM, Simionescu AA, Caravia L, et al. Complex effects of imatinib on spontaneous and oxytocin-induced contractions in human non-pregnant myometrium. Acta Physiol Hung 2011;98:329–38.
[43] Cretoiu SM, Cretoiu D, Simionescu AA, Popescu LM. Telocytes in human fallopian tube and uterus express estrogen and progesterone receptors. In: Kahn SM, editor. Sex Steroids. Rijeka: InTech; 2012. pp. 91–114.
[44] Popescu LM, Ciontea SM, Cretoiu D. Interstitial Cajal-like cells in human uterus and fallopian tube. Ann NY Acad Sci 2007;1101:139–65.
[45] Hatta K, Huang ML, Weisel RD, Li RK. Culture of rat endometrial telocytes. J Cell Mol Med 2012. doi:10.1111/j.1582-4934.2012.01583.x.
[46] Cretoiu SM, Cretoiu D, Suciu L, Popescu LM. Interstitial Cajal-like cells of human Fallopian tube express estrogen and progesterone receptors. J Mol Histol 2009;40:387–94.
[47] Suciu L, Popescu LM, Gherghiceanu M, et al. Telocytes in human term placenta: morphology and phenotype. Cells Tissue Organ 2010;192:325–39.
[48] Rusu MC, Mirancea N, Manoiu VS, et al. Skin telocytes. Ann Anat 2012;194:359–67.
[49] Ceafalan L, Gherghiceanu M, Popescu LM and Simionescu O. Telocytes in human skin - are they involved in skin regeneration? J Cell Mol Med 2012. doi:10.1111/j.1582-4934.2012.01580.x.
[50] Hinescu ME, Popescu LM, Gherghiceanu M, Faussone-Pellegrini MS. Interstitial Cajal-like cells in rat mesentery: an ultrastructural and immunohistochemical approach. J Cell Mol Med 2008;12: 260–70.
[51] Gherghiceanu M, Hinescu ME, Andrei F, et al. Interstitial Cajal-like cells (ICLC) in myocardial sleeves of human pulmonary veins. J Cell Mol Med 2008;12:1777–81.
[52] Rusu MC, Pop F, Hostiuc S, et al. Extrahepatic and intrahepatic human portal interstitial Cajal cells. Anat Rec 2011;294:1382–92.

[53] Cantarero I, Luesma MJ, Junquera C. The primary cilium of telocytes in the vasculature: electron microscope imaging. J Cell Mol Med 2011;15:2594–600.

[54] Lionetti V. How resident stem cells communicate with cardiac cells in beating heart? J Stem Cell Res Ther 2011;1:e104. doi:10.4172/2157-7633.1000e104.

[55] Popescu LM, Gherghiceanu M, Kostin S, Faussone-Pellegrini MS. Telocytes and heart renewing. In: Wang P, Kuo CH, Takeda N, Singal PK, editors. Adaptation Biology and Medicine. New Delhi: Narosa; 2011. p. 17–39.

[56] Gherghiceanu M, Popescu LM. Heterocellular communication in the heart: electron tomography of telocyte-myocyte junctions. J Cell Mol Med 2011;15:1005–11.

[57] Popescu LM, Gherghiceanu M, Manole CG, Faussone-Pellegrini MS. Cardiac renewing: interstitial Cajal-like cells nurse cardiomyocyte progenitors in epicardial stem cell niches. J Cell Mol Med 2009;13:866–86.

[58] Popescu LM, Gherghiceanu M, Cretoiu D, Radu E. The connective connection: interstitial cells of Cajal (ICC) and ICC-like cells establish synapses with immunoreactive cells. Electron microscope study in situ. J Cell Mol Med 2005;9:714–30.

[59] Liu JJ, Shen XT, Zheng X, et al. Distribution of telocytes in the rat heart. J Clin Rehabil Tiss Eng Res 2011;15:3546–8.

[60] Nicolescu MI, Bucur A, Dinca O, et al. Telocytes in parotid glands. Anat Rec 2012;295:spc1.

[61] Chen Y, Gelfond JA, McManus LM, Shireman PK. Reproducibility of quantitative RT-PCR array in miRNA expression profiling and comparison with microarray analysis. BMC Genomics 2009; 10:407.

[62] Cismasiu VB, Radu E, Popescu LM. miR-193 expression differentiates telocytes from other stromal cells. J Cell Mol Med 2011;15:1071–4.

[63] Manole CG, Cismasiu V, Gherghiceanu M, Popescu LM. Experimental acute myocardial infarction: telocytes involvement in neo-angiogenesis. J Cell Mol Med 2011;15:2284–96.

[64] Mitchell KJ, Pannerec A, Cadot B, et al. Identification and characterization of a non-satellite cell muscle resident progenitor during postnatal development. Nat Cell Biol 2010;12:257–66.

[65] Mandache E, Gherghiceanu M, Macarie C, et al. Telocytes in human isolated atrial amyloidosis: ultrastructural remodelling. J Cell Mol Med 2010;14:2739–47.

[66] Ardeleanu C, Bussolati G. Telocytes are the common cell of origin of both PEComas and GISTs: an evidence-supported hypothesis. J Cell Mol Med 2011;15:2569–74.

Chapter 12

Pericytes as the Source of Mesenchymal Stem Cells

Lindolfo da Silva Meirelles*, Arnold I. Caplan†, Nance Beyer Nardi‡

* *University of São Paulo, Ribeirão Preto, São Paulo, Brazil*
† *Skeletal Research Center, Department of Biology, Case Western Reserve University, Cleveland, Ohio, USA*
‡ *Laboratory of Stem Cells and Cell Therapy, Universidade Luterana do Brazil, Canoas, Rio Grande do Sul, Brazil*

Pericytes

Pericytes may be defined as a subset of perivascular cells with cellular processes that wrap endothelial cells in blood vessels (Figure 12.1A). The precise identification of pericytes in tissue sections, however, is not always easy because they are included in a continuum of perivascular cells called mural cells [1]. So far, the only widely accepted criteria for the identification of pericytes are their anatomic location relative to endothelial cells and their morphology as seen in tissue sections. Stricter definitions consider that pericytes are present on capillaries but not in larger vessels and, in addition, that pericytes are embedded in the basement membrane located around blood vessels [2]. Some authors are less strict and consider that pericytes are not restricted to capillaries [3]. Indeed, there are studies that provide evidence that cells with the characteristics of pericytes are present in larger blood vessels [4,5].

Resident Stem Cells and Regenerative Therapy. http://dx.doi.org/10.1016/B978-0-12-416012-5.00012-8

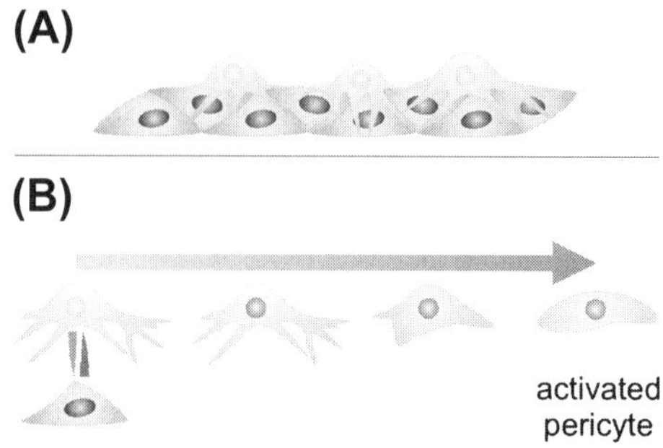

Figure 12.1 Schematic view of pericytes. **(A)** Pericytes are represented as light blue cells with projections that literally embrace a capillary formed by endothelial cells, represented in green. **(B)** Under steady-state conditions, pericytes maintain a complex bidirectional communication with endothelial cells as represented by the arrows on the left portion of the panel. Alterations in this intricate relationship induce gene expression changes in the pericyte, which may then undergo a series of changes to become activated. In this activated state, pericytes are postulated to be similar to cultured MSCs—that is, they exhibit characteristics such as proliferation, migratory capabilities, and secretion of various bioactive molecules that are trophic to the surrounding cells and exert diverse effects on immune system cells. **See Plate 28.**

In some cases, the presence of a surrounding basement membrane does not define a pericyte. Perisinusoidal cells such as hepatic stellate cells, for example, are not surrounded by a basement membrane but are considered a type of pericyte [6]; bone marrow adventitial reticular cells, another type of perisinusoidal cell, also exhibit pericytic characteristics [7]. Both hepatic stellate cells and bone marrow adventitial reticular cells can be viewed as tissue-specific forms of pericytes. Another tissue-specific pericyte is the mesangial cell [8]. The apparent heterogeneity intrinsic to the pericyte compartment may result from both tissue-specific environments and developmental differences.

The developmental origins of pericytes are still poorly defined. Pericytes are generally assumed to be mesodermal in origin [9], even though this assumption leaves room for discussion. The debate regarding the origin of hepatic stellate cells, for instance, has been ongoing for decades [10], and recent studies indicate that these cells originate from the mesenchyme [11,12]. It is clear, however, that pericytes are ultimately derived from neural crest cells in the head [13,14] and in the thymus [15], which suggests they originate from the ectomesenchyme in these areas. Another relevant question regarding the developmental origins of pericytes is whether they descend from committed precursors (as proposed by the "angioblast theory") or result from the recruitment of adjacent cells by endothelial cells (as proposed by the "local origin theory") [16]. These theories will not be discussed in detail in this chapter; however, it should be noted that they may be not mutually exclusive.

So far, only a few molecules are accepted as pericyte markers and only in particular contexts, with no single specific pericyte marker recognized. Alkaline phosphatase has been shown to be present in pericytes of human muscle [17] and in mouse bone marrow adventitial reticular cells [18]. Liver stellate cells (Ito cells) accumulate vitamin A droplets in their cytoplasm, which can be used as a parameter for cell sorting [19]. The nerve/glial antigen 2 (NG2) may be used to identify pericytes, although its expression is associated with a proliferative rather than a mature phenotype [20,21]. CD146 has been used in some studies as a pericyte marker [22,23], but its expression by endothelial cells [24] and other perivascular cells such as vascular smooth muscle cells [25] precludes its use for the isolation of highly purified pericytes. In mice, the regulator of G protein signaling 5 (RGS5) has been shown to be preferentially expressed in pericytes and vascular smooth muscle cells [26]; however, its expression in pericytes may not be restricted to a mature phenotype [27], similarly to the NG2 antigen. The antigen defined by the STRO-1 antibody, which has been shown to enrich for bone marrow stromal cells [28], has been reported to stain pericytes in bone marrow [29] but it also stains endothelial cells [30]. Furthermore, a recent study questions the usefulness of the STRO-1 antibody to identify pericytes, as it was found to stain primarily endothelial cells and only some perivascular cells located in microanatomic positions not compatible with a pericytic phenotype [31]. Bone marrow adventitial reticular cells have been shown to selectively express the low affinity nerve growth factor receptor, CD271 [32]. The antigen defined by the 3G5 antibody has been shown to selectively stain pericytes in the microvasculature [33,34]; however, some hematopoietic cells also express this molecule [30,35]. Alpha-smooth muscle actin (αSMA), which is a marker for smooth muscle cells [36], has been shown to identify some pericytes in tissue sections [33]; however, it is possible that most pericytes, with the exception of those near arterioles, are negative for αSMA in situ [16]. Rat mesangial cells, for example, do not express αSMA in situ even though αSMA expression is detected after culture [37]. Expression of αSMA by pericytes may be associated with activation because αSMA-negative mesangial cells become αSMA-positive after kidney injury [38].

In adult tissues, most pericytes are mature cells in a quiescent state. Under particular circumstances, however, pericytes may undergo changes that allow them to proliferate and migrate locally. In this state, pericytes are said to be activated (Figure 12.1B). Pericyte activation is associated with blood vessel development, wound healing, inflammation, and tissue remodeling.

During active angiogenesis, activated pericytes proliferate and follow the lead of angiogenic endothelial cells. One of the key signaling axes which coordinate pericyte migration alongside endothelial cells during angiogenesis or vasculogenesis is the platelet-derived growth factor B (PDGFB)/platelet-derived growth factor receptor beta (PDGFRβ) pathway [39]. In humans, activated pericytes express the high molecular weight melanoma-associated antigen (HMW-MAA) [40]. The homolog molecule to HMW-MAA is the nerve/glial antigen 2 (NG2) in rats and AN2 in mice [41]. In spite of these species-specific designations, most authors currently refer to this molecule as NG2. NG2 expression in pericytes is up-regulated during angiogenesis or vascular remodeling [20,40]. Another molecule reported to be up-regulated in activated pericytes is aminopeptidase A [42]. In mice, RGS5 has been associated with pericyte activation [27]. Recently, rat pericytes were shown to express beta-3 tubulin upon activation in vivo [5].

During wound healing, the physical association of activated pericytes or pericyte precursors with sprouting vessels is associated with waning of endothelial cell proliferation [43]. It is likely that pericytes play an anti-angiogenic role in vivo, as they inhibit endothelial cell proliferation in vitro [44]. Sato and Rifkin [44] found that this effect was dependent on activation of transforming growth factor beta-1 (TGFβ1), which is produced in a latent form by endothelial cells, pericytes, and vascular smooth muscle cells. Later, other groups published data indicating that formation of gap junctions between endothelial cells and pericyte precursor cells is a requirement for production of active TGFβ [45]. Therefore, the "marriage" between endothelial cells and cells able to acquire a mature pericytic phenotype ultimately instructs endothelial cells to stop proliferation, which allows consequent vessel stabilization. Blood vessel stabilization by pericytes is important not only in wound healing but also during development: loss of pericytes due to lack of PDGFB causes microaneurysms in mice [46]. It is clear, therefore, that stabilizing blood vessels is a primordial function of pericytes. In the brain, pericyte deficiency increases blood vessel permeability, which indicates that pericytes are important regulators of the blood-brain barrier [47,48]. Another function attributed to pericytes is blood flow control by means of contraction/relaxation. Direct electrical stimulation of brain capillary pericytes was found to elicit contraction and consequent decrease in vessel luminal diameter; on the other hand, stimulation with glutamate was found to promote relaxation [49].

Relationship between Pericytes and MSCs

In addition to their importance in the structural maintenance of blood vessel integrity, pericytes have gained further attention from researchers owing to their resemblance to mesenchymal stem cells (MSCs). In theory, an MSC may be defined as a cell that is able to give rise to connective tissue cell types while remaining a stem cell [50,51]. In practical terms, MSC biology has been studied through the analysis of cultured adherent cells that are operationally defined as "MSCs" if they contain cells able to exhibit traits of bone, cartilage, and adipose tissue after culture in specific conditions (Figure 12.2). These cultured cells also express a number of characteristic surface molecules (e.g., CD13, CD29, CD44, CD73, and CD105). Interestingly, cultured MSCs and pericytes share the expression of a number of surface markers (see [52] for a review of MSC and pericyte surface markers).

The establishment of MSC cultures involves tissue disaggregation (mechanical or enzymatic) and seeding of the resulting cell suspension in cell culture-treated plastic dishes. After a few days in culture, the nonadherent cells are removed along with medium during medium change; the adherent cell fraction is then expanded, immunophenotyped, and assayed to assess their differentiation potential [53]. If the cultured cell populations contain cells able to differentiate along the osteogenic, chondrogenic, and adipogenic pathways, they are operationally defined as MSCs. Some authors prefer that these cultures be called "multipotent mesenchymal stromal cells" rather than "mesenchymal stem cells" [54].

Even though MSCs were formerly hailed for their ability to differentiate into various cell types, their most prominent feature was later found to be the secretion of bioactive molecules [55,56]. Indeed, MSCs have been reported to release a range of molecules, which are responsible for trophic, antiapoptotic, immunomodulatory, angiogenic, and antiscarring effects [57,58].

Multiple cell types (e.g., pericytes, fibroblasts, adipocytes, vascular smooth muscle cells, endothelial cells, monocytes, and other blood cells) are generally present in the cell suspension used for the establishment of MSC cultures. This initial cellular heterogeneity does not allow immediate identification of the cell type or cell types that give rise to MSC cultures. It is also possible that various cell types give rise to heterogeneous cultured populations that contain bona fide MSCs.

In 1999, Bianco and Cossu proposed that bone marrow stromal cell cultures are derived from adventitial reticular cells [85] because the latter express alkaline phosphatase and are able to differentiate into adipocytes in vivo [18]. A relationship between cultured pericytes and MSCs is also postulated on the basis of studies that highlight their differentiation capabilities. Some reports showed that pericytes can deposit a mineralized matrix when cultured under appropriate conditions [59–61]. The ability of pericytes to differentiate along the adipogenic and chondrogenic pathways in vitro has been reported [17,22,62]. The osteogenic differentiation of pericytes in vivo has also been described [60]. Finally, at least two groups have reported differentiation of pericytes into skeletal muscle cells in vitro and in vivo [17,22]. These facts indicate that pericytes can give rise to cultures that contain cells that fit the operational definition of MSC. Further evidence that links pericytes to MSCs is the establishment of MSC cultures from virtually any vascularized tissue, including decapsulated glomeruli, but not from circulating blood [63]. Furthermore, the number of fibroblastic colonies, used as an indication of the number of MSCs in a given tissue, strongly correlates with blood vessel density [64].

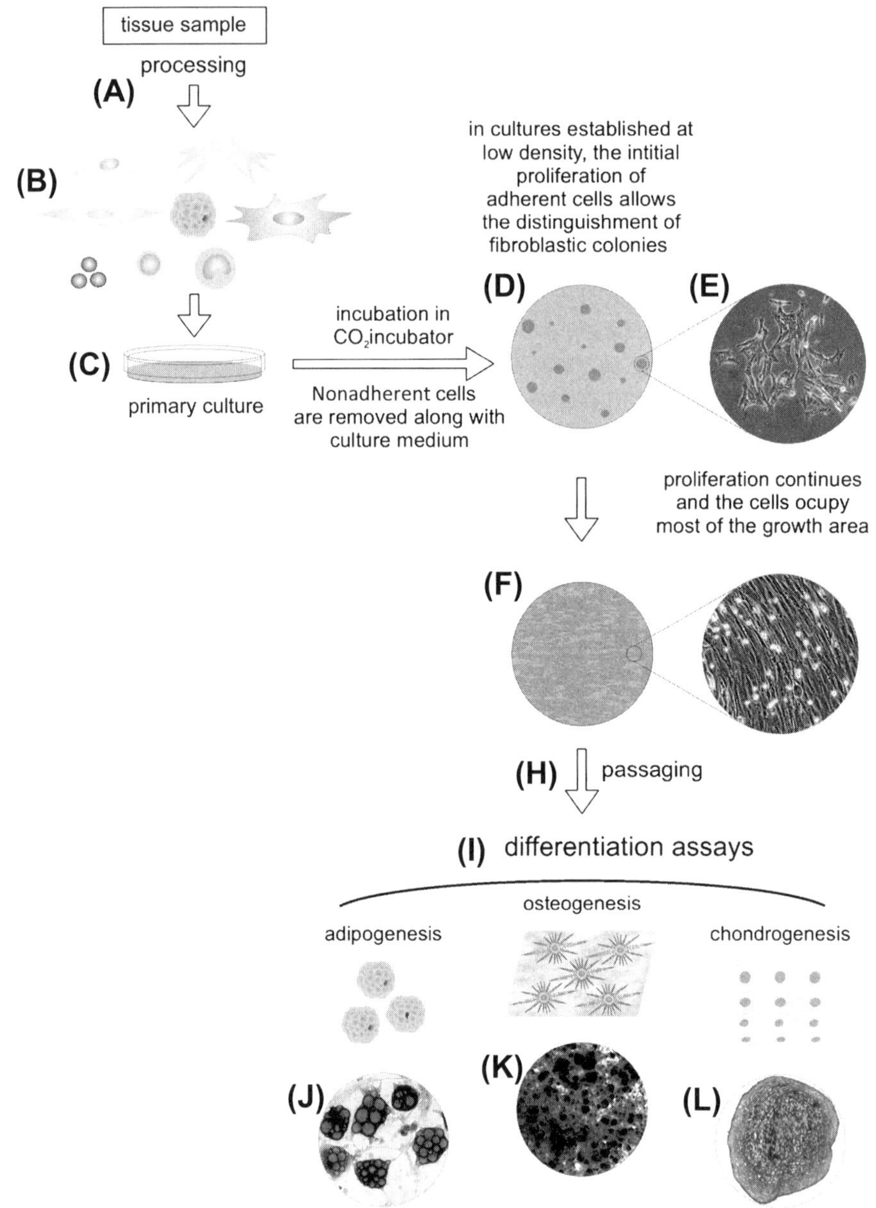

Figure 12.2 Schematic overview of the establishment of MSC cultures. Initially, a tissue sample is processed by methods such as mechanical or enzymatic disaggregation or fractionation by centrifugation on a density gradient **(A)**. Usually the cell population obtained comprises endothelial cells, fibroblasts, vascular smooth muscle cells, pericytes, adipocytes, monocytes, lymphocytes, and some red blood cells **(B)**, even though

In view of the intrinsic relationship between pericytes and MSCs shown here, we proposed a model in which activated pericytes that arise as a result of tissue injury are similar to the cultured cells defined as MSCs [52]. The implication of this assumption is that activated pericytes are MSC-like cells, which exert trophic and immunomodulatory effects in their surroundings during the course of wound healing.

The Plasticity Issue

It is important to mention that, in most instances, pericytes have been cultured before induction of differentiation. The evidence that pericytes correspond to MSCs based on differentiation of cultured cells is challenged by studies that demonstrated that not only pericytes, but also other mature cells such as chondrocytes [65], adipocytes [66], monocytes [67], and even endothelial cells [68] exhibit potential to differentiate into various mesenchymal lineages after dedifferentiation is induced by in vitro culture (Figure 12.3). Among the difficulties concerning the interpretation of these results are the possibility of introduction of artificial changes in the cells by in vitro cultivation and, in some cases, uncertainties regarding the purity of the cells used for the establishment of cultures and the proportion of the various cells that survive in primary culture. The simplest pericyte isolation protocols, for example, use blood vessels for the establishment of cultures [69,70]; in these cases, it is not readily possible to formally exclude the chance that vascular smooth muscle cells, fibroblasts, and other blood vessel-associated cells also contribute to the establishment of cultures that contain cells able to differentiate. Even complex

the cell composition may vary depending on the organ of origin and methods used for processing. These cells are dispensed in a culture vessel (such as a treated Petri dish) containing basal medium with 10% to 20% fetal bovine serum **(C)**. The culture is then left in an incubator at 37° C in a humidified atmosphere that contains 5% to 10% CO_2. Three days later, the medium is replaced, which depletes most nonadherent cells from the culture. The culture is further incubated, and the medium is changed twice a week. Depending on the initial seeding density, discrete fibroblastic cell colonies can be observed as cells proliferate **(D)**. A micrograph of an initial fibroblastic colony is shown in **(E)**. As the adherent cells proliferate, they eventually occupy most of the area available for growth **(F)**. At this point, the culture is said to be confluent. A micrograph of a confluent culture is shown in **(G)**. Cells in confluent cultures are removed from the dish with an enzymatic treatment and seeded into new dishes at a lower density to allow further expansion **(H)**. At such passages, the adherent cells can be subjected to differentiation assays **(I)**. If the cultures contain cells able to differentiate along the adipogenic, osteogenic, and chondrogenic pathways, they can be operationally referred to as mesenchymal stem cells. Differentiation can be detected as development of lipid-laden vacuoles (stained orange in **J**), deposition of calcium-rich extracellular matrix (stained red in **K**), or changes in morphology accompanied by deposition of a sulfated glycosaminoglycan extracellular matrix (stained light purple in **L**). **See Plate 29.**

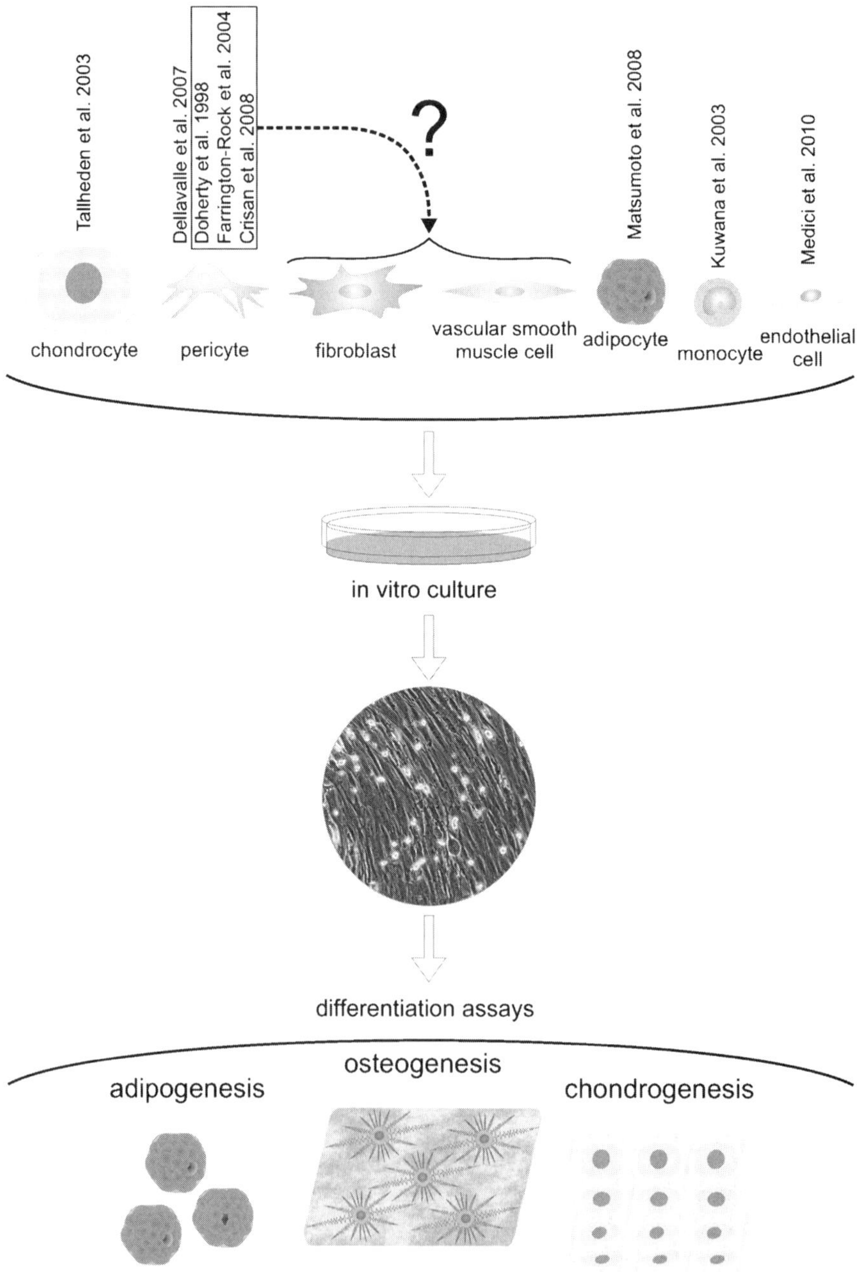

Figure 12.3 Phenotypic plasticity of various terminally differentiated cells. Mature cell types such as chondrocytes, pericytes, adipocytes, monocytes, and even endothelial cells have been shown to be able to establish cultures that contain cells capable of differentiation along the adipogenic, osteogenic, and chondrogenic pathways. In the case of pericytes, some isolation protocols (studies indicated in the box) may provide fibroblasts and vascular smooth muscle cells in addition to pericytes.

fluorescence-activated cell sorting protocols may not provide a "universal" pericyte isolation solution. The markers used by Crisan et al. [22] to isolate perivascular cells from various tissues, for example, seem to be imprecise to provide highly purified pericytes from adipose tissue in view of the markers these cells express in this particular tissue [25].

Therefore, in vitro cultivation introduces variables that cloud the identification of which cell type or cell types can effectively be regarded as MSCs in vivo. The fact that various cell types can be operationally defined as MSCs after culture could be viewed as an argument that undermines the hypothesis that pericytes correspond to MSCs in vivo. The limitations of in vitro studies require, thus, different approaches to determine which cell or cells correspond to the MSC in the body.

Pericytes as Stem Cells in the Body

The alternative to the problems brought up in the previous section is the study of putative MSCs in vivo. To date, there is no evidence that terminally differentiated cells such as osteoblasts or adipocytes, for example, can revert to a more primitive state in vivo. On the other hand, several studies provide evidence that pericytes can do so. In situ studies indicate that pericytes give rise to adipocytes, fibroblasts, chondrocytes, osteocytes, Leydig cells, endometrial stromal cells, odontoblasts, and skeletal muscle fibers in vivo (summarized in Figure 12.4). The results from these studies, which will be further discussed later in this chapter, define pericytes as stem or progenitor cells for the tissues to which they belong.

Evidence that pericytes give rise to adipocytes in situ derive from a study that focused on the analysis of histologic sections of portions of rat adipose tissue harvested at various time points after focal thermal injury induced by an incandescent wire [71]. The appearance of some signs of activation such as increase in volume and amount of rough endoplasmic reticulum could be detected in pericytes 10 hours after induction of the lesion. Pericyte activation was evident 2 days after injury. At this time point, fibroblastic cells were also observable. At subsequent time points, fibroblastic cells that accumulated lipid vacuolae were observed. Based on these results, the authors proposed that pericytes (in addition to "typical fibroblasts") give rise to fibroblastic cells that, in turn, become adipocytes. Nonetheless, these data do not constitute definitive proof that pericytes give rise to adipocytes in adipose tissue. However, more recently, Tang et al. used an inducible genetic lineage tracing system which allowed the detection of cells that express a master adipogenesis gene, PPAR-γ, in the adipose tissue of mice [72]. With this system, the progeny of the cells that express PPAR-γ is also labeled. With the help of another genetic lineage tracing system, the authors determined that the cells that express PPAR-γ are present in the perivascular compartment and proliferate in postnatal life. The location of these cells relative to

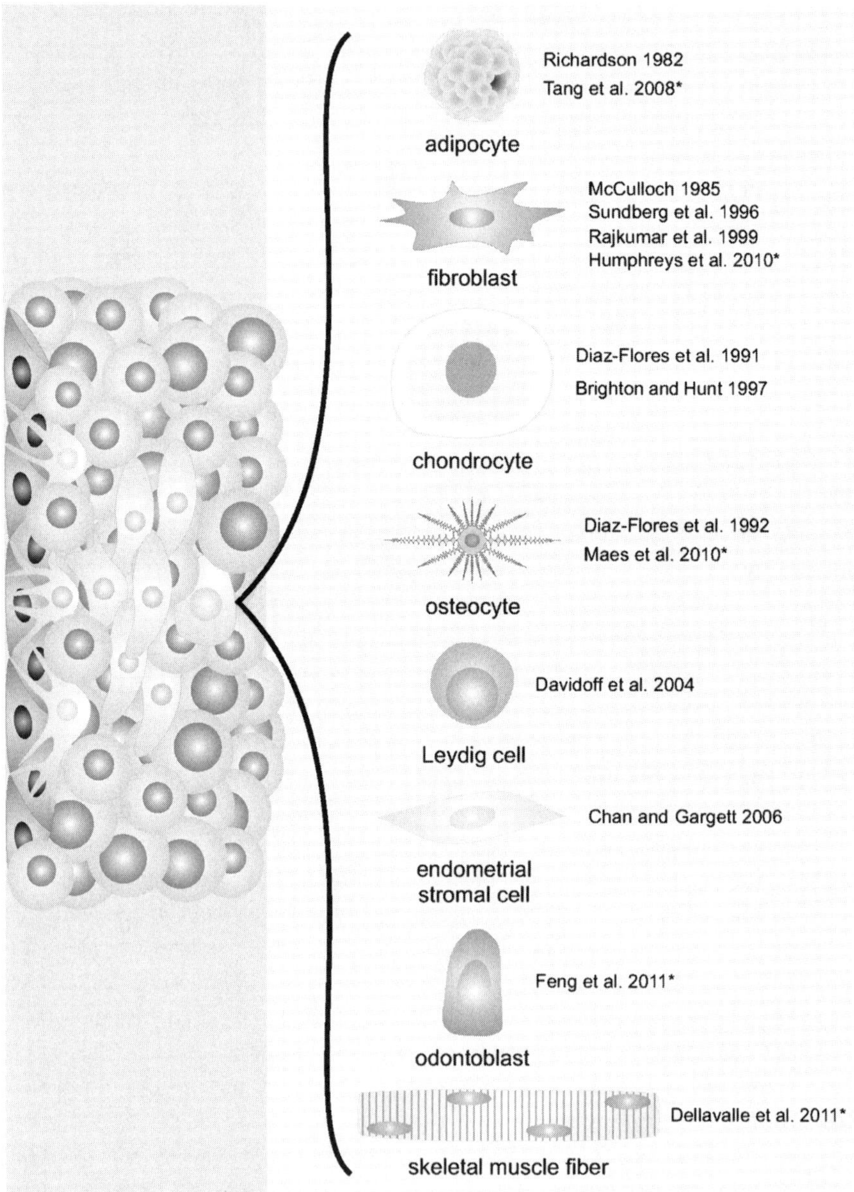

Figure 12.4 Differentiation of pericytes in situ. The left portion of the figure depicts a blood vessel and its associated pericytes inside a tissue composed of mature cells (represented in pink). Environmental cues stimulate the activation of one of the pericytes. The resulting activated pericytes can then enter a differentiation pathway in order to give rise to mature, tissue-specific cells. This process is inferred on the basis of studies (indicated on the right side of the figure) that found evidence that pericytes give rise to adipocytes, fibroblasts, chondrocytes, osteocytes, Leydig cells, endometrial stromal cells, odontoblasts, and skeletal muscle fibers in situ. References marked with an asterisk indicate studies in which genetic labeling was used. **See Plate 30.**

endothelial cells and the expression of molecules such as platelet-derived growth factor receptor β (PDGFRβ), neural/glial antigen 2 (NG2), and α-smooth muscle actin (αSMA) characterize these cells as pericytes, although the authors preferred to use the term "mural cells." Another interesting finding by Tang et al. is the absence of PPAR-γ expression in perivascular cells from other tissues (muscle and retina), which indicates the existence of differences between pericytes from different tissues.

The contribution of pericytes to the formation of cartilage and bone cells in situ was investigated by Dias-Flores et al. in rats by means of experiments with cartilage grafts [73] and periosteal lesion [74]. After injury on the surface of bone by lifting a periosteal flap (which contained microscopic bone fragments) or after transplantation of a cartilage graft, an increase in vascular permeability was induced by histamine and the Monastral Blue stain delivered intravascularly. Histologic analyses performed shortly after administration of the stain revealed the presence of Monastral Blue inclusions in perivascular cells. When tissue samples were harvested and analyzed at various subsequent time points, inclusions were observed in cells with mature phenotypes, which indicated that they were derived from pericytes.

Brighton and Hunt studied the morphologic alterations that occur in blood vessels during healing of bone fracture in rabbits by transmission electron microscopy [75]. Five days after the fracture, the authors found chondroblasts in perivascular areas in the proximal region of the inner layer of the periosteal callus. These cells were embedded in a cartilaginous matrix in which remaining fragments of the basal lamina were observable, which suggested that they were derived from pericytes—although the authors did not discard the hypothesis that these cells could be derived from endothelial cells. Six days after the fracture, hypertrophic chondrocytes could be seen in these areas. Subsequently, endochondral bone formation was observed in the proximal portion of the inner layer of the callus. This does not mean that pericyte-derived chondrocytes further differentiated into bone cells. Chondrocytes formed during bone healing do not turn into osteocytes; rather, they provide a provisional matrix for bone formation. Recently, Maes et al. used two genetic lineage tracing systems to visualize cells of the osteoblastic lineage and study the origin of bone cells during endochondral formation of the long bones in mice [76]. As a result, the authors found that the cells that give rise to osteoblasts have a pericytic habitat and invade the hypertrophic cartilage along with vascular elements. Similar results were found, in the same study, in experiments in which bone fracture was induced in adult mice. Taken together, these results indicate that the bone healing process depends on pericytes located inside the lesion for the formation of hypertrophic cartilage and also on pericytes surrounding the injury site, which invade as a second wave of cells into the hypertrophic cartilage along with blood vessels and form new osseous tissue.

Data indicating that pericytes give rise to fibroblasts in situ are also available in the literature [77–80]. McCulloch [78] used a DNA labeling technique (incorporation of tritiated thymidine) to study the development of dental ligament in rats. He found that some perivascular cells incorporated the DNA label; some time after labeling, labeled cell clumps were also observable at areas adjacent to blood vessels, which suggested that pericytes gave rise to proliferative cells that differentiate into mature cells (fibroblasts and ligamentocytes) of the ligament. Sundberg et al. [80] stained histologic sections from cases of excessive scarring with antibodies which recognize the high molecular weight-melanoma-associated antigen (HMW-MAA), a marker for activated pericytes [40] also referred to as NG2, and prolyl-4-hydroxylase beta-subunit, which is expressed by cells that actively synthesize collagen, along with antibodies to other markers. Cell subpopulations defined by the markers used led the authors to suggest that a population of pericytes migrate into the perivascular space and become fibroblasts responsible for fibrosis. Rajkumar et al. [79] performed immunohistologic analyses on skin biopsies from patients with autoimmune Raynaud's phenomenon or systemic sclerosis in search of cells that expressed HMW-MAA and the platelet-derived growth factor receptor β (PDGFRβ), which is expressed not only by pericytes but also by fibroblasts. The authors found that about two thirds of the PDGFRβ-positive cells also expressed the HMW-MAA in the patients studied; in contrast, only about 20% of cells in normal skin samples co-expressed these two markers. The authors interpreted these results as evidence that activated pericytes contribute to fibrosis. More recently, a study by Humphreys et al. [77] provided stronger evidence that pericytes give rise to myofibroblasts, which are contractile fibroblasts that develop during wound repair, by genetically tracing the fate of pericytes during the course of kidney fibrosis induced by ischemia/reperfusion in mice. The authors identified pericytes as descendants from cells that express the transcription factor FoxD1 during kidney development. As a result, fibroblasts and myofibroblasts were found to derive from pericytes during kidney fibrosis. The authors also used two other lineage tracing systems to exclude the possible origin of myofibroblasts from other cell types.

The differentiation of pericytes (and vascular smooth muscle cells) into Leydig cells in situ has been suggested by Davidoff et al. [81] after chemical ablation of Leydig cells using ethane dimethane sulfonate in rats. To visualize proliferative cells, bromodeoxyuridine (BrdU) was administered to the animals 2 hours prior to euthanasia. Leydig cells were undetectable 3 days after the treatment. Leydig cells were observable again about 2 weeks after administration of the drug as cell clusters in the vicinity of intertubular blood vessels. Three weeks after the treatment, these cells exhibited a Leydig cell marker, cytochrome P450 side chain cleavage enzyme, and most of them were located around the seminiferous tubules. Thirty days after the chemical lesion, the new Leydig cells were localized in the areas between the tubules. Immunohistologic analyses showed that the newly generated Leydig cells expressed the pericyte markers PDGFRβ and NG2. These data indicate that pericytes have the ability to (re)generate Leydig cells.

Chan and Gargett [82] used a label-retaining methodology to search for stem cells in the endometrium of mice. In that study, 3-day-old mice received BrdU for 3 days in order to label proliferative cells, which are abundant at this age. Using this approach, only cells that remain in a quiescent state retain BrdU as the animal grows. In a tissue that undergoes cyclic regeneration such as the endometrium, the frequency of quiescent, BrdU-positive cells diminishes with each regeneration cycle. The analysis of the anatomic sites that contain BrdU-positive cells whose frequency decreases cycle after cycle allows the identification of stem cell niches. Using this rationale, the authors were able to identify the existence of epithelial and stromal label-retaining cells. One third of the stromal label-retaining cells were located in perivascular locations, which suggests that pericytic cells behave as stem cells in the endometrium.

Recently, the contribution of pericytes to odontoblasts has been demonstrated using a genetic lineage tracing approach [83]. In that study, the authors used genetically engineered mice to allow the observation of the progeny of cells that expressed the pericyte marker NG2. In one of the mouse strains used, the genetic tracing system was active throughout life; in a second mouse strain, this system was inactive until induction by injection of tamoxifen. As a result, pericytes were found to give rise to odontoblasts both during physiologic tooth growth and under injury conditions. Another dental cell population was also found to give rise to odontoblasts in addition to pericytes, which adds evidence to the proposition that pericytes are not the only stem/precursor cells in vascularized tissues.

Finally, pericytes were also shown to give rise to skeletal muscle fibers. Dellavalle et al. [84] used genetically engineered mice to observe the contribution of tissue nonspecific alkaline phosphatase-positive cells to muscle fibers during normal development in a mouse model of muscular dystrophy and, also, after muscle injury with cardiotoxin. These authors had previously found that alkaline phosphatase is present in pericytes but absent in satellite cells in human skeletal muscle [17]. In mice, however, endothelial cells also express alkaline phosphatase, which led the authors to use animals engineered to allow the observation of the fate of VE-cadherin-positive (endothelial) cells. The authors found that pericytes, but not endothelial cells, give rise to muscle fibers and, to some extent, to satellite cells. They also determined that the contribution of pericytes to skeletal muscle formation is highest during the first month of murine postnatal life and becomes minimal during adulthood even after cardiotoxin-induced muscle injury. This finding indicates that the possible use of pericytes for skeletal muscle regeneration in humans may be dependent on the proper timing of cell administration.

Conclusion

One of the consequences of the intense efforts to unravel the identity of MSCs in vivo is the realization that pericytes are stem or progenitor cells in all vascularized tissues, a fact that is corroborated by genetic labeling strategies in animal models today.

Pericytes are not expected to be the only stem/progenitor cells present in the body; nonetheless, the finding that they behave as such in various scenarios makes it clear that they have a great therapeutic potential not only as tools for cell-based therapies but also as targets for pharmacologic approaches. Exploring this potential in preclinical animal models is key for the development of new therapies or improvement of the existing ones.

Acknowledgments

The authors are indebted to Conselho National de Desenvolvimento Científico e Tecnológico (CNPq), Fundação de Amparo à Pesquisa do Estado do Rio Grande do sul, Fundação de Amparo à Pesquisa do Estado de São Paulo (FAPESP), and the Virginia and David Baldwin Research Fund for funding. Discussions regarding pericyte anatomy and ontogeny with Dr. Radovan Borejevic are gratefully acknowledged. L. da S. M. holds a postdoctoral fellowship from FAPESP.

References

[1] Hirschi KK, D'Amore PA. Pericytes in the microvasculature. Cardiovasc Res 1996 Oct;32(4): 687–98.

[2] Armulik A, Genové G, Betsholtz C. Pericytes: developmental, physiological, and pathological perspectives, problems, and promises. Dev Cell 2011 Aug 16;21(2):193–215.

[3] Sims DE. Diversity within pericytes. Clin Exp Pharmacol Physiol 2000 Oct;27(10):842–6.

[4] Andreeva ER, Pugach IM, Gordon D, Orekhov AN. Continuous subendothelial network formed by pericyte-like cells in human vascular bed. Tissue Cell 1998 Feb;30(1):127–35.

[5] Stapor PC, Murfee WL. Identification of class III β-tubulin as a marker of angiogenic perivascular cells. Microvasc Res 2011 Sep 17 [Epub ahead of print].

[6] Pinzani M. Hepatic stellate (ITO) cells: expanding roles for a liver-specific pericyte. J Hepatol 1995 Jun;22(6):700–6.

[7] Díaz-Flores L, Gutiérrez R, Madrid JF, Varela H, Valladares F, Acosta E, et al. Pericytes. Morphofunction, interactions and pathology in a quiescent and activated mesenchymal cell niche. Histol Histopathol 2009 Jul;24(7):909–69.

[8] Schlondorff D. The glomerular mesangial cell: an expanding role for a specialized pericyte. FASEB J 1987 Oct;1(4):272–81.

[9] Sims DE. The pericyte–a review. Tissue Cell 1986;18(2):153–74.

[10] Geerts A. On the origin of stellate cells: mesodermal, endodermal or neuro-ectodermal? J Hepatol 2004 Feb;40(2):331–4.

[11] Asahina K, Tsai SY, Li P, Ishii M, Maxson Jr RE, Sucov HM, et al. Mesenchymal origin of hepatic stellate cells, submesothelial cells, and perivascular mesenchymal cells during mouse liver development. Hepatology 2009 Mar;49(3):998–1011.

[12] Asahina K, Zhou B, Pu WT, Tsukamoto H. Septum transversum-derived mesothelium gives rise to hepatic stellate cells and perivascular mesenchymal cells in developing mouse liver. Hepatology 2011 Mar;53(3):983–95.

[13] Etchevers HC, Vincent C, Le Douarin NM, Couly GF. The cephalic neural crest provides pericytes and smooth muscle cells to all blood vessels of the face and forebrain. Development 2001 Apr;128(7):1059–68.

[14] Korn J, Christ B, Kurz H. Neuroectodermal origin of brain pericytes and vascular smooth muscle cells. J Comp Neurol 2002 Jan 1;442(1):78–88.

[15] Foster K, Sheridan J, Veiga-Fernandes H, Roderick K, Pachnis V, Adams R, et al. Contribution of neural crest-derived cells in the embryonic and adult thymus. J Immunol 2008 Mar 1;180(5): 3183–9.

[16] Dore-Duffy P. Pericytes: pluripotent cells of the blood brain barrier. Curr Pharm Des 2008;14(16):1581–93.

[17] Dellavalle A, Sampaolesi M, Tonlorenzi R, Tagliafico E, Sacchetti B, Perani L, et al. Pericytes of human skeletal muscle are myogenic precursors distinct from satellite cells. Nat Cell Biol 2007 Mar;9(3):255–67.

[18] Bianco P, Costantini M, Dearden LC, Bonucci E. Alkaline phosphatase positive precursors of adipocytes in the human bone marrow. Br J Haematol 1988 Apr;68(4):401–3.

[19] Kubota H, Yao HL, Reid LM. Identification and characterization of vitamin A-storing cells in fetal liver: implications for functional importance of hepatic stellate cells in liver development and hematopoiesis. Stem Cells 2007 Sep;25(9):2339–49.

[20] Murfee WL, Rehorn MR, Peirce SM, Skalak TC. Perivascular cells along venules upregulate NG2 expression during microvascular remodeling. Microcirculation 2006 Apr-May;13(3):261–73.

[21] Ozerdem U, Grako KA, Dahlin-Huppe K, Monosov E, Stallcup WB. NG2 proteoglycan is expressed exclusively by mural cells during vascular morphogenesis. Dev Dyn 2001 Oct;222(2):218–27.

[22] Crisan M, Yap S, Casteilla L, Chen CW, Corselli M, Park TS, et al. A perivascular origin for mesenchymal stem cells in multiple human organs. Cell Stem Cell 2008 Sep 11;3(3):301–13.

[23] Sacchetti B, Funari A, Michienzi S, Di Cesare S, Piersanti S, Saggio I, et al. Self-renewing osteoprogenitors in bone marrow sinusoids can organize a hematopoietic microenvironment. Cell 2007 Oct 19;131(2):324–36.

[24] Mutin M, Dignat-George F, Sampol J. Immunologic phenotype of cultured endothelial cells: quantitative analysis of cell surface molecules. Tissue Antigens 1997 Nov;50(5):449–58.

[25] Zimmerlin L, Donnenberg VS, Pfeifer ME, Meyer EM, Péault B, Rubin JP, et al. Stromal vascular progenitors in adult human adipose tissue. Cytometry A 2010 Jan;77(1):22–30.

[26] Bondjers C, Kalén M, Hellström M, Scheidl SJ, Abramsson A, Renner O, et al. Transcription profiling of platelet-derived growth factor-B-deficient mouse embryos identifies RGS5 as a novel marker for pericytes and vascular smooth muscle cells. Am J Pathol 2003 Mar;162(3):721–9.

[27] Berger M, Bergers G, Arnold B, Hämmerling GJ, Ganss R. Regulator of G-protein signaling-5 induction in pericytes coincides with active vessel remodeling during neovascularization. Blood 2005 Feb 1;105(3):1094–101.

[28] Simmons PJ, Torok-Storb B. Identification of stromal cell precursors in human bone marrow by a novel monoclonal antibody, STRO-1. Blood 1991 Jul 1;78(1):55–62.

[29] Bianco P, Riminucci M, Gronthos S, Robey PG. Bone marrow stromal stem cells: nature, biology, and potential applications. Stem Cells 2001;19(3):180–92.

[30] Shi S, Gronthos S. Perivascular niche of postnatal mesenchymal stem cells in human bone marrow and dental pulp. J Bone Miner Res 2003 Apr;18(4):696–704.

[31] Lin G, Liu G, Banie L, Wang G, Ning H, Lue TF, et al. Tissue distribution of mesenchymal stem cell marker stro-1. Stem Cells Dev 2011 Oct;20(10):1747–52.

[32] Cattoretti G, Schiró R, Orazi A, Soligo D, Colombo MP. Bone marrow stroma in humans: anti-nerve growth factor receptor antibodies selectively stain reticular cells in vivo and in vitro. Blood 1993 Apr 1;81(7):1726–38.

[33] Helmbold P, Wohlrab J, Marsch WC, Nayak RC. Human dermal pericytes express 3G5 ganglioside–a new approach for microvessel histology in the skin. J Cutan Pathol 2001 Apr;28(4):206–10.

[34] Nayak RC, Berman AB, George KL, Eisenbarth GS, King GL. A monoclonal antibody (3G5)-defined ganglioside antigen is expressed on the cell surface of microvascular pericytes. J Exp Med 1988 Mar 1;167(3):1003–15.

[35] Rabinowe SL, Nayak RC, Krisch K, George KL, Eisenbarth GS. Aging in man. Linear increase of a novel T cell subset defined by antiganglioside monoclonal antibody 3G5. J Exp Med 1987 May 1;165(5):1436–41.

[36] Owens GK, Kumar MS, Wamhoff BR. Molecular regulation of vascular smooth muscle cell differentiation in development and disease. Physiol Rev 2004 Jul;84(3):767–801.

[37] Elger M, Drenckhahn D, Nobiling R, Mundel P, Kriz W. Cultured rat mesangial cells contain smooth muscle alpha-actin not found in vivo. Am J Pathol 1993 Feb;142(2):497–509.

[38] Johnson RJ, Iida H, Alpers CE, Majesky MW, Schwartz SM, Pritzi P, et al. Expression of smooth muscle cell phenotype by rat mesangial cells in immune complex nephritis. Alpha-smooth muscle actin is a marker of mesangial cell proliferation. J Clin Invest 1991 Mar;87(3):847–58.

[39] Hellström M, Kalén M, Lindahl P, Abramsson A, Betsholtz C. Role of PDGF-B and PDGFR-beta in recruitment of vascular smooth muscle cells and pericytes during embryonic blood vessel formation in the mouse. Development 1999 Jun;126(14):3047–55.

[40] Schlingemann RO, Rietveld FJ, de Waal RM, Ferrone S, Ruiter DJ. Expression of the high molecular weight melanoma-associated antigen by pericytes during angiogenesis in tumors and in healing wounds. Am J Pathol 1990 Jun;136(6):1393–405.

[41] Stallcup WB. The NG2 proteoglycan: past insights and future prospects. J Neurocytol 2002 Jul-Aug;31(6-7):423–35.

[42] Schlingemann RO, Oosterwijk E, Wesseling P, Rietveld FJ, Ruiter DJ. Aminopeptidase a is a constituent of activated pericytes in angiogenesis. J Pathol 1996 Aug;179(4):436–42.

[43] Crocker DJ, Murad TM, Geer JC. Role of the pericyte in wound healing. An ultrastructural study. Exp Mol Pathol 1970 Aug;13(1):51–65.

[44] Sato Y, Rifkin DB. Inhibition of endothelial cell movement by pericytes and smooth muscle cells: activation of a latent transforming growth factor-beta 1-like molecule by plasmin during co-culture. J Cell Biol 1989 Jul;109(1):309–15.

[45] Hirschi KK, Burt JM, Hirschi KD, Dai C. Gap junction communication mediates transforming growth factor-beta activation and endothelial-induced mural cell differentiation. Circ Res 2003 Sep 5;93(5):429–37.

[46] Lindahl P, Johansson BR, Levéen P, Betsholtz C. Pericyte loss and microaneurysm formation in PDGF-B-deficient mice. Science 1997 Jul 11;277(5323):242–5.

[47] Armulik A, Genové G, Mäe M, Nisancioglu MH, Wallgard E, Niaudet C, et al. Pericytes regulate the blood-brain barrier. Nature 2010 Nov 25;468(7323):557–61.

[48] Daneman R, Zhou L, Kebede AA, Barres BA. Pericytes are required for blood-brain barrier integrity during embryogenesis. Nature 2010 Nov 25;468(7323):562–6.

[49] Peppiatt CM, Howarth C, Mobbs P, Attwell D. Bidirectional control of CNS capillary diameter by pericytes. Nature 2006 Oct 12;443(7112):700–4.

[50] Caplan AI. Mesenchymal stem cells. J Orthop Res 1991 Sep;9(5):641–50.

[51] Caplan AI. The mesengenic process. Clin Plast Surg 1994 Jul;21(3):429–35.

[52] da Silva Meirelles L, Caplan AI, Nardi NB. In search of the in vivo identity of mesenchymal stem cells. Stem Cells 2008 Sep;26(9):2287–99.

[53] Meirelles Lda S, Nardi NB. Methodology, biology and clinical applications of mesenchymal stem cells. Front Biosci 2009 Jan 1;14:4281–98.

[54] Horwitz EM, Le Blanc K, Dominici M, Mueller I, Slaper-Cortenbach I, Marini FC, et al. International Society for Cellular Therapy. Clarification of the nomenclature for MSC: The International Society for Cellular Therapy position statement. Cytotherapy 2005;7(5):393–5.

[55] Caplan AI, Dennis JE. Mesenchymal stem cells as trophic mediators. J Cell Biochem 2006 Aug 1;98(5):1076–84.

[56] Caplan AI. Adult mesenchymal stem cells for tissue engineering versus regenerative medicine. J Cell Physiol 2007 Nov;213(2):341–7.

[57] Caplan AI, Correa D. The MSC: an injury drugstore. Cell Stem Cell 2011 Jul 8;9(1):11–5.

[58] Meirelles Lda S, Fontes AM, Covas DT, Caplan AI. Mechanisms involved in the therapeutic properties of mesenchymal stem cells. Cytokine Growth Factor Rev 2009 Oct-Dec;20(5-6): 419–27.

[59] Brighton CT, Lorich DG, Kupcha R, Reilly TM, Jones AR, Woodbury 2nd RA. The pericyte as a possible osteoblast progenitor cell. Clin Orthop Relat Res 1992 Feb;(275):287–99.

[60] Doherty MJ, Ashton BA, Walsh S, Beresford JN, Grant ME, Canfield AE. Vascular pericytes express osteogenic potential in vitro and in vivo. J Bone Miner Res 1998 May;13(5):828–38.

[61] Schor AM, Allen TD, Canfield AE, Sloan P, Schor SL. Pericytes derived from the retinal microvasculature undergo calcification in vitro. J Cell Sci 1990 Nov;97(Pt 3):449–61.

[62] Farrington-Rock C, Crofts NJ, Doherty MJ, Ashton BA, Griffin-Jones C, Canfield AE. Chondrogenic and adipogenic potential of microvascular pericytes. Circulation 2004 Oct 12;110(15): 2226–32.

[63] da Silva Meirelles L, Chagastelles PC, Nardi NB. Mesenchymal stem cells reside in virtually all post-natal organs and tissues. J Cell Sci 2006 Jun 1;119(Pt 11):2204–13.

[64] da Silva Meirelles L, Sand TT, Harman RJ, Lennon DP, Caplan AI. MSC frequency correlates with blood vessel density in equine adipose tissue. Tissue Eng Part A 2009 Feb;15(2):221–9.

[65] Tallheden T, Dennis JE, Lennon DP, Sjögren-Jansson E, Caplan AI, Lindahl A. Phenotypic plasticity of human articular chondrocytes. J Bone Joint Surg Am 2003;85-A(Suppl. 2):93–100.

[66] Matsumoto T, Kano K, Kondo D, Fukuda N, Iribe Y, Tanaka N, et al. Mature adipocyte-derived dedifferentiated fat cells exhibit multilineage potential. J Cell Physiol 2008 Apr;215(1):210–22.

[67] Kuwana M, Okazaki Y, Kodama H, Izumi K, Yasuoka H, Ogawa Y, et al. Human circulating CD14+ monocytes as a source of progenitors that exhibit mesenchymal cell differentiation. J Leukoc Biol 2003 Nov;74(5):833–45.

[68] Medici D, Shore EM, Lounev VY, Kaplan FS, Kalluri R, Olsen BR, et al. Conversion of vascular endothelial cells into multipotent stem-like cells. Nat Med 2010 Dec;16(12):1400–6.

[69] Balabanov R, Washington R, Wagnerova J, Dore-Duffy P. CNS microvascular pericytes express macrophage-like function, cell surface integrin alpha M, and macrophage marker ED-2. Microvasc Res 1996 Sep;52(2):127–42.

[70] Maier CL, Shepherd BR, Yi T, Pober JS, et al. Explant outgrowth, propagation and characterization of human pericytes. Microcirculation 2010 Jul;17(5):367–80.

[71] Richardson RL, Hausman GJ, Campion DR. Response of pericytes to thermal lesion in the inguinal fat pad of 10-day-old rats. Acta Anat (Basel) 1982;114(1):41–57.

[72] Tang W, Zeve D, Suh JM, Bosnakovski D, Kyba M, Hammer RE, et al. White fat progenitor cells reside in the adipose vasculature. Science 2008 Oct 24;322(5901):583–6.

[73] Diaz-Flores L, Gutierrez R, Gonzalez P, Varela H. Inducible perivascular cells contribute to the neochondrogenesis in grafted perichondrium. Anat Rec 1991 Jan;229(1):1–8.

[74] Diaz-Flores L, Gutierrez R, Lopez-Alonso A, Gonzalez R, Varela H. Pericytes as a supplementary source of osteoblasts in periosteal osteogenesis. Clin Orthop Relat Res 1992 Feb;(275):280–6.

[75] Brighton CT, Hunt RM. Early histologic and ultrastructural changes in microvessels of periosteal callus. J Orthop Trauma 1997 May;11(4):244–53.

[76] Maes C, Kobayashi T, Selig MK, Torrekens S, Roth SI, Mackem S, et al. Osteoblast precursors, but not mature osteoblasts, move into developing and fractured bones along with invading blood vessels. Dev Cell 2010 Aug 17;19(2):329–44.

[77] Humphreys BD, Lin SL, Kobayashi A, Hudson TE, Nowlin BT, Bonventre JV, et al. Fate tracing reveals the pericyte and not epithelial origin of myofibroblasts in kidney fibrosis. Am J Pathol 2010 Jan;176(1):85–97.

[78] McCulloch CA. Progenitor cell populations in the periodontal ligament of mice. Anat Rec 1985 Mar;211(3):258–62.

[79] Rajkumar VS, Sundberg C, Abraham DJ, Rubin K, Black CM. Activation of microvascular pericytes in autoimmune Raynaud's phenomenon and systemic sclerosis. Arthritis Rheum 1999 May;42(5):930–41.

[80] Sundberg C, Ivarsson M, Gerdin B, Rubin K. Pericytes as collagen-producing cells in excessive dermal scarring. Lab Invest 1996 Feb;74(2):452–66.

[81] Davidoff MS, Middendorff R, Enikolopov G, Riethmacher D, Holstein AF, Müller D. Progenitor cells of the testosterone-producing Leydig cells revealed. J Cell Biol 2004 Dec 6;167(5):935–44.

[82] Chan RW, Gargett CE. Identification of label-retaining cells in mouse endometrium. Stem Cells 2006 Jun;24(6):1529–38.

[83] Feng J, Mantesso A, De Bari C, Nishiyama A, Sharpe PT. Dual origin of mesenchymal stem cells contributing to organ growth and repair. Proc Natl Acad Sci USA 2011 Apr 4 [Epub ahead of print].

[84] Dellavalle A, Maroli G, Covarello D, Azzoni E, Innocenzi A, Perani L, et al. Pericytes resident in postnatal skeletal muscle differentiate into muscle fibres and generate satellite cells. Nat Commun 2011 Oct 11;2:499. doi:10.1038/ncomms1508.

[85] Bianco P, Cossu G. Uno, nessuno e centomila: searching for the identity of mesodermal progenitors. Exp Cell Res 1999 Sep 15;251(2):257–63.

Index

Note: Page numbers followed by "f" indicate figures and "t" indicate tables.

D

E

F

G

H

I

M

O

P

R

U

V

Color Plates

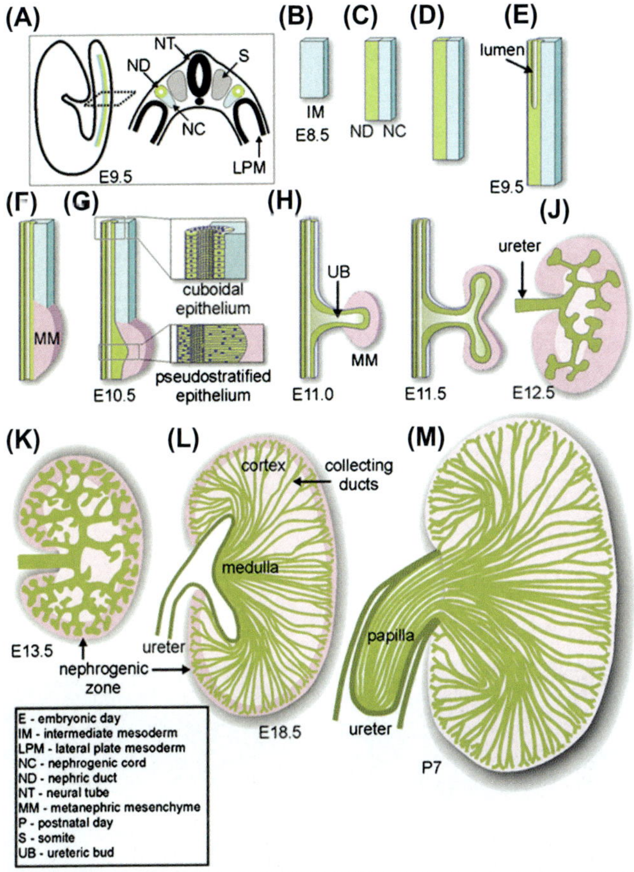

Plate 1 Renal organogenesis: origins of metanephric mesenchyme and ureteric bud. For full caption see Chapter 1, page 6.

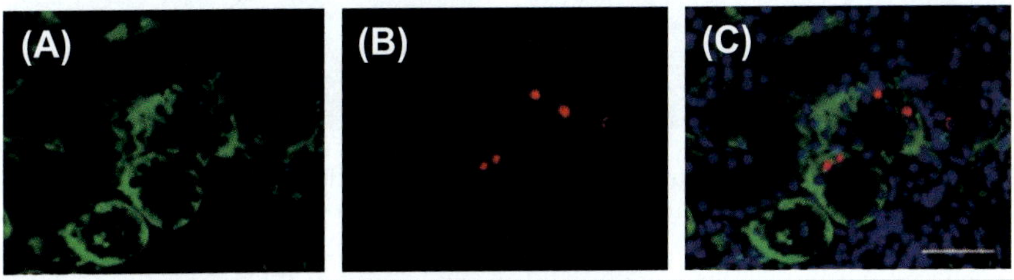

Plate 2 BrdU-positive cells in renal tubules of normal rat kidneys. For full caption see Chapter 1, page 11.

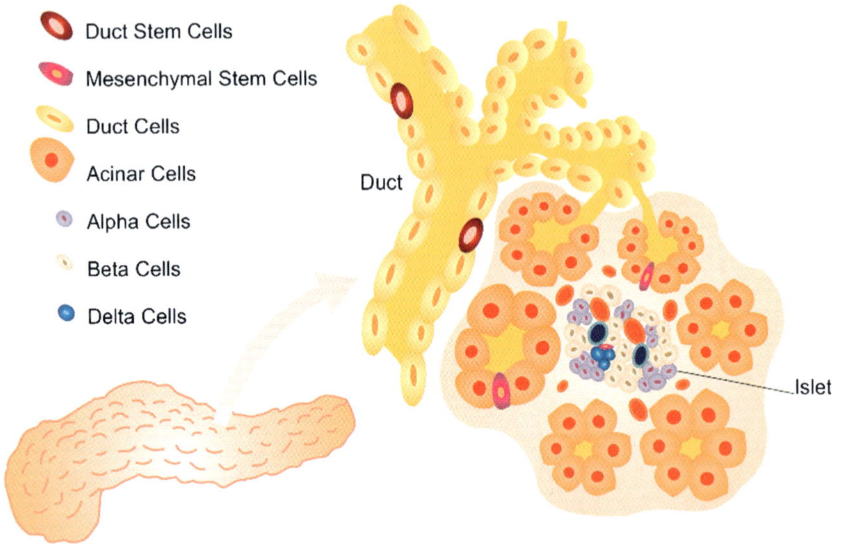

Plate 3 Different cell types and putative stem cell populations residing inside the pancreatic tissue.

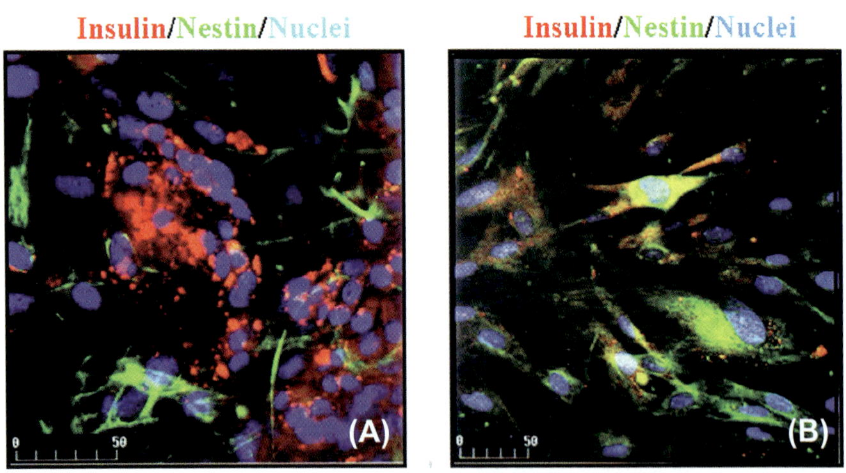

Plate 4 Confocal microscopy of triple-labeled immunofluorescence for insulin (red), nestin (green), and DAPI for nuclei (blue). Human pancreatic islet cell cultures were maintained for 24 **(A)** and 60 **(B)** days on uncoated plastic. Whereas the insulin signal tended to decrease with time, nestin-positive cells constitute the great majority of cells in long-term culture and co-localization of insulin and nestin is observed (B). Scale bars in μm. *(From [12].)*

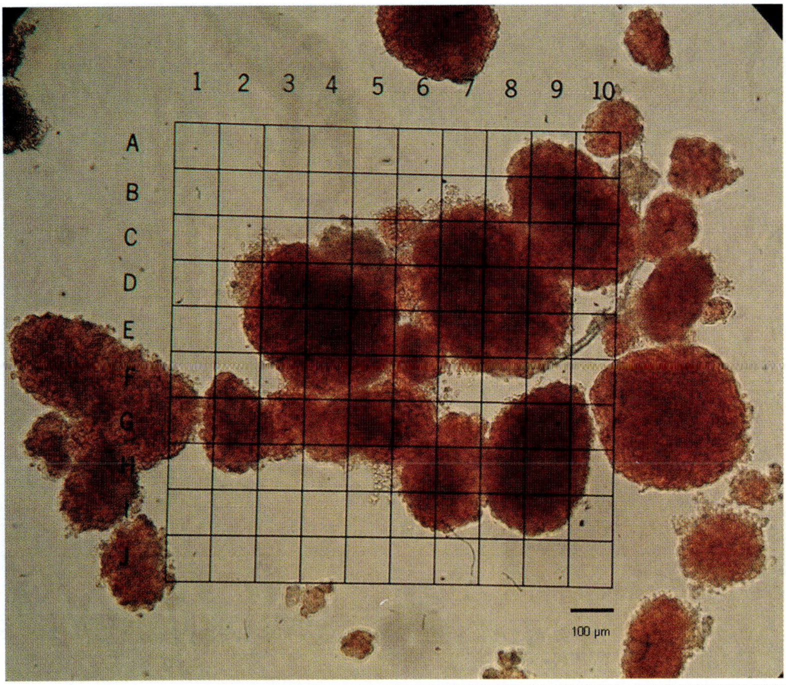

Plate 5 Purified human pancreatic islet. For full caption see Chapter 2, page 40.

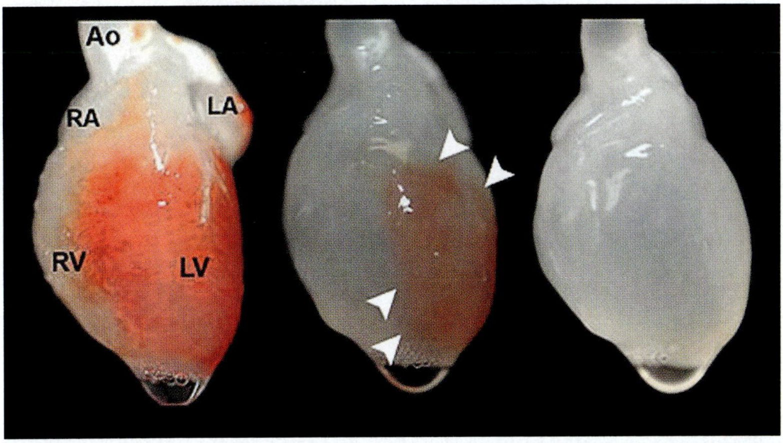

Plate 6 Whole heart decellularization. For full caption see Chapter 3, page 47.

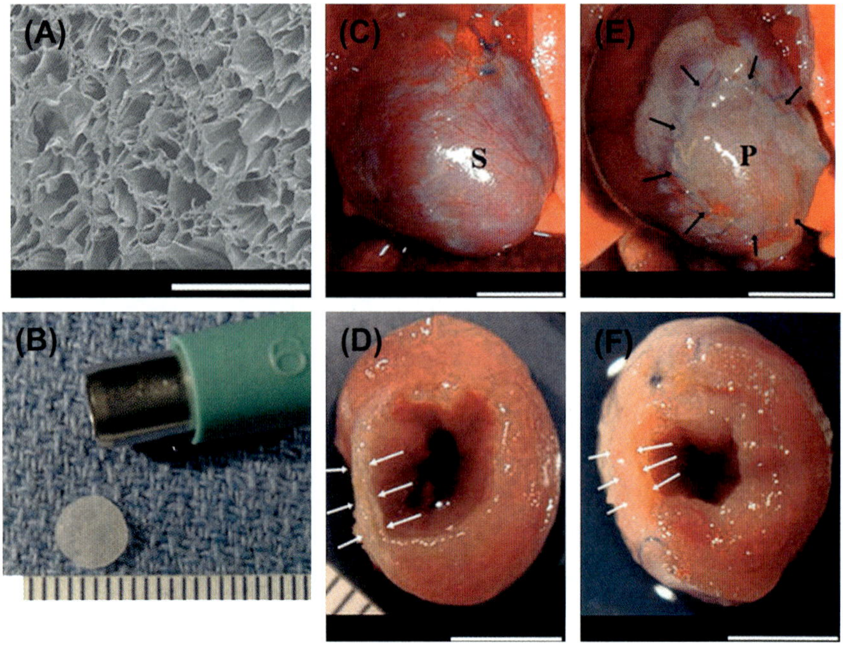

Plate 7 Cardiac patch application. For full caption see Chapter 3, page 56.

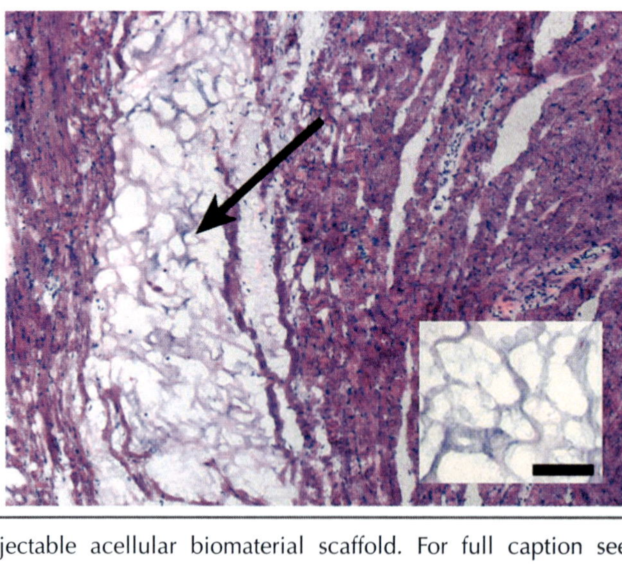

Plate 8 Injectable acellular biomaterial scaffold. For full caption see Chapter 3, page 59.

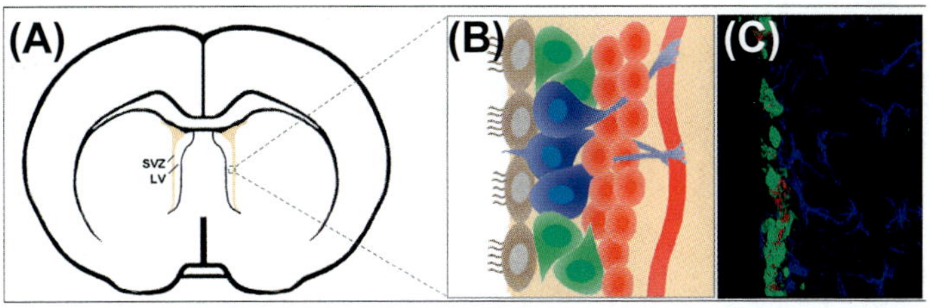

Plate 9 Subventricular zone of adult rodents. For full caption see Chapter 4, page 76.

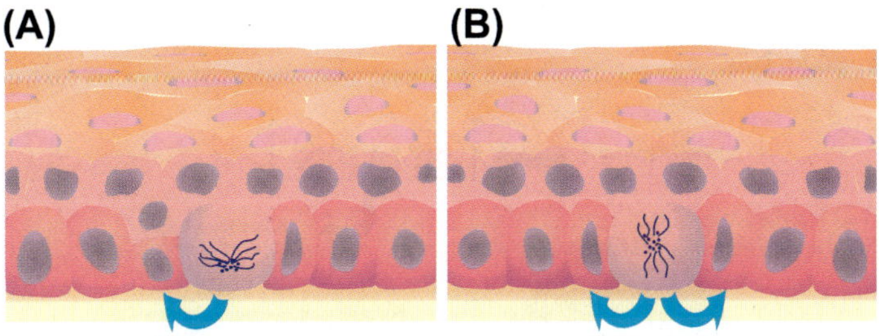

Plate 10 The position of the mitotic spindle is determinant of the destiny of epithelial cells after stem cell division. For full caption see Chapter 5, page 91.

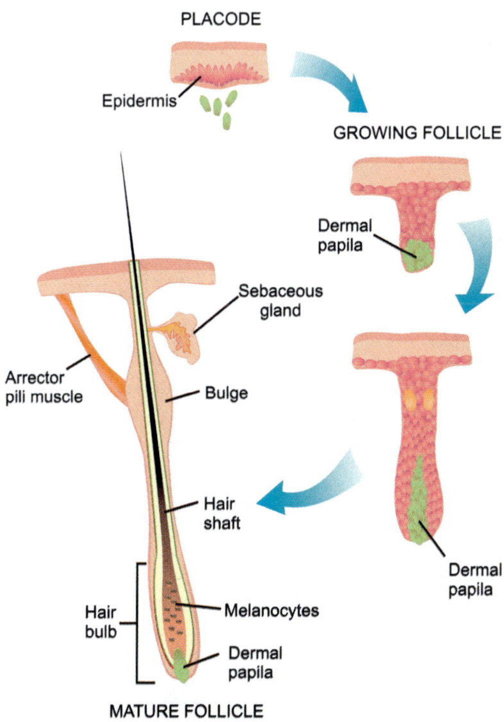

Plate 11 For caption see Chapter 5, page 93.

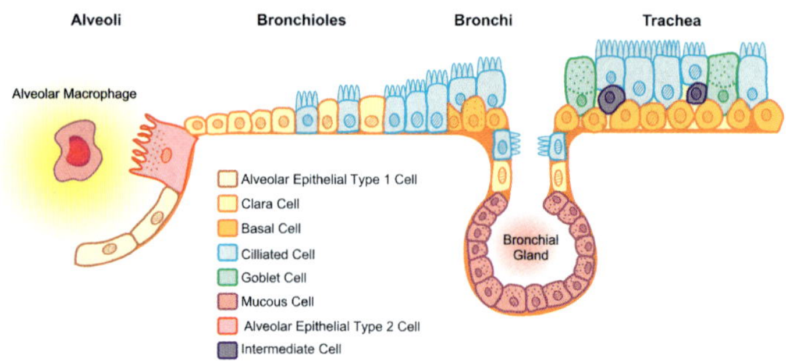

Plate 12 Schematic representation of pulmonary cell types. For full caption see Chapter 6, page 106.

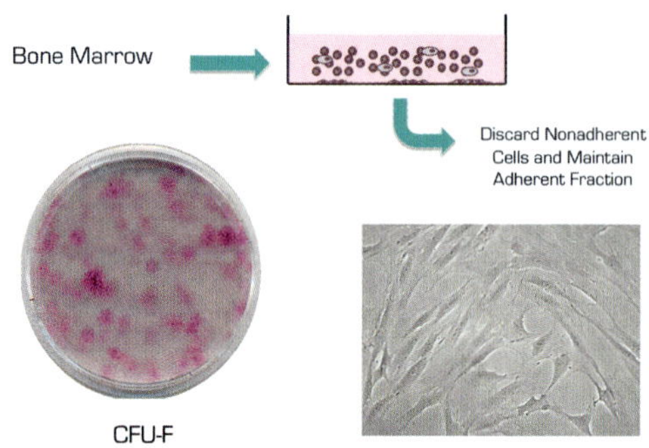

Plate 13 CFU-F colonies (Crystal violet stain).

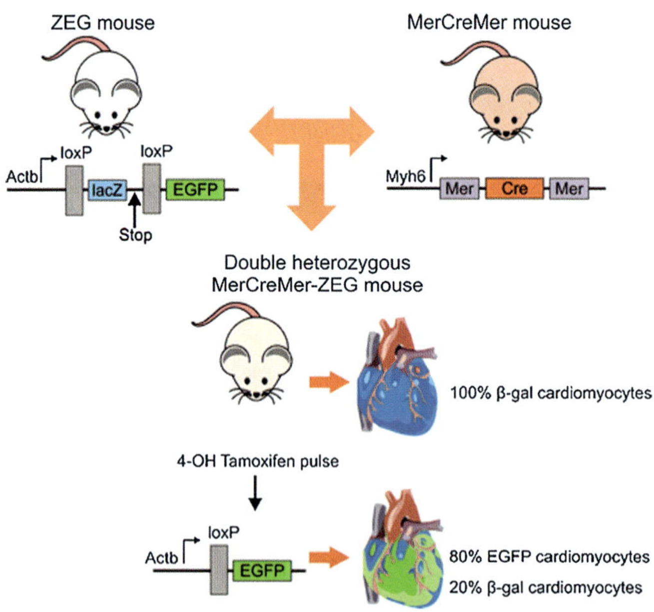

Plate 14 For caption see Chapter 8, page 143.

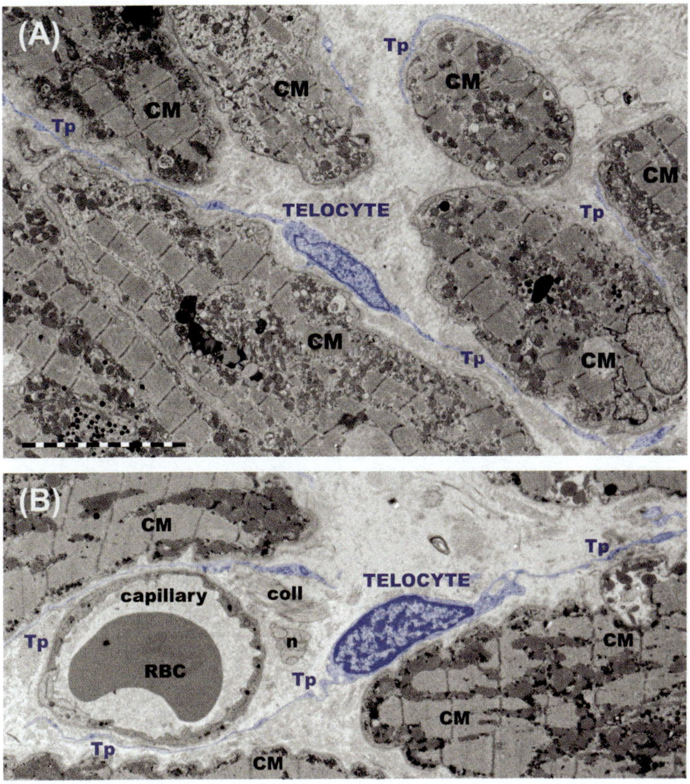

Plate 15 Human atrial myocardium, transmission electron microscopy. For full caption see Chapter 11, page 206.

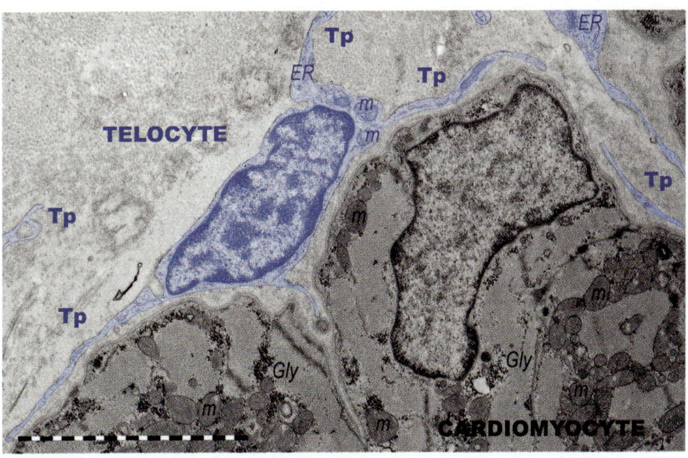

Plate 16 Human myocardium. Transmission electron microscopy. For full caption see Chapter 11, page 207.

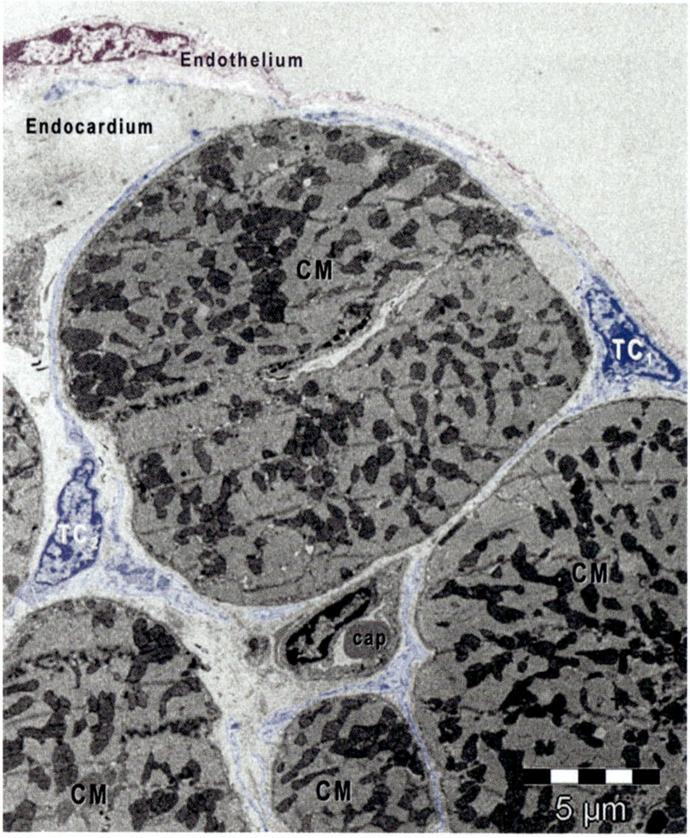

Plate 17 For caption see Chapter 11, page 207.

(A)

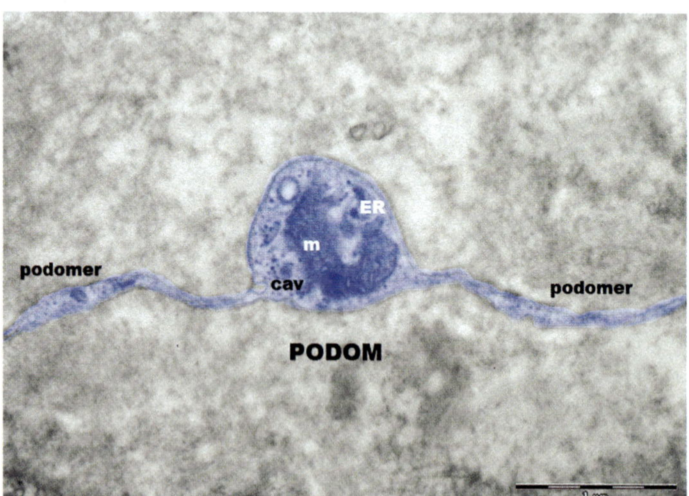

Plate 18 For caption see Chapter 11, page 208.

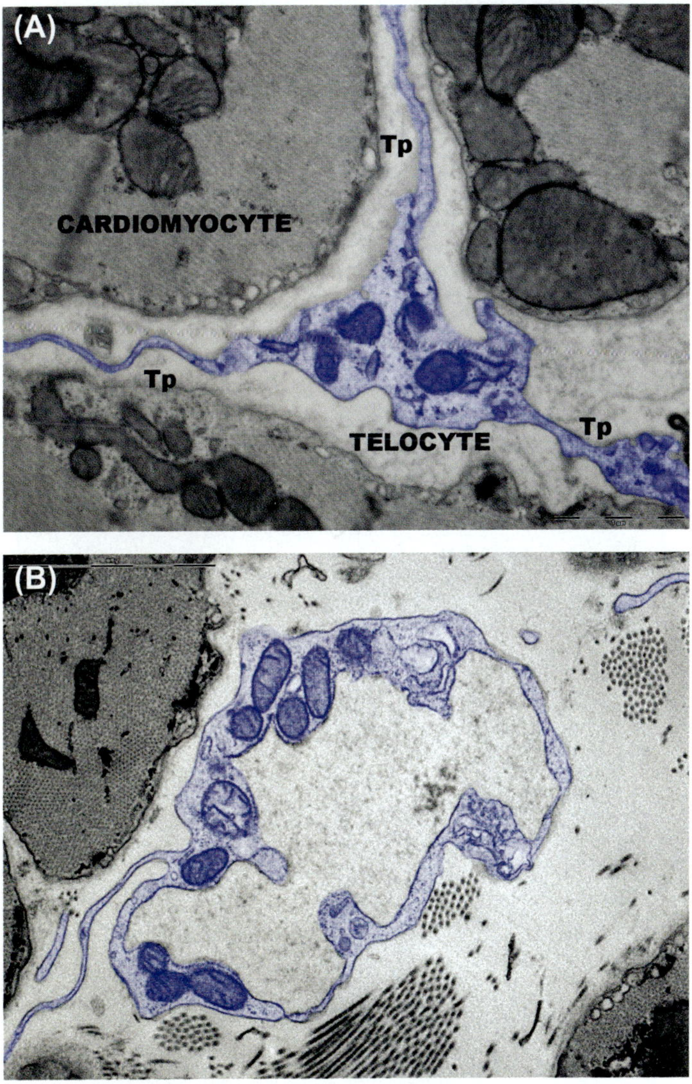

Plate 19 **(A)** A triangular telocyte (digitally colored in blue) with three emerging telopodes (Tp). Rat myocardium. **(B)** A particular circular appearance of a telopode, with both podoms and podomers, in rat heart interstitium.

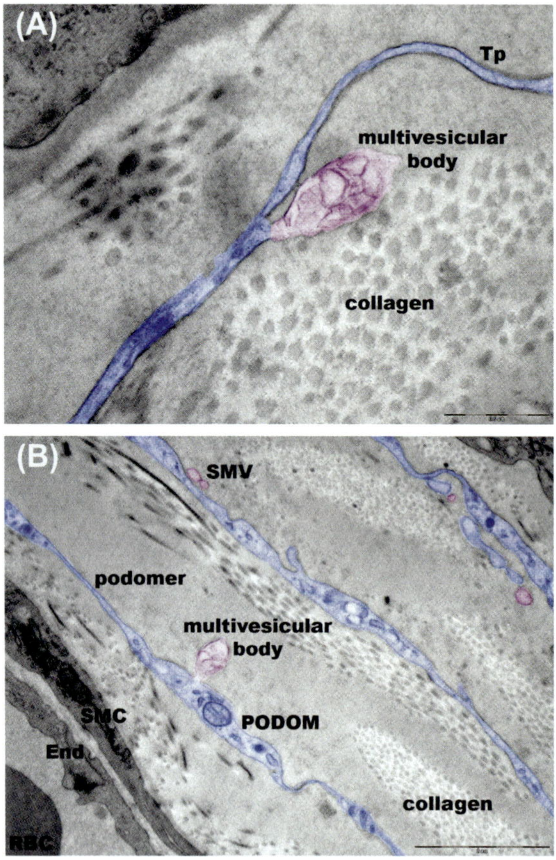

Plate 20 For caption see Chapter 11, page 214.

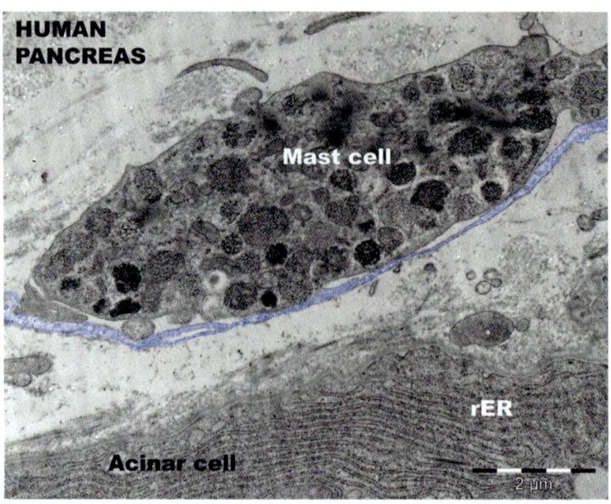

Plate 21 Human exocrine pancreas. For full caption see Chapter 11, page 217.

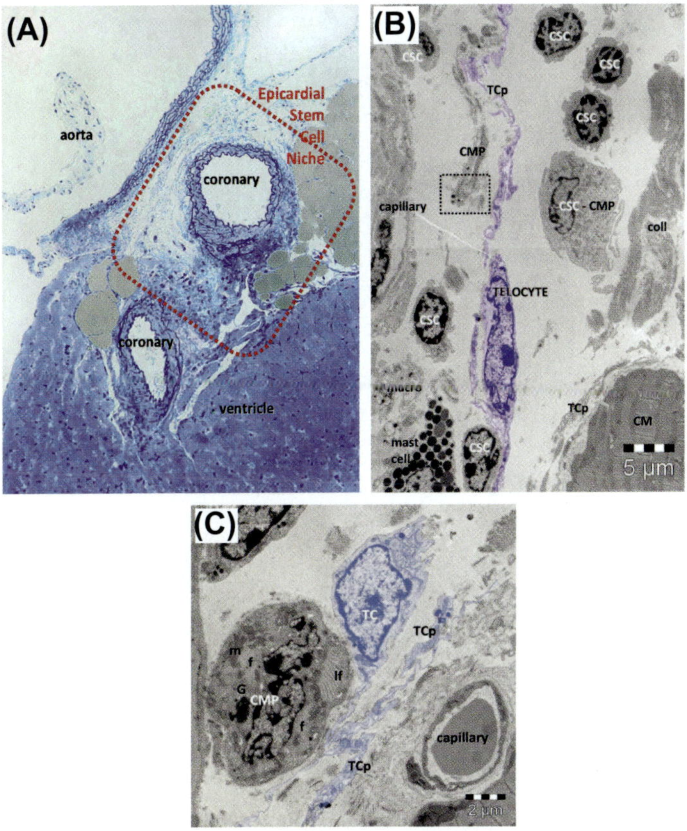

Plate 22 For caption see Chapter 11, page 220.

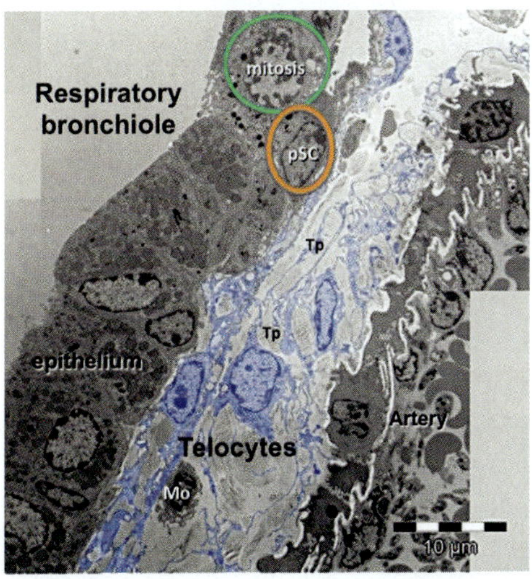

Plate 23 Transmission electron microscopy of mouse respiratory bronchiole. For full caption see Chapter 11, page 221.

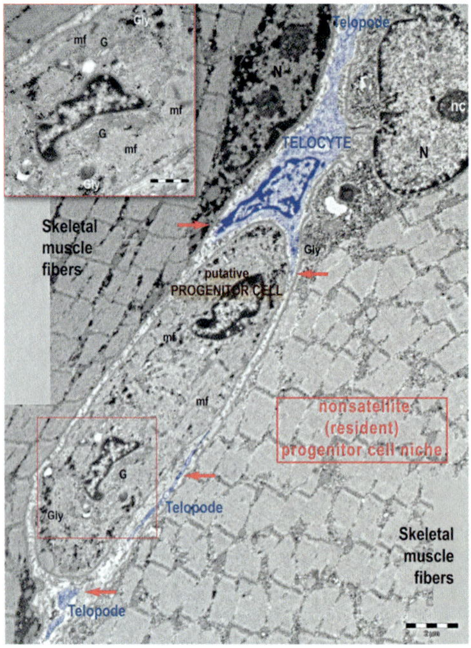

Plate 24 For caption see Chapter 11, page 222.

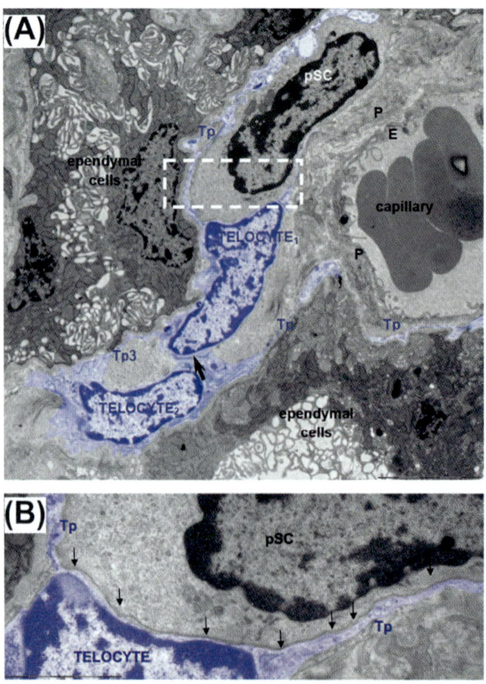

Plate 25 EM of adult rat choroid plexus. For full caption see Chapter 11, page 223.

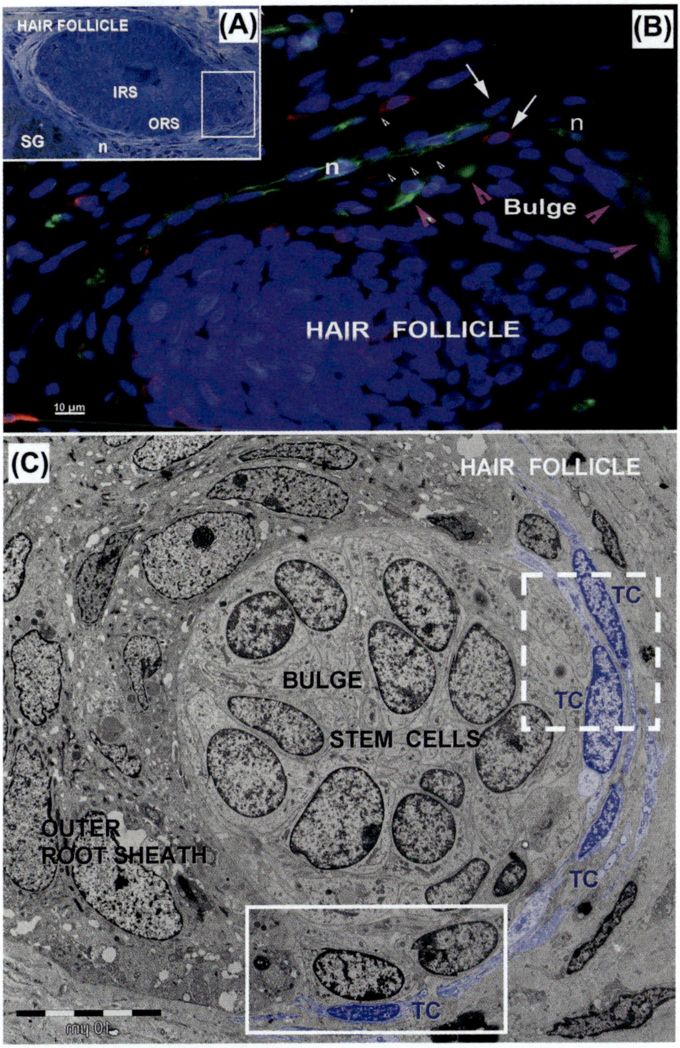

Plate 26 Telocytes in reticular dermis. **(A)** Light microscopy image (stained by toluidine-blue) showing a hair follicle and an adjacent sebaceous gland (SG). IRS, inner root sheath; ORS, outer root sheath; n, perifollicular nerve fibers. **(B)** Epifluorescence microscopy: double labelling revealing nestin positive cells (green), stem cells from the bulge area of hair follicle (magenta arrowheads), and perifollicular nerve fibers (N), surrounded by CD117 positive TCs (red, white arrows) with long Tps (white arrowhead). Nuclei were counterstained with DAPI (blue). Original magnification 400x. **(C)** EM of the boxed area in **(A)** shows a cluster of stem cells in the outer root sheath of a hair follicle. The stem cells are bordered by TCs with Tps, as well as other cells from the outer root sheath. *Reproduced with permission after [49].*

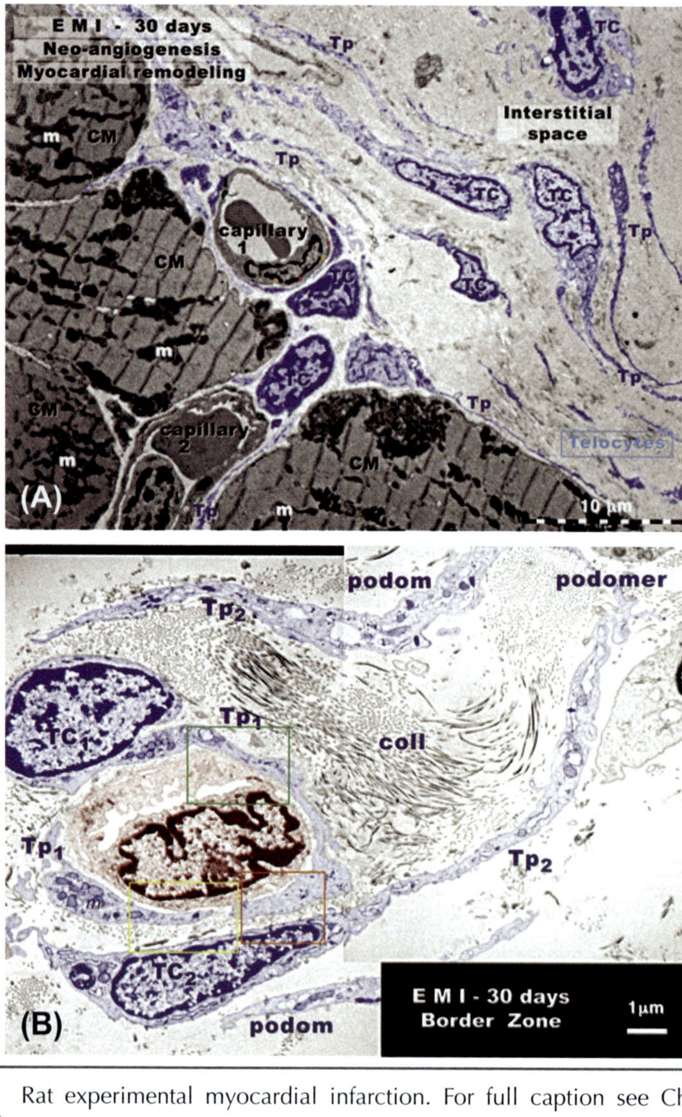

Plate 27 Rat experimental myocardial infarction. For full caption see Chapter 11, page 226.

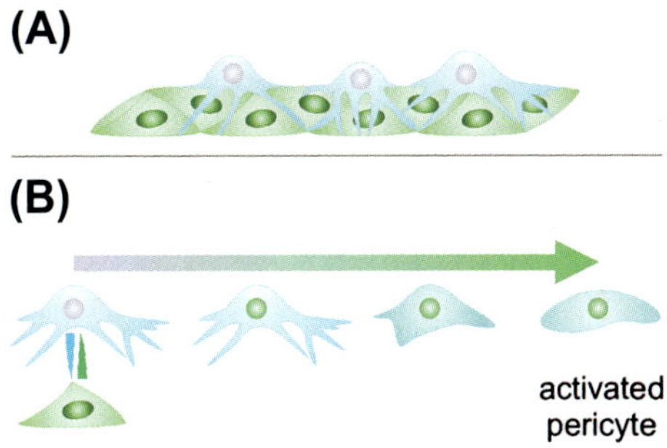

Plate 28 Schematic view of pericytes. For full caption see Chapter 12, page 234.

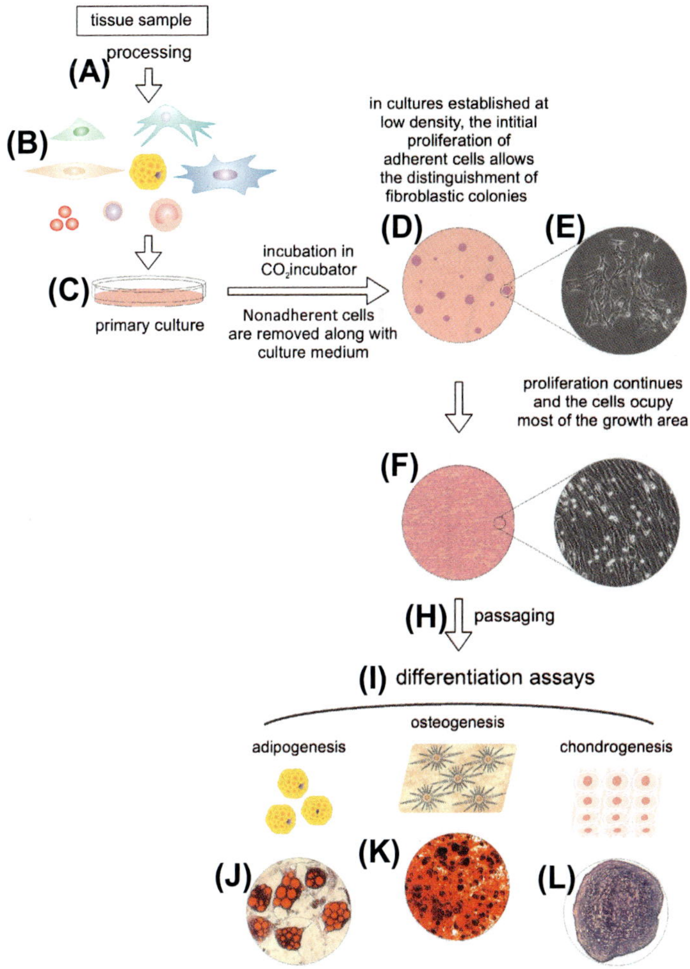

Plate 29 Schematic overview of the establishment of MSC cultures. For full caption see Chapter 12, page 238.

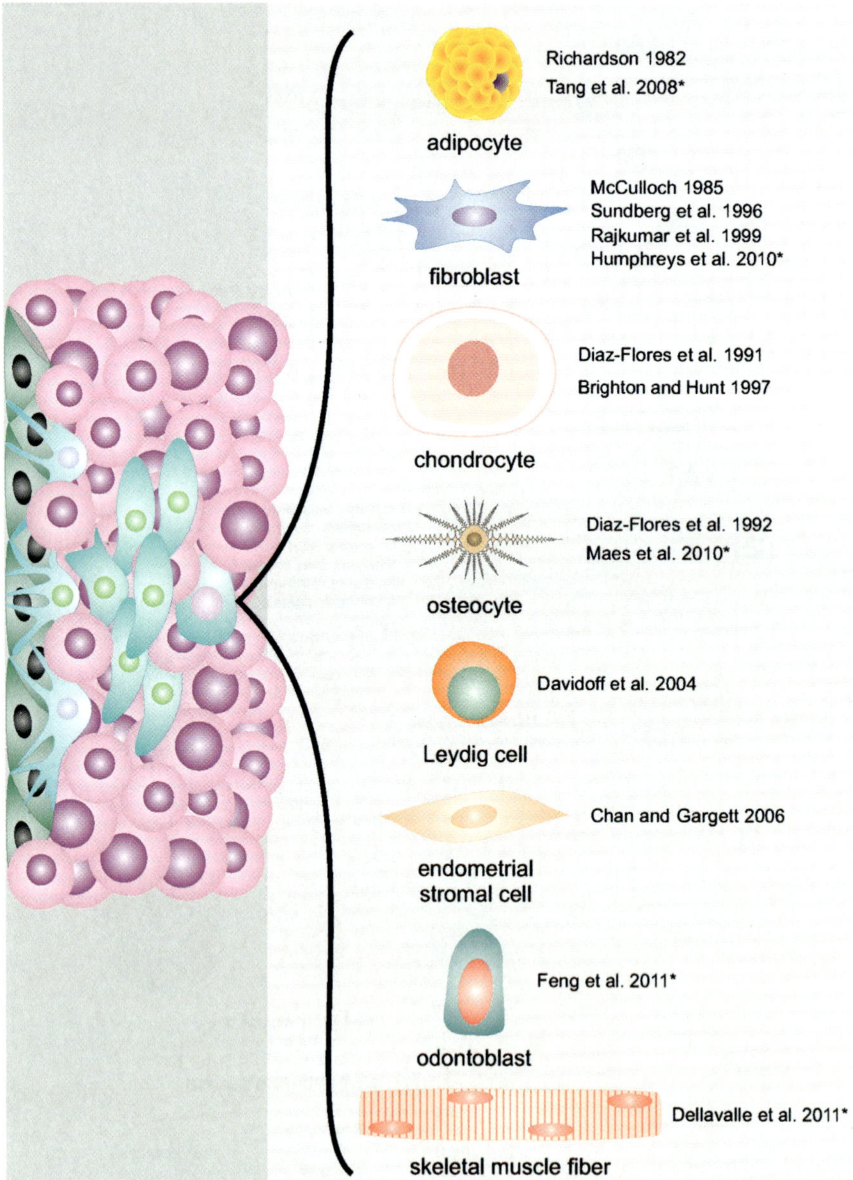

Plate 30 Differentiation of pericytes in situ. The left portion of the figure depicts a blood vessel and its associated pericytes inside a tissue composed of mature cells (represented in pink). Environmental cues stimulate the activation of one of the pericytes. The resulting activated pericytes can then enter a differentiation pathway in order to give rise to mature, tissue-specific cells. This process is inferred on the basis of studies (indicated on the right side of the figure) that found evidence that pericytes give rise to adipocytes, fibroblasts, chondrocytes, osteocytes, Leydig cells, endometrial stromal cells, odontoblasts, and skeletal muscle fibers in situ. References marked with an asterisk indicate studies in which genetic labeling was used.